Covariance Analysis and Beyond

Wei Lan • Chih-Ling Tsai

Covariance Analysis and Beyond

Springer

Wei Lan
School of Statistics and Data Science and Center of Statistical Research
Southwestern University of Finance and Economics
Chengdu, Sichuan, China

Chih-Ling Tsai
Graduate School of Management
University of California - Davis
Davis, CA, USA

ISBN 978-3-032-08795-9 ISBN 978-3-032-08796-6 (eBook)
https://doi.org/10.1007/978-3-032-08796-6

This Springer imprint is published by the registered company Springer Nature Switzerland AG
The registered company address is: Gewerbestrasse 11, 6330 Cham, Switzerland

To our parents:
Lige Lan and Lijun Zhang
Liang-Chih Tsai and Chin-Lin Lu Tsai

and our families:
Yue Ding, Jingling Lan, and Yuning Lan
Ching-Ju Liao, Yu-Yen Ho Tsai, Wen-Lin Tsai, and Wen-Ting Tsai

Preface

The Fundamental Principle of Statistics, Data Science, AI, AGI, and Life is Weighting.

The great Chinese philosopher Lao-Tzu wrote, in Chapter 42 of his *Tao-Te-Ching* around the sixth century BC, that "The Tao generates One; One generates Two; Two generates Three; and Three generates Universe." According to an interpretation of this principal, humans naturally generate variables (or features or attributes or data) with one dimension, two dimensions, and three or higher dimensions across all sorts of areas. Among those generated variables, we tend to study the relationship between two related variables such as "high and low,"[1] "great and infinity,"[2] and "great and sparse,"[3] interpreted from the *Tao-Te-Ching*. It is not surprising that such an examination of relationships becomes more pronounced when considering contemporary variables induced by technological advancements and emerging needs such as genes and proteins, stock returns and financial behaviors, social networks and media, and signal, image, and speech processes. In sum, from ancient to modern times, assessing the relationship between two variables is a common and intrinsic practice.

In the real world, we often collect (or create or observe) more than two variables. For the sake of simplicity, we can represent a collection of variables in the form of a vector. From the data analysts' or statisticians' perspective, covariance or correlation is a common mean for assessing the relationship between any two variables. In fact, the concept of correlation and its associated covariance has a long history in data analyses. The four eminent pioneers, Bravais (1844), Galton (1889), Edgeworth (1893), and Pearson (1900), have made fundamental contributions to this critical subject, and they have also opened a new era for bivariate and multivariate analyses; see, e.g., the seminal book and innovative article of Fisher (1925), Wishart (1928), respectively. After collecting the covariances obtained from all possible pairs of variables in a given vector, together with each variable's own variation, we have

[1] "High and low are contrastingly related to each other."

[2] " Great proceeds far, far approaches infinite."

[3] "The greatest sound is almost soundless and sparse."

the covariance matrix, which is an extractive summarization of the dispersions and dependencies across the multivariate variables. In the past 100 years, a lot of theoretical properties and numerous real examples across various quantitative disciplines have been developed and exploited through utilizing covariance matrices (or correlation matrices). For example, artificial intelligence, big data, high-dimensional data, machine learning, network analysis, and tensor analysis have grown rapidly and led to pioneering research and real-world applications in the last two decades. One of the key drivers of this rapid progress has been the role of the covariance matrix in exploring theoretical foundations, dimension reductions, inferences, and forecasts via algorithms, models, datasets, and variables, among others. This motivates us to write a book to investigate covariance matrices and their applications.

The main focus of this book is studying covariance matrices used in the innovative emerging fields mentioned above. To this end, we employ accessible and synthetic approaches to introduce purposes, methods, and theories. In addition, this book not only presents systematic reviews of cutting age researches but also sheds light on avenues for future research. Moreover, it provides examples to concisely illustrate practical applications. As a result, this book is suitable for graduate students and research fellows across various quantitative disciplines such as biological science, business analytics, data science, economics, engineering, finance, geography, marketing, mathematics, medicine, psychology, and statistics. We believe this novel book makes a valuable contribution to covariance data analysis and beyond. In writing this book, we appreciate Amanda Kimball, Tianyi Zhu, Xuan Yang, Bo Lei, and Dongxue Zhang for their valuable assistance. The first author gratefully acknowledges financial support from the National Natural Science Foundation of China (72422020, 72333001, 12171395, 71991472) and the Southwestern University of Finance and Economics. The second author also expresses his great gratitude for strong support from the Robert W. Glock Chair endowment established at the University of California-Davis. Finally, we would like to thank our families for their strong encouragement and emotional support.

Chengdu, China
Davis, CA, USA
August 2025

Wei Lan
Chih-Ling Tsai

Declarations

Competing Interests The authors have no competing interests to declare that are relevant to the content of this manuscript.

Contents

Chapter 1
Introduction

In data analysis, the covariance matrix plays a pivotal role in studying the relationships between a set of variables. It has been commonly and widely used across various fields regardless of the data being dense or sparse, low-dimension or high-dimension, discrete or continuous, fixed or random, time series or non-time series, structured or non-structured, and so forth. Hence, the main theme of this book is "covariance." To study this subject, we first introduce multivariate data and its expressions, along with two examples. We next state the scope and the contributions of the book and then briefly describe each of the next nine chapters' key points. Finally, we present some notation, norms, operations, decompositions, order operations, asymptotic properties, and inequalities that will be directly and indirectly used in this book and its associated references.

1.1 Multivariate Data

Advanced technologies have significantly impacted fields such as agriculture, biology, business, communications, economics, engineering, finance, marketing, mathematics, medicine, science, and social science. Technological advancements have made it easier to generate, store, and analyze large amounts of data, often multivariate in nature. Multivariate refers to data that contain observations (or objects) with more than one variable (or feature). Each individual variable can be classified as a dependent (or response) variable or an independent variable (or covariate) based on its role in the data analysis. Although this book focuses on multivariate data, the single variable dataset is also discussed. Each variable can be continuous or discrete. In addition, the observations can be repeated over time. Moreover, the number of variables can exceed the number of observations. Useful reference books of multivariate data analysis can be found in Mardia et al. (1979),

W. Lan, C.-L. Tsai, *Covariance Analysis and Beyond*,
https://doi.org/10.1007/978-3-032-08796-6_1

Timm (2002), Anderson (2003), Johnson and Wichern (2007), Izenman (2008), Everitt and Hothorn (2011), Härdle and Simar (2015), and Hair et al. (2019).

1.2 Data Expression

To describe multivariate data, let $Y = (y_1, \cdots, y_p)'$ denote a $p \times 1$ vector, where the symbol " $'$ " represents the transpose. After collecting n samples, let $Y_i = (y_{i1}, \cdots, y_{ip})' \in \mathbb{R}^p$ denote the i-th observation associated with the p variables, for $i = 1, \cdots, n$, where $\mathbb{R}^p$ is the set of the p-tuples of real numbers. In addition, for $j = 1, \cdots, p$, let $Y_{(j)} = (y_{1j}, \cdots, y_{nj})'$ be the $n \times 1$ vector that denotes the n observations on the j-th variable. Thus, the data with n observations and p variables can be displayed as an $n \times p$ matrix, $\mathbf{Y} = (Y_1, \cdots, Y_n)' = (Y_{(1)}, \cdots, Y_{(p)})$. In practice, Y_i $(i = 1, \cdots, n)$ can depend on a covariate vector $X_i = (x_{i1}, \cdots, x_{iq})'$, where q is the number of covariates. For $j = 1, \cdots, q$, let $X_{(j)} = (x_{1j}, \cdots, x_{nj})'$ be the $n \times 1$ vector. Thus, $\mathbf{X} = (X_1, \cdots, X_n)' = (X_{(1)}, \cdots, X_{(q)})$ is a $n \times q$ matrix.

In practice, X_i can be a matrix. To this end, denote $\mathbf{X}^{(i)} = (X_1^{(i)}, \cdots, X_q^{(i)})$, where $X_k^{(i)} = (X_{k,1}^{(i)}, \cdots, X_{k,p}^{(i)})' \in \mathbb{R}^p$ and $k = 1, \cdots, q$. As a result, $\mathbf{X}^{(i)}$ is a $p \times q$ matrix for $i = 1, \cdots, n$. As for the data expression for multidimensional arrays, we will discuss it in Chap. 10. The next section employs two examples to illustrate the above notation.

1.3 Examples

Example 1.1 (Concrete Slump and Flow) This dataset was created by Yeh (2007, 2008), which includes 103 observations and three dependent variables, slump (cm), flow (cm), and 28-day compressive strength (mpa). As a result, $n = 103$, $p = 3$, and $\mathbf{Y}$ is a 103×3 matrix. Specifically, $\mathbf{Y} = (Y_1, \cdots, Y_{103})' = (Y_{(1)}, \cdots, Y_{(3)})$, where $Y_i = (y_{i1}, \cdots, y_{i3})'$ and $Y_{(j)} = (y_{1j}, \cdots, y_{103,j})'$. In addition, seven covariates are collected, and they are (i) cement (kg/m^3); (ii) blast furnace slag (kg/m^3); (iii) fly ash (kg/m^3); (iv) water (kg/m^3); (v) super-plasticizer (kg/m^3); (vi) coarse aggregate (kg/m^3); and (vii) fine aggregate (kg/m^3). Thus, $\mathbf{X}$ is a 103×7 matrix, and $\mathbf{X} = (X_1, \cdots, X_{103})' = (X_{(1)}, \cdots, X_{(7)})$, where $X_i = (x_{i1}, \cdots, x_{i7})'$ and $X_{(j)} = (x_{1j}, \cdots, x_{103,j})'$.

Example 1.2 (US Stock Market) This example has been studied by Fan et al. (2024a). They collected the monthly stock returns of 400 stocks in the US stock market from January 2016 to July 2018, where the dataset was downloaded from Wharton Research Data Services. The response variable consists of $p = 400$ stocks' monthly returns, and there are $n = 31$ months in total. Accordingly, $\mathbf{Y}$ is a 31×400 matrix, and $\mathbf{Y} = (Y_1, \cdots, Y_{31})' = (Y_{(1)}, \cdots, Y_{(400)})$, where $Y_i = (y_{i1}, \cdots, y_{i,400})'$ and $Y_{(j)} = (y_{1j}, \cdots, y_{31,j})'$ for $i = 1, \cdots, 31$ and $j = 1, \cdots, 400$. In addition, they collected $q = 22$ covariates, which are associated with stock returns. Although the covariate data can be displayed in the three-dimensional array, we follow the approach of Fan et al. (2024a) and present it in matrix form. Specifically, for $i = 1, \cdots, 31$, denote $\mathbf{X}^{(i)} = (X_1^{(i)}, \cdots, X_{22}^{(i)})$, where $X_k^{(i)} = (X_{k,1}^{(i)}, \cdots, X_{k,400}^{(i)})'$ and $k = 1, \cdots, 22$. As a result, $\mathbf{X}^{(i)}$ is a 400×22 matrix for $i = 1, \cdots, 31$. The detailed descriptions of the 22 covariates are presented in Fan et al. (2024a).

1.4 Scope and Contributions

After collecting a set of multivariate data, it is natural to seek the relationship between the variables (if any). Based on the data structure and the goal of the study, one can further employ various techniques to conduct dimension reduction, separation, identification, decisions, and predictions. These techniques include regression, principal component analysis, factor analysis, canonical correlation, discrimination, classification, and clustering, among others. All of these techniques are related to covariance matrices that have been widely used in machine learning and data science; see, e.g., Clarke et al. (2009), Hastie et al. (2009), Alpaydin (2020), Kuo and Madni (2023), and Tian and Feng (2023). Furthermore, large covariance matrices have in turn been incorporated into the techniques in modern multivariate analysis; see, e.g, Pourahmadi (2013), Yao et al. (2015), Fan et al. (2016), and Bose and Bhattacharjee (2018).

It is of interest to note that covariance regression models have paid considerable attention to linking the covariance matrix with the explanatory variables; see, e.g., Anderson (2003), Hoff and Niu (2012) and Zou et al. (2017). In particular, those explanatory variables can be either fixed or random (Zou et al. 2025). Recently, Zou et al. (2022) extended covariance regression models to covariance-mean regression models, which link both the mean and the covariance to the explanatory variables. One such model is the spatial autoregressive (or network autocorrelation) model that takes into account the dependence and spatial heterogeneity of responses by leveraging spatial (or network) information; see, e.g., Lee (2004), LeSage and Pace (2009), and Zou et al. (2021).

In addition to the above topics, factor analysis has been widely used to reduce a large number of variables into fewer factors. Accordingly, the number of parameters involved in covariance matrix is reduced. The classical factor models, together with various modifications and extensions, have been developed into popular modern factor models such as approximate factor models (Chamberlain and Rothschild 1983), factor-augmented regression models (Bai and Ng 2002; Stock and Watson 2002a), dynamic factor models (Geweke 1977; Sargent and Sims 1977), matrix factor models (Wang et al. 2019; Chen and Fan 2023), and tensor factor models (Chen et al. 2022; Chen et al. 2024b).

In the last decade, machine learning has played a critical role in the development and advancement of artificial intelligence (AI). In the field of machine learning, deep learning is a powerful tool to learn from data and make predictions or classifications using artificial neural networks, especially on complex data. Hence, it seems natural that covariance matrices enable improvements in deep learning (see, e.g., El-Nouby et al. 2021; Wang et al. 2022; Sihag et al. 2022), while deep learning can also increase the accuracy of covariance matrix estimation (see, e.g., Cai and Kim 2024). Besides machine learning, tensor analysis has become a popular and important field for handling multidimensional data and solving complex problems. Hence, the extensions of factor models and covariance matrices to tensor factor models and tensor covariance matrices, respectively, have proved essential for the extraction of insightful information from data; see, e.g., Li and Sun (2011), Manceur and Dutilleul (2013), Leng and Pan (2018), Chen et al. (2022), Chang et al. (2023), Deng et al. (2023), Chen and Lam (2024), Chen et al. (2024a,b), and Han et al. (2024a,b).

As discussed above, the covariance matrix plays a fundamental role in both classical and modern data analysis. This motivates us to write this book on the way we express how variables relate to one another (i.e., the covariance matrix). It not only reviews the fundamentals of multivariate analysis but also encompasses recent developments in covariance models across various subjects. As a result, this book makes the following contributions. First, it integrates covariance-related subjects together in a way that includes structured covariance matrices, covariance regression models and their extensions, fixed and random covariance models, spatial autoregressive (or network autocorrelation) models, covariance factor models, machine learning, and tensor analysis. Second, it employs cutting-edge methods to study covariances and their practical applications. These methods include activation, banding, blocking, dimension reduction, filtering, penalizing, selection, shrinking, tapering, thresholding, matrix factor, tensor factor, tensor decomposition, deep learning, and transfer learning, among others.

Third, it allows readers to integrate two or more concepts from different subjects to explore possible future research avenues such as deep covariance regression analysis, tensor covariance regression analysis, tensor structured covariance models, tensor random regression models, tensor regression and factor models, domain generalizations (see, e.g., Wu 2021; Nguyen et al. 2022; Wang et al. 2023; Zhou et al. 2023), sentiment and text analyses (see, e.g., Liu 2020; Jockers and Thalken 2020; Kumar et al. 2023; Ke et al. 2024), and Transformers (see, e.g., El-Nouby et al. 2021; Kamath et al. 2022; Lin et al. 2022; Islam et al. 2024; Esmaeili Nezhad and Samimi 2024; Noci et al. 2023).

1.5 Contents

To accomplish our goals, the rest of this book is organized as follows. Chapter 2 introduces the covariance and precision matrices and then employs the covariance matrix to obtain the mean, variance, and covariance of quadratic forms. In addition, the four covariance estimations (moment estimation, maximum likelihood estimation, penalized estimation, and robust estimation) are presented.

Chapter 3 studies structured covariance matrices, which include some known matrices such as compound symmetry, autoregressive with order one (AR(1)), moving average process with order one (MA(1)), symmetric Toeplitz matrices, banding, thresholding, and linear covariance matrices. In addition, it introduces linear covariance models that link the covariance matrix with a set of known symmetric matrices. To assure the covariance matrix is positive definite, unconstrained parameterizations are discussed.

Chapter 4 introduces covariance regression models and shows theoretical properties of regression parameters. The non-Gaussian linear covariance models and covariance model with general linear structure are also analyzed. Chapter 5 reviews the four known mean regression models with non-identity covariance matrices: weighted regression, generalized least squares estimator, longitudinal data, and multivariate regression. These models motivate us to introduce covariance-mean regression models. We then present the theoretical properties of regression parameter estimators. We further provide a test statistic to analyze the necessity of the auxiliary information and propose a test to assess the adequacy of the covariance-mean regression model. To broaden practical applications, Chap. 6 generalizes random mean regression models to random covariance regression models. It proposes the covariance estimators under the Random-X setting and then shows the theoretical properties of estimators. Furthermore, the model assessment theory of covariance regression has been developed. The theoretical results demonstrate that the Random-X setting can yield larger expected test errors than those of the Fixed-X setting.

Chapter 7 extends classical spatial autoregressive (SAR) models to network influence regression (NIR) models. Accordingly, the influential nodes (or actors) of a network can be identified. Note that NIR can be considered as an application of covariance regression. In addition, an inward-and-outward network influence model is considered to examine nodal heterogeneity between inward influence and outward influence. Moreover, a procedure to screen out portal nodes for large-scale social networks is investigated, and a multivariate SAR model for large-scale social networks is studied. Then, four additional types of network autocorrelation models are briefly discussed. Chapter 8 first reviews classical factor models. Then it studies approximate factor models, factor-augmented regression models, dynamic factor models, and matrix factor models. Parameter estimation and factor selection across models are also investigated.

Chapter 9 introduces machine learning (which includes supervised learning, unsupervised learning, and semi-supervised learning) and briefly discusses reinforcement learning. It also studies the applications of covariance matrices in machine learning, such as classification and clustering. Moreover, it explores three deep learning methods, convolution neural networks (CNNs), graph convolutional networks (GCNs), and Transformer and discusses their relationships with covariance matrices. Subsequently, transfer learning and covariance matrix modeling are investigated. It is worth noting that, unlike traditional machine learning/deep learning books, Chap. 9 focuses on exploring the connection between machine learning/deep learning and covariance matrices.

Finally, Chap. 10 extends the matrix product, decomposition, correlation, and canonical correlation to higher-order tensors. Furthermore, it discusses separable covariance models, core shrinkage covariance models, and tensor factor models. Two applications via the tensor-based covariance approach are presented.

For the sake of convenience, we have compiled links to the articles with most of the examples used in this book at the following link: https://github.com/Covariance-Analysis-and-Beyond/Data-Link. We have compiled these resources based on their locations and availability before publishing this book, and we make no promises about their accessibility or future availability. We are not responsible for any future removal, renaming, or changes that may be made to any of the individual links or data resources herein. In addition, we are not responsible for providing any data via the above link.

1.6 Notation, Norms, Operations, and Decompositions

In Sect. 1.3, we have used some notation to describe the data in two examples. For the sake of completeness, we add more notation as well as introduce the norms, operations, and decompositions. This allows us to present the model structure, theoretical properties, and examples that will be used in this book and in the referenced literature.

1.6.1 Notation

Denote $\mathbf{a} = (a_1, \cdots, a_m)'$ and $\mathbf{b} = (b_1, \cdots, b_m)'$ as $m \times 1$ generic vectors, where a_i and b_i are real numbers for $i = 1, \cdots, m$. Note that the symbol "$\,'\,$" means the transpose, as mentioned in Sect. 1.2. Analogously, denote $\mathbf{c} = (c_1, \cdots, c_n)'$ as an $n \times 1$ generic vector, where c_i is a real number for $i = 1, \cdots, n$. Subsequently, denote $\mathbf{A} = (a_{ij})$ and $\mathbf{B} = (b_{ij})$ as $m \times n$ generic matrices, where a_{ij} and b_{ij} are real numbers for $i = 1, \cdots, m$ and $j = 1, \cdots, n$. In addition, denote $\mathbf{C} = (c_{ij})$ and $\mathbf{D} = (d_{kl})$ as $n \times p$ and $q \times r$ generic matrices, respectively, where c_{ij} and d_{kl} are real numbers for $i = 1, \cdots, n$, $j = 1, \cdots, p$, $k = 1, \cdots, q$ and $l = 1, \cdots, r$. For the sake of convenience, we sometimes denote matrices by non-boldface letters such as A, B, C, and D. To show asymptotic properties, we use $\rightarrow$, $\stackrel{d}{\rightarrow}$, and $\stackrel{p}{\rightarrow}$ to denote the convergence, the convergence in distribution, and the convergence in probability, respectively. Let $\mathbf{G}_1$ and $\mathbf{G}_2$ be two generic symmetric matrices. Then $\mathbf{G}_1 \succ \mathbf{G}_2$ denotes that the difference between $\mathbf{G}_1$ and $\mathbf{G}_2$ is positive definite. In addition, $\mathbf{G}_1 \triangleq \mathbf{G}_2$ (or $\mathbf{G}_1 := \mathbf{G}_2$ or $\mathbf{G}_1 \equiv \mathbf{G}_2$) denotes that $\mathbf{G}_1$ is defined to be equal to $\mathbf{G}_2$, and $\mathbf{G}_1 := \mathbf{G}_2$ also denotes that $\mathbf{G}_1$ is to be defined as $\mathbf{G}_2$. Moreover, $\mathbf{G}_1 \approx \mathbf{G}_2$ indicates that $\mathbf{G}_1$ is approximate to $\mathbf{G}_2$, and $\mathbf{G}_1 \sim \mathbf{G}_2$ (or $\mathbf{G}_1 \asymp \mathbf{G}_2$) denotes that $\mathbf{G}_1$ is asymptotically equivalent to $\mathbf{G}_2$. Suppose that A could be a subset of a set X. Define $\mathbf{1}_A$ (or $\mathbf{1}(A)$) to be the following indicator function: $\mathbf{1}_A(x) = 1$ if $x \in A$, and $\mathbf{1}_A(x) = 0$ otherwise. On the other hand, $1_n = (1, \cdots, 1)'$ denotes an $n \times 1$ vector. For the sake of simplicity, we sometimes denote it $\mathbf{1}$. Finally, for any real number x, let $[x]$ (or $\lfloor x \rfloor$) represent the greatest integer not exceeding x, and denote $(x)_+ = x$ if $x > 0$, and 0 otherwise.

1.6.2 Norms

The L_k-norm of any vector $\mathbf{a}$ is defined as $||\mathbf{a}||_k = \{\sum_{i=1}^m |a_i|^k\}^{1/k}$ for $k = 1, \cdots, \infty$. Thus, the L_1-norm is $||\mathbf{a}||_1 = \sum_{i=1}^m |a_i|$, and the L_2-norm is $||\mathbf{a}||_2 = \{\sum_{i=1}^m a_i^2\}^{1/2}$. For the sake of convenience, we sometimes denote $|| \cdot ||_2$ by $|| \cdot ||$. In addition, the L_∞-norm is defined as $||\mathbf{a}||_\infty = \max_{1 \le i \le m} |a_i|$. Analogously, the L_k entry-wise norm of any matrix $\mathbf{A}$ is defined as $||\mathbf{A}||_k = ||\text{vec}(\mathbf{A})||_k = \{\sum_{i=1}^m \sum_{j=1}^n |a_{ij}|^k\}^{1/k}$ for $k = 1, \cdots, \infty$, where vec($\mathbf{A}$) represents an $mn \times 1$ column vector by stacking the n-column vectors of the matrix $\mathbf{A}$ on top of one another. Thus, the L_1-norm is $||\mathbf{A}||_1 = ||\text{vec}(\mathbf{A})||_1 = \sum_{i=1}^m \sum_{j=1}^n |a_{ij}|$ and the L_2-norm is $||\mathbf{A}||_2 = ||\text{vec}(\mathbf{A})||_2 = \{\sum_{i=1}^m \sum_{j=1}^n a_{ij}^2\}^{1/2}$, which is also called the Frobenius norm and is denoted by $\|\mathbf{A}\|_F$. For the sake of convenience, let $|| \cdot ||_1^-$ denote the L_1-norm with all off-diagonal elements so that $||\mathbf{A}||_1^- = \sum_{i=1}^m \sum_{j \ne i}^n |a_{ij}|$. In addition, the L_∞-norm is defined as $||\mathbf{A}||_\infty = \|\text{vec}(\mathbf{A})\|_\infty = \max_{1 \le i \le m, 1 \le j \le n} |a_{ij}|$. Note that the above norms are defined via the element-wise of a matrix. In contrast, the p-norm of a matrix $\mathbf{A}$ can be defined as follows:

$$||\mathbf{A}||_p = \max_{\mathbf{x} \ne 0} \frac{||\mathbf{Ax}||_p}{||\mathbf{x}||_p},$$

where $\mathbf{x}$ is any $n \times 1$ vector and the p-norm of any vector $\mathbf{v} = (v_1, \cdots, v_n)'$ is $||v||_p = (|v_1|^p + \cdots + |v_n|^p)^{1/p}$. The three commonly used matrix norms with p=1, 2, and ∞ are, respectively, given by (i) $||\mathbf{A}||_1 = \max_{1 \le j \le n} \sum_{i=1}^m |a_{ij}|$; (ii) the spectral norm, $||\mathbf{A}||_2 = \sqrt{\gamma_{max}(\mathbf{A}'\mathbf{A})}$, where $\gamma_{max}(\mathbf{A}'\mathbf{A})$ is the largest eigenvalue of $\mathbf{A}'\mathbf{A}$; and (iii) $||\mathbf{A}||_\infty = \max_{1 \le i \le m} \sum_{j=1}^n |a_{ij}|$.

1.6.3 Operations

We first define vector addition, subtraction, inner product, and outer product as follows: $\mathbf{a} + \mathbf{b} = (a_1 + b_1, \cdots, a_m + b_m)'$; $\mathbf{a} - \mathbf{b} = (a_1 - b_1, \cdots, a_m - b_m)'$; $\mathbf{a} \cdot \mathbf{b} = \mathbf{a}'\mathbf{b} = \mathbf{b}'\mathbf{a} = \sum_{i=1}^m a_i b_i$; and $\mathbf{a} \otimes \mathbf{c}' = \mathbf{c}' \otimes \mathbf{a} = \mathbf{ac}'$, which is the $m \times n$ matrix such that $(\mathbf{a} \otimes \mathbf{c}')_{ij} = a_i c_j$ for $i = 1, \cdots, m$ and $j = 1, \cdots, n$. The outer product $\otimes$ is sometimes denoted by $\circ$. We then define matrix addition, subtraction, and product given below. $\mathbf{A}+\mathbf{B} = (a_{ij} + b_{ij})$, $\mathbf{A}-\mathbf{B} = (a_{ij} - b_{ij})$, and $\mathbf{AC} = (g_{ij})$

with dimension $m \times p$, where $g_{ij} = \sum_{k=1}^{n} a_{ik} c_{kj}$. In addition, the Hadamard (Schur) product and Kronecker product are defined as follows: $\mathbf{A} * \mathbf{B} = (a_{ij} b_{ij})$ with dimension $m \times n$; and $\mathbf{A} \otimes \mathbf{D} = (a_{ij} \mathbf{D})$ with dimension $mq \times nr$. Note that the Kronecker product is a generalization of the outer product.

We next introduce a few terms often used in matrix operations. The rank of matrix **A** is denoted by rank(**A**), which is the maximum number of linearly independent columns of **A**. Suppose **A** is a square matrix of dimension m. It is symmetric if $\mathbf{A} = \mathbf{A}'$, and it is idempotent if $\mathbf{A}^2 = A$. The diagonal of **A** is denoted as $\text{diag}(\mathbf{A}) = (a_{11}, a_{22}, \cdots, a_{mm})'$. Conversely, diag(**a**) refers to the $m \times m$ square matrix, which has **a** on the diagonal and zero elsewhere. In addition, the trace of **A** is denoted by $\text{tr}(\mathbf{A}) = \sum_{i=1}^{m} a_{ii}$. We further employ the Leibniz formula to define the determinant: $|\mathbf{A}| = \det(\mathbf{A}) = \sum_{\sigma \in S_m} \mathbf{sgn}(\sigma) \prod_{i=1}^{m} a_{i,\sigma(i)}$, where S_m is the set of permutations σ of the integers $1, \cdots, m$, and $\mathbf{sgn}(\sigma)$ is the sign of the permutation σ, which is determined by the number of pairwise exchanges required to transform $\sigma(1), \cdots, \sigma(m)$ into $1, \cdots, m$. Note that the products from an odd permutation receive a minus sign, while the products from an even permutation receive a plus sign. If $|\mathbf{A}| \neq 0$, then **A** is non-singular. As a result, it has a unique inverse matrix $\mathbf{A}^{-1}$, which satisfies $\mathbf{A}\mathbf{A}^{-1} = \mathbf{A}^{-1}\mathbf{A} = \mathbf{I}_m$, where $\mathbf{I}_m$ is an identity matrix of dimension m. For the sake of convenience, we sometimes denote $\mathbf{I}_m$ by **I**. Moreover, **A** is orthogonal if $\mathbf{A}^{-1} = \mathbf{A}'$. Accordingly, $\mathbf{A}\mathbf{A}' = \mathbf{A}'\mathbf{A} = \mathbf{I}_m$. Note that if **A** is an $m \times n$ rectangular matrix, we can define a Moore–Penrose inverse (i.e., pseudo inverse or generalized inverse) of **A**. Denote by $\mathbf{A}^+$ the Moore-Penrose inverse, defined as the matrix that satisfies (i) $\mathbf{A}\mathbf{A}^+\mathbf{A} = \mathbf{A}$; (ii) $\mathbf{A}^+\mathbf{A}\mathbf{A}^+ = \mathbf{A}^+$; (iii) $(\mathbf{A}\mathbf{A}^+)' = \mathbf{A}\mathbf{A}^+$; and (iv) $(\mathbf{A}^+\mathbf{A})' = \mathbf{A}^+\mathbf{A}$. Obviously, $\mathbf{A}^{-1}$ satisfies these conditions.

Subsequently, we present a few operations of Kronecker products since they can be useful in multi-dimensional analysis. Let **A** and **B** be $m \times n$ matrices, **C** be an $n \times p$ matrix, and **D** be a $q \times r$ matrix as defined above. In addition, let **E** be a $q \times r$ matrix, **F** be an $r \times s$ matrix, and **G** be an $s \times t$ matrix. Then (i) $(\mathbf{A} \otimes \mathbf{D})' = \mathbf{A}' \otimes \mathbf{D}'$; (ii) $(\mathbf{A} \otimes \mathbf{D}) \otimes \mathbf{G} = \mathbf{A} \otimes (\mathbf{D} \otimes \mathbf{G})$; (iii) $(\mathbf{AC}) \otimes (\mathbf{EF}) = (\mathbf{A} \otimes \mathbf{E})(\mathbf{C} \otimes \mathbf{F})$; (iv) $(\mathbf{A} + \mathbf{B}) \otimes (\mathbf{D} + \mathbf{E}) = (\mathbf{A} \otimes \mathbf{D}) + (\mathbf{A} \otimes \mathbf{E}) + (\mathbf{B} \otimes \mathbf{D}) + (\mathbf{B} \otimes \mathbf{E})$; (v) $\text{rank}(\mathbf{A} \otimes \mathbf{D}) = \text{rank}(\mathbf{A})\text{rank}(\mathbf{D})$; and (vi) $\text{vec}(\mathbf{DFG}) = (\mathbf{G}' \otimes \mathbf{D})\text{vec}(\mathbf{F})$. Suppose that $m = n$ and $q = r$, and A and D are invertible. Then, (vii.) $(\mathbf{A} \otimes \mathbf{D})^{-1} = \mathbf{A}^{-1} \otimes \mathbf{D}^{-1}$ and (viii.) $|\mathbf{A} \otimes \mathbf{D}| = |\mathbf{A}|^q |\mathbf{D}|^m$.

Finally, we provide some vector and matrix differentiations, and useful references can be found in Kollo and Rosen (2005), Seber (2008), and Magnus and Neudecker (2019). Let y be a scalar, and let $Y = (y_1, \cdots, y_p)'$, $X = (x_1, \cdots, x_n)'$, $\mathbf{c} = (c_1, \cdots, c_n)'$ and $\mathbf{d} = (d_1, \cdots, d_q)'$ be $p \times 1$, $n \times 1$, $n \times 1$ and $q \times 1$ vectors, respectively. In addition, let $\mathbf{Y} = (y_{ij})$, $\mathbf{A} = (a_{kl})$, $\mathbf{X} = (x_{st})$ and $\mathbf{B} = (b_{uv})$ be $N \times p$, $m \times n$, $n \times q$, and $q \times r$ matrices, respectively, where $i = 1, \cdots, N$, $j = 1, \cdots, p$, $k = 1, \cdots, m$, $l = 1, \cdots, n$, $s = 1, \cdots, n$, $t = 1, \cdots, q$, $u = 1, \cdots, q$, and $v = 1, \cdots, r$. Define the derivative of **Y** with respect to $\mathbf{X}'$, the derivative of Y with respect to X', the derivative of y with respect to X', and the derivative of y with respect to $\mathbf{X}'$ as follows: $\frac{\partial \mathbf{Y}}{\partial \mathbf{X}'} = \partial vec(\mathbf{Y}) / \partial vec(\mathbf{X})'$ with dimension $Np \times nq$; $\frac{\partial Y}{\partial X'} = (\partial y_i/\partial x_j)$ with dimension $p \times n$, $\frac{\partial y}{\partial X'} = (\partial y/\partial x_j)$ with dimension $1 \times n$, and $\frac{\partial y}{\partial \mathbf{X}'} = (\partial y/\partial x_{ts})$ with dimension $q \times n$. Note that the above definitions are sometimes denoted by $\frac{\partial \mathbf{Y}}{\partial \mathbf{X}}$, $\frac{\partial Y}{\partial X}$, $\frac{\partial y}{\partial X}$ and $\frac{\partial y}{\partial \mathbf{X}}$.

By the above definitions, we have (i) $\frac{\partial(\mathbf{c}'X)}{\partial X'} = \mathbf{c}'$; (ii) $\frac{\partial(\mathbf{A}X)}{\partial X'} = \mathbf{A}$; (iii) if $\mathbf{A}$ is symmetric with $m = n$, then $\frac{\partial(X'\mathbf{A}X)}{\partial X'} = 2X'\mathbf{A}$ and $\frac{\partial^2(X'\mathbf{A}X)}{\partial X \partial X'} = 2\mathbf{A}$; (iv) if $\mathbf{A}$ is symmetric, $\mathbf{X}$ is invertible, and $m = n = q$, then $\frac{\partial tr(\mathbf{AX})}{\partial \mathbf{X}'} = \mathbf{A}$, $\frac{\partial tr(\mathbf{X}'\mathbf{AX})}{\partial \mathbf{X}'} = 2\mathbf{X}'\mathbf{A}$, $\frac{\partial tr(\mathbf{X}^{-1}\mathbf{A})}{\partial \mathbf{X}'} = -\mathbf{X}^{-1}\mathbf{A}\mathbf{X}^{-1}$, $\frac{\partial |\mathbf{X}|}{\partial \mathbf{X}'} = |\mathbf{X}|\mathbf{X}^{-1}$, $\frac{\partial \log |\mathbf{X}|}{\partial \mathbf{X}'} = \mathbf{X}^{-1}$, and $\frac{\partial |\mathbf{X}'\mathbf{AX}|}{\partial \mathbf{X}'} = 2|\mathbf{X}'\mathbf{AX}|\mathbf{X}^{-1}$; (v) $\frac{\partial(\mathbf{c}'\mathbf{Xd})}{\partial \mathbf{X}'} = \mathbf{dc}'$; and (vi) $\frac{\partial(\mathbf{AXB})}{\partial \mathbf{X}'} = \mathbf{B}' \otimes \mathbf{A}$.

1.6.4 Decompositions

In this subsection, we consider three decompositions: singular value decomposition, spectral decomposition, and Cholesky decomposition. The singular value decomposition (SVD) of an $m \times n$ matrix $\mathbf{A}$ is $\mathbf{A} = \tilde{\mathbf{U}}\tilde{\mathbf{\Lambda}}\tilde{\mathbf{V}}'$, where $\tilde{\mathbf{U}}$ and $\tilde{\mathbf{V}}$ are $m \times m$ and $n \times n$ orthogonal matrices, respectively, and $\tilde{\mathbf{\Lambda}} = (\tilde{\lambda}_{ij})$ is an $m \times n$ matrix such that $\tilde{\lambda}_{ii} > 0$ for $i = 1, \cdots, r$ and 0 otherwise, where r is the rank of matrix $\mathbf{A}$. Let $r \leq \min(m, n)$. Then, an alternative definition of SVD is $\mathbf{A} = \mathbf{U}\mathbf{\Lambda}\mathbf{V}'$, where $\mathbf{U}$ and $\mathbf{V}$ are $m \times r$ and $n \times r$ matrices, respectively, $\mathbf{U}'\mathbf{U} = \mathbf{I}_r$, $\mathbf{V}'\mathbf{V} = \mathbf{I}_r$, and $\mathbf{\Lambda} = (\lambda_{ii})$ is an $r \times r$ diagonal matrix with positive diagonal elements.

We next introduce spectral decomposition. Define eigenvalue (or characteristic) and eigenvector (or characteristic vector) as follows. Let $\mathbf{A}$ be a square matrix with dimension m. Then, any value γ satisfying $|\mathbf{A} - \gamma \mathbf{I}| = 0$ is called the eigenvalue of $\mathbf{A}$. For the given eigenvalue γ, any $m \times 1$ vector $\mathbf{u} \neq \mathbf{0}$ satisfying $\mathbf{Au} = \gamma \mathbf{u}$ is called the eigenvector of $\mathbf{A}$ associated with γ. Assume that $\mathbf{A}$ is also a symmetric matrix. Then the spectral decomposition of $\mathbf{A}$ is defined as $\mathbf{A} = \mathbf{U}\mathbf{\Gamma}\mathbf{U}'$, where $\mathbf{\Gamma} = \text{diag}(\gamma_1, \cdots, \gamma_m)$, and γ_is are the eigenvalues of $\mathbf{A}$. We say that $\mathbf{A}$ is a positive definite matrix iff (i.e., if and only if) all $\gamma_i > 0$, and $\mathbf{A}$ is a positive semidefinite matrix iff all $\gamma_i \geq 0$. For the sake of convenience, we sometimes denote the smallest and largest eigenvalues of A by $\gamma_{min}(\mathbf{A})$ and $\gamma_{max}(\mathbf{A})$, respectively. It is worth noting that the spectral decomposition is a special case of the singular value decomposition.

Before presenting the Cholesky decomposition, we define a generic matrix $\mathbf{G}$ as an upper triangular matrix for which all of the elements below the diagonal of $\mathbf{G}$ are zeros. In contrast, $\mathbf{G}$ is a lower triangular matrix if all the elements above its main diagonal are zeros. Then the Cholesky decomposition of a positive semidefinite matrix $\mathbf{A}$ is defined by $\mathbf{A} = \mathbf{G}\mathbf{G}'$, where $\mathbf{G}$ is a lower (or upper) triangular matrix. If $\mathbf{A}$ is a positive definite matrix, then $\mathbf{G}$ is unique.

To save space, we only present the matrix information that is frequently used in this book. More matrix notation, operations, and decompositions can be found in the books of Kollo and Rosen (2005), Seber (2008), and Schott (2017).

1.7 Orders, Asymptotic Properties, and Inequalities

This section introduces three subjects: non-stochastic and stochastic orders, asymptotic properties, and inequalities. Some useful references are Rao (1965), Bishop et al. (1975), Lehmann (1998), Chung (2001), and White (2001).

1.7.1 Non-stochastic Orders (Orders of Big O and Small o)

Suppose that a_n, b_n, c_n, and d_n are four sequences of real numbers. Denote $a_n = O(b_n)$ if $|a_n/b_n|$ is bounded for all sufficiently large n. That is, there exists a finite real number $K > 0$ and an integer $m > 0$ such that $|a_n| \leq K|b_n|$ for all $n > m$. For example, $a_n = O(1)$ means that $|a_n| \leq K$ for all $n > m$. The above definition indicates that the sequence a_n has approximately the same order of magnitude as the sequence b_n. Specifically, a_n and b_n have the same order (i.e., $a_n \asymp b_n$) if $|a_n/b_n|$ is bounded away from both 0 and ∞. In other words, $a_n \asymp b_n$ iff $a_n = O(b_n)$ and $b_n = O(a_n)$.

In contrast to big O, we define small o by denoting $c_n = o(d_n)$ if $|c_n/d_n|$ converges to zero. That is, for any $\varepsilon > 0$, there exists an integer m such that $|c_n| < \varepsilon|d_n|$ for any $n > m$ (i.e., $c_n/d_n \to 0$). This indicates that c_n is of a smaller order than d_n. For example, $c_n = o(1)$ means that $c_n \to 0$. In addition, if $c_n = n^\gamma$ and $d_n = n^\delta$, then $c_n = o(d_n)$ provided $\gamma < \delta$. It is obvious that small-o implies big-O.

Suppose that $\mathbf{a}_n$ and $\mathbf{c}_n$ are two vectors. Then $\mathbf{a}_n = O(b_n)$ and $\mathbf{c}_n = o(d_n)$ mean that $||\mathbf{a}_n||_2 = O(b_n)$ and $||\mathbf{c}_n||_2 = o(d_n)$, respectively. In addition, let $\mathbf{A}_n$ and $\mathbf{C}_n$ be two matrices. Then $\mathbf{A}_n = O(b_n)$ and $\mathbf{C}_n = o(d_n)$ mean that $||\mathbf{A}_n||_2 = O(b_n)$ and $||\mathbf{C}_n||_2 = o(d_n)$, respectively, where $\|\cdot\|_2$ denotes the matrix norm (i.e., the spectral norm).

A few operations of big O and small o are given below. (i) $O(a_n)O(b_n) = O(a_nb_n)$; (ii) $k_1O(a_n) = O(a_n)$, where k_1 is a non-zero constant; (iii) $O(a_n) + O(b_n) = O(\max\{|a_n|, |b_n|\})$; (iv) $o(c_n)o(d_n) = o(c_nd_n)$; (v) $k_2o(c_n) = o(c_n)$, where k_2 is a constant; (vi) $o(c_n) + o(d_n) = o(\max\{|c_n|, |d_n|\})$; (vii) $O(a_n)o(c_n) = o(a_nc_n)$; (viii) $O(a_n) + o(c_n) = O(\max\{|a_n|, |c_n|\})$; (ix) $o(O(a_n)) = o(a_n)$; and (x) $O(o(c_n)) = o(c_n)$. Some illustrations are as follows. Let $a_n = O(n^\alpha)$, $b_n = O(n^\beta)$, $c_n = o(n^\gamma)$, and $d_n = O(n^\delta)$. Based on the above operation rules, we have that (i) $a_nb_n = O(n^{\alpha+\beta})$; (ii) $c_nd_n = o(n^{\gamma+\delta})$; (iii) $a_n + b_n = O(n^\kappa)$, where $\kappa = \max(\alpha, \beta)$; (iv) $c_n + d_n = o(n^\lambda)$, where $\lambda = \max(\gamma, \delta)$; (v) $a_n = O(n^{\alpha+\nu})$ for any $\nu > 0$; (vi) $a_n = o(n^{\alpha+\xi})$ for any $\xi > 0$; and (vii) $c_n = O(n^\gamma)$. Note that if every element of a finite-dimensional vector or matrix is of order $O(n^\alpha)$ or $o(n^\gamma)$, then that vector or matrix is of order $O(n^\alpha)$ or $o(n^\gamma)$.

1.7.2 Stochastic Orders (Orders of Big O_p and Small o_p)

Let $\{y_n\} = \{y_n : n = 1, 2, \cdots\}$ be a sequence of random variables, where $y_n \in \mathbb{R}^1$. Define $y_n = O_p(1)$ if, for every $\varepsilon > 0$, then there exists $M > 0$ such that $P\{|y_n| > M\} < \varepsilon$. Accordingly, $y_n = O_p(a_n)$ if $y_n/a_n = O_p(1)$. For example, if $\{y_n\}$ is a sequence of random variables with mean μ_n and variance σ_n^2, then $y_n - \mu_n = O_p(\sigma_n)$. Suppose that $\{y_n\}$ has a distribution function F_n and y has a distribution function F. Then $\{y_n\}$ converges in distribution to y if $\lim_{n\to\infty} F_n(y^*) = \lim_{n\to\infty} P(y_n \leq y^*) = F(y^*)$ for all continuity points y^* of F. For the sake of simplicity, it can be denoted by $y_n \xrightarrow{d} y$. Note that if $\{y_n\}$ converges in distribution to a random variable y, then $y_n = O_p(1)$.

In contrast to big O_p, define $z_n = o_p(1)$ if, for every $\varepsilon > 0$, then $\lim_{n\to\infty} P\{|z_n| > \varepsilon\} = 0$ (i.e., $\lim_{n\to\infty} P\{|z_n| \leq \varepsilon\} = 1$ or z_n converges in probability to 0). As a result, $z_n = o_p(c_n)$ if $z_n/c_n = o_p(1)$. For example, if σ_n converges in probability to σ, then $\sigma_n = \sigma + o_p(1)$ (i.e., $\sigma_n/\sigma = 1 + o_p(1)$ if $\sigma \neq 0$). Note that z_n converges in probability to z can be denoted by $z_n \xrightarrow{p} z$, e.g., $\sigma_n \xrightarrow{p} \sigma$.

Suppose that $Y_n \in \mathbb{R}^p$ and $Z_n \in \mathbb{R}^p$ are two sequences of random vectors, where $p > 1$. In addition, let a_n and d_n be defined as above. Then $Y_n = O_p(a_n)$ and $Z_n = o_p(d_n)$ mean that $||Y_n||_2 = O_p(b_n)$ and $||Z_n||_2 = o_p(d_n)$, respectively. In addition, let $\mathbf{Y}_n$ and $\mathbf{Z}_n$ be two sequence of random matrices. Then $\mathbf{Y}_n = O_p(b_n)$ and $\mathbf{Z}_n = o_p(d_n)$ mean that the spectral norms of $\mathbf{Y}_n$ and $\mathbf{Z}_n$ satisfy that $||\mathbf{Y}_n||_2 = O_p(b_n)$ and $||\mathbf{Z}_n||_2 = o_p(d_n)$, respectively. Finally, the operations of big O_p and small o_p are the same as those of big O and small o as described above.

1.7.3 Asymptotic Properties

1.7.3.1 Law of Large Numbers (LLN)

(i) Weak Law of Large Numbers (WLLN):

Let $\{y_i\}$ be a sequence of independent and identically distributed (i.i.d.) random variables with mean $\mu < \infty$ for $i = 1, \cdots, n$. Then the sample mean $\bar{y} = n^{-1}\sum_{i=1}^n y_i$ converges to μ in probability, and it is denoted by $\bar{y} \xrightarrow{p} \mu$. That is, $\lim_{n\to\infty} P\{|\bar{y} - \mu| > \varepsilon\} = 0$ for every $\varepsilon > 0$.

(ii) Strong Law of Large Numbers (SLLN):

Let $\{y_i\}$ be a sequence of i.i.d. random variables with mean μ for $i = 1, \cdots, n$, and $E(|y_i|) < \infty$. Then the sample mean $\bar{y} = n^{-1}\sum_{i=1}^n y_i$ converges almost surely to μ, and it is denoted by $\bar{y} \xrightarrow{a.s.} \mu$. That is, $P\{\lim_{n\to\infty} \bar{y} = \mu\} = 1$.

1.7.3.2 Consistency

A parameter estimator $\widehat{\theta}_n$ is consistent if it converges in probability to the true parameter θ as the sample size increases. That is, $\lim_{n\to\infty} P\{|\widehat{\theta}_n - \theta| > \varepsilon\} = 0$ for every $\varepsilon > 0$ (i.e., $\widehat{\theta}_n = \theta + o_p(1)$ or $\widehat{\theta}_n \xrightarrow{p} \theta$). This result is also applicable in a vector setting with fixed dimension.

1.7.3.3 Slutsky's Theorem

Suppose $\widehat{\theta}_n \xrightarrow{d} \theta$ and $a_n \xrightarrow{p} a$ and $b_n \xrightarrow{p} b$, respectively, where a and b are constants. Then $a_n + b_n\widehat{\theta}_n \xrightarrow{d} a + b\theta$. This theorem is also applicable in a vector setting with fixed dimension.

1.7.3.4 Central Limit Theorem (CLT)

(i) Univariate case. Let $\{y_i\}$ be a sequence of i.i.d. random variables with mean $\mu < \infty$ and variance $\sigma^2 < \infty$ for $i = 1, \cdots, n$. Then $(\bar{y} - E(\bar{y}))/\sqrt{\text{var}(\bar{y})} = (\bar{y} - \mu)/(\sigma/\sqrt{n}) \xrightarrow{d} N(0, 1)$, where $N(0, 1)$ represents the standard normal with mean 0 and variance 1.

(ii) Multivariate case. Let $Y_i = (y_{i1}, \cdots, y_{ip})'$ be a sequence of i.i.d. random vectors with mean $\boldsymbol{\mu} = (\mu_1, \cdots, \mu_p)'$ and finite covariance matrix $\boldsymbol{\Sigma}$ for $i = 1, \cdots, n$, where $\boldsymbol{\Sigma}$ is positive definite with dimension $p \times p$. In addition, denote the sample mean by $\bar{Y} = (\bar{y}_1, \cdots, \bar{y}_p)'$, where $\bar{y}_j = n^{-1}\sum_{i=1}^n y_{ij}$ for $j = 1, \cdots, p$. Then $\sqrt{n}(\bar{Y} - \boldsymbol{\mu}) \xrightarrow{d} N(\mathbf{0}, \boldsymbol{\Sigma})$, where $\mathbf{0}$ is a $p \times 1$ vector of zeros.

1.7.3.5 Lindeberg Central Limit Theorem (Lindeberg-CLT)

(i.a) Univariate case with the single array:
Let $\{y_i : i = 1, \cdots, n\}$ be a collection of independent random variables with $E(y_i) = \mu_i < \infty$ and $0 < \text{var}(y_i) = \sigma_i^2 < \infty$ for $i = 1, \cdots, n$. Define $B_n^2 = \sum_{i=1}^n \sigma_i^2$. In addition, assume that the following version of Lindeberg's condition holds. For every $\varepsilon > 0$,

$$\frac{1}{B_n^2}\sum_{i=1}^n E\{(y_i - \mu_i)^2 \mathbf{1}_{\{|y_i - \mu_i| > \varepsilon B_n\}}\} \to 0$$

as $n \to \infty$, where $\mathbf{1}(\cdot)$ is the indicator function. Then $B_n^{-1}\sum_{i=1}^n (y_i - \mu_i) \xrightarrow{d} N(0, 1)$.

(i.b) Univariate case with the double array:
Let $\{y_{ni} : i = 1, \cdots, k_n\}$ be a collection of independent random variables with $E(y_{ni}) = \mu_{ni} < \infty$ and $0 < \text{var}(y_{ni}) = \sigma_{ni}^2 < \infty$ for $i = 1, \cdots, k_n$, where $k_n \to \infty$ as $n \to \infty$. Note that the n in the first subscript refers to the n-th row. Thus, if $k_n = n$, then it becomes a triangular array; see, e.g., Chung (2001).

Define $B_n^2 = \sum_{i=1}^{k_n} \sigma_{ni}^2$. In addition, assume that the following version of Lindeberg's condition holds. For every $\varepsilon > 0$,

$$\frac{1}{B_n^2} \sum_{i=1}^{k_n} E\{(y_{ni} - \mu_{ni})^2 \mathbf{1}_{\{|y_{ni} - \mu_{ni}| > \varepsilon B_n\}}\} \to 0$$

as $n \to \infty$. Then $B_n^{-1} \sum_{i=1}^{k_n} (y_{ni} - \mu_{ni}) \xrightarrow{d} N(0, 1)$. Note that if $k_n = n$ and $y_{ni} = y_i$, then it reduces to the single array case.

(ii) Multivariate case with the triangular array:
Let $\{Y_{ni} = (y_{ni1}, \cdots, y_{nip})' : i = 1, \cdots, n\}$ be a collection of independent random vectors with mean $E(Y_{ni}) = \boldsymbol{\mu}_{ni} = (\mu_{ni1}, \cdots, \mu_{nip})' < \infty$ and finite covariance matrix $\boldsymbol{\Sigma}_{ni}$ for $i = 1, \cdots, n$, where $\boldsymbol{\Sigma}_{ni}$ is positive definite with dimension $p \times p$ $(p < \infty)$. In addition, let $\bar{Y}_n = n^{-1} \sum_{i=1}^{n} Y_{ni}$, $\bar{\boldsymbol{\mu}}_n = n^{-1} \sum_{i=1}^{n} \boldsymbol{\mu}_{ni}$, and $\bar{\boldsymbol{\Sigma}}_n = n^{-1} \sum_{i=1}^{n} \boldsymbol{\Sigma}_{ni} \to \boldsymbol{\Sigma}$, where $\boldsymbol{\Sigma}$ is positive definite. Assume that, for every $\varepsilon > 0$,

$$\frac{1}{n} \sum_{i=1}^{n} E\{||(Y_{ni} - \boldsymbol{\mu}_{ni})|_2^2 \mathbf{1}_{\{||Y_{ni} - \boldsymbol{\mu}_{ni}||_2 > \varepsilon\sqrt{n}\}}\} \to 0$$

as $n \to \infty$. Then $\sqrt{n}\boldsymbol{\Sigma}^{-1/2}(\bar{Y}_n - \bar{\boldsymbol{\mu}}_n) \xrightarrow{d} N(\mathbf{0}, \boldsymbol{I}_p)$.

1.7.3.6 Martingale Central Limit Theorem (Martingale-CLT, Hall and Heyde 1980)

Let $\{s_{ni}, 1 \le i \le k_n\}$ be a zero-mean and square-integrable martingale array with respect to the σ-field $\mathcal{F}_{ni}$. That is, $E(s_{ni}) = 0$, $\sup_{n,i} E(s_{ni}^2) < \infty$, $E(s_{ni}|\mathcal{F}_{mi}) = s_{mi}$ for any $m \le n$. In addition, assume that $\mathcal{F}_{ni} \subseteq \mathcal{F}_{n+1,i}$ for any $1 \le i \le k_n$ and $n \ge 1$, and k_n tends to infinity as $n \to \infty$. Define the martingale difference $x_{ni} = s_{ni} - s_{n,i-1}$ and let η^2 be an almost surely (a.s.) finite random variable and $\eta > 0$ almost surely. Suppose that

(i) for any $\epsilon > 0$, $\sum_{i=1}^{k_n} E\{x_{ni}^2 \mathbf{1}_{\{|x_{ni}| > \epsilon\}} | \mathcal{F}_{n,i-1}\} \xrightarrow{p} 0$; and
(ii) $v_{nk_n}^2 = \sum_{i=1}^{k_n} E(x_{ni}^2 | \mathcal{F}_{n,i-1}) \xrightarrow{p} \eta^2$.

Then, we have $s_{nk_n}/\eta \xrightarrow{d} N(0, 1)$.

1.7.3.7 Delta Method

(i) Univariate case. Suppose that $\sqrt{n}(\widehat{\theta_n} - \theta) \xrightarrow{d} N(0, \sigma^2)$, where $\theta_n \in \mathbb{R}^1$. Then $\sqrt{n}(g(\widehat{\theta_n}) - g(\theta)) \xrightarrow{d} N(0, \sigma^2\{g'(\theta)\}^2)$, where g is a continuously differentiable function and $g'(\theta) = \partial g/\partial\theta \neq 0$.

(ii) Multivariate case. Suppose that $\sqrt{n}(\widehat{\boldsymbol{\theta}}_n - \boldsymbol{\theta}) \xrightarrow{d} N(0, \boldsymbol{\Sigma}(\boldsymbol{\theta}))$, where $\widehat{\boldsymbol{\theta}}_n = (\widehat{\theta_1}, \cdots, \widehat{\theta_p})'$, $\boldsymbol{\theta} = (\theta_1, \cdots, \theta_p)'$, and $\boldsymbol{\Sigma}(\boldsymbol{\theta})$ is the $p \times p$ asymptotic covariance matrix of $\widehat{\boldsymbol{\theta}}_n$, and it is positive definite. In addition, assume that $\mathbf{g}$ is a continuously differential function, and $\mathbf{g} : \mathbb{R}^p \rightarrow \mathbb{R}^q$ such that $\mathbf{g}(\boldsymbol{\theta}) = (g_1(\boldsymbol{\theta}), \cdots, g_q(\boldsymbol{\theta}))'$. Then $\sqrt{n}(\mathbf{g}(\widehat{\boldsymbol{\theta}}_n) - \mathbf{g}(\boldsymbol{\theta})) \xrightarrow{d} N(0, \{\partial\mathbf{g}/\partial\boldsymbol{\theta}\}\boldsymbol{\Sigma}(\boldsymbol{\theta})\{\partial\mathbf{g}/\partial\boldsymbol{\theta}\}')$, where $\partial\mathbf{g}/\partial\boldsymbol{\theta} \neq \mathbf{0}$ is a $q \times p$ matrix whose (i, j)-th element is $\partial g_i(\boldsymbol{\theta})/\partial\theta_j$ for $i = 1, \cdots, q$ and $j = 1, \cdots, p$.

1.7.4 Inequalities

1.7.4.1 Chebyshev's Inequality

Let y be a random variable with mean μ and variance $0 < \sigma^2 < \infty$. Then, for any $k > 0$, $P(|y - \mu| \geq k) \leq \frac{\sigma^2}{k^2}$.

1.7.4.2 Markov's Inequality

Let y be a non-negative random variable. Then, for any $k > 0$, $P(y \geq k) \leq \frac{E(y)}{k}$.

1.7.4.3 Jensen's Inequality

Let $f : \mathbb{R} \rightarrow \mathbb{R}$ be a convex function on an interval $U \in \mathbb{R}$. Suppose that $P(y \in U) = 1$, and $E(f(y))$ and $E(y)$ exist. Then $E(f(y)) \geq f(E(y))$.

1.7.4.4 Hölder's Inequality

Let y and z be two random variables. Suppose that $r > 1, s > 1$ and $1/r + 1/s = 1$. If $E|y|^r < \infty$ and $E|z|^s < \infty$, then $E(|yz|) \leq (E|y|^r)^{1/r}(E|z|^s)^{1/s}$. Note that $r = s = 2$ leads to Cauchy–Schwarz inequality.

1.7.4.5 Minkowski's Inequality

Let y and z be two random variables. Suppose that $r \geq 1$. If $E|y|^r < \infty$ and $E|z|^r < \infty$, then $\{E(|y+z|^r)\}^{1/r} \leq (E|y|^r)^{1/r} + (E|z|^r)^{1/r}$.

1.7.4.6 Bernstein's Inequality

Let $y_1, \cdots, y_n$ be a sequence of independent random variables such that $E(y_i) = 0$, $E(y_i^2) = b_i$ and $E(|y_i|^d) \leq b_i h^{d-2} d!/2$, where $d > 2$ and h is a constant independent of i and d. Define $s_n = y_1 + \cdots + y_n$ and $m_n = \sum_{i=1}^n b_i$, then for any $k > 0$,

$$P(|s_n| > k) \leq 2\exp\Big\{ - k^2/(2m_n + 2kh)\Big\}.$$

1.7.4.7 Hoeffding's Inequality (Hoeffding 1963)

Let $y_1, \cdots, y_n$ be a sequence of independent random variables and satisfy $a_i \leq y_i \leq b_i$ with probability 1 for any i, where $-\infty < a_i < b_i < \infty$. Define $s_n = y_1 + \cdots + y_n$, then for any $k > 0$,

$$P\Big\{|s_n - E(s_n)| > k\Big\} \leq 2\exp\Big\{ - 2k^2/\sum_{i=1}^{n}(b_i - a_i)^2\Big\}.$$

1.7.4.8 Hanson-Wright's Inequality (Hanson and Wright 1971)

Let $Y = (y_1, \cdots, y_n)'$ be a random vector with independent components y_is, which satisfy $E(y_i) = 0$ and $\sup_{d>0} d^{-1/2}\{E(|y_i|^d)\}^{1/d} \leq K$ for a finite constant K. Let $\mathbf{A}$ be an $n \times n$ matrix. Then, for any $k > 0$, there exists a finite positive constant c, independent of k, such that

$$P\Big\{|Y'\mathbf{A}Y - E(Y'\mathbf{A}Y)| > k\Big\} \leq 2\exp\Big\{ - c\min\left(\frac{k^2}{K^4\|\mathbf{A}\|_F^2}, \frac{k}{K^2\|\mathbf{A}\|_2}\right)\Big\},$$

where $\|\mathbf{A}\|_2$ is calculated by matrix norm and $\|\mathbf{A}\|_F^2$ is calculated by entry-wise norm.

Chapter 2
Covariance Matrices, Precision (Concentration) Matrices, Estimation, and Tests

This chapter introduces covariance matrices and precision (concentration) matrices. We then demonstrate the application of covariance matrices for obtaining the mean, variance, and covariance of quadratic forms. In addition, four types of covariance estimation (moment estimation, maximum likelihood estimation, penalized estimation, and robust estimation) are presented. Afterward, covariance estimations via eigenvector thresholding and eigenvalue shrinkage are explored. Subsequently, precision matrix estimation, time-dependent covariance matrices, positive definiteness, and the tests of covariance matrices are studied. Finally, empirical examples of the applications of covariance matrices are briefly discussed.

2.1 Covariance Matrices

2.1.1 Random Vectors

A vector $Y = (y_1, \cdots, y_p)'$ of size p is a random vector if the elements of Y are random variables whose value is based on a specific distribution (or probability) function. For the sake of convenience, we use the term "distribution function" in this book for both continuous and discrete random variables. The p continuous (or discrete) random variables $y_1, \cdots, y_p$ are defined as mutually independent if their joint probability density function (or joint probability distribution function) is the same as the product of their marginal density functions (or marginal probability distribution functions). Suppose that $\mathbf{y}_1, \cdots, \mathbf{y}_n$ are i.i.d. realizations of random vector Y. For the sake of convenience, we often use the symbols $Y_1, \cdots, Y_n$ in this book, rather than $\mathbf{y}_1, \cdots, \mathbf{y}_n$, with a slight abuse of notation.

W. Lan, C.-L. Tsai, *Covariance Analysis and Beyond*,
https://doi.org/10.1007/978-3-032-08796-6_2

2.1.2 Covariance and Precision (Concentration) Matrices

For a given distribution function, the expectation (or mean) vector of Y is denoted by $E(Y) = (E(y_1), \cdots, E(y_p))' = \boldsymbol{\mu} = (\mu_1, \cdots, \mu_p)'$. In addition, the covariance matrix of Y is $\text{Cov}(Y) = \boldsymbol{\Sigma} = (\sigma_{ij}) = E\{(Y - \boldsymbol{\mu})(Y - \boldsymbol{\mu})'\}$ of dimension $p \times p$, where $\sigma_{ij} = E\{(y_i - \mu_i)(y_j - \mu_j)\}$ for $i = 1, \cdots, p$ and $j = 1, \cdots, p$. The resulting correlation matrix is $\boldsymbol{\rho} = (\rho_{ij}) = (\sigma_{ij}/\sqrt{\sigma_{ii}\sigma_{jj}})$, where $\rho_{ij} = \text{corr}(y_i, y_j)$, i.e., the correlation between y_i and y_j, and it is unit-free. The covariance, σ_{ij} (or correlation ρ_{ij}), measures the linear relationship between two random variables, y_i and y_j. If y_i and y_j are independent, then $\sigma_{ij} = \rho_{ij} = 0$.

Before defining the precision matrix, we introduce some related terms. Let $Y_{\bullet\backslash\{i,j\}}$ be the random vector excluding the components y_i and y_j of Y. Then the conditional covariance of y_i and y_j for the given $Y_{\bullet\backslash\{i,j\}}$ is defined as $\sigma_{ij_\bullet} = \text{cov}(y_i, y_j|Y_{\bullet\backslash\{i,j\}}) = \text{cov}(y_i, y_j) - \text{Cov}(y_i, Y_{\bullet\backslash\{i,j\}})\{\text{Cov}(Y_{\bullet\backslash\{i,j\}})\}^{-1}\text{Cov}(Y_{\bullet\backslash\{i,j\}}, y_j)$ and it is also called the partial covariance. It can be shown that $\text{cov}(y_i, y_j|Y_{\bullet\backslash\{i,j\}}) = \text{cov}(e_{i\backslash\{i,j\}}, e_{j\backslash\{i,j\}})$, where $e_{i\backslash\{i,j\}}$ and $e_{j\backslash\{i,j\}}$ are the residuals from fitting the linear regressions of y_i versus $Y_{\bullet\backslash\{i,j\}}$ and y_j versus $Y_{\bullet\backslash\{i,j\}}$, respectively.

Analogously, the partial correlation $\rho_{ij_\bullet}$ is the correlation between $e_{i\backslash\{i,j\}}$ and $e_{j\backslash\{i,j\}}$, which is the conditional correlation between y_i and y_j given $Y_{\bullet\backslash\{i,j\}}$. If $\rho_{ij_\bullet} = 0$, then y_i and y_j are conditionally independent, provided that Y follows a multivariate normal distribution. We next define the precision matrix, which is the inverse of the covariance matrix. It can be shown that $\boldsymbol{\Sigma}^{-1} = (\Theta_{ij})$, where Θ_{ij} measures the linear relationship between y_i and y_j after removing the contribution from the rest of variables in Y. Accordingly, the partial correlation is $\rho_{ij_\bullet} = -\Theta_{ij}/\sqrt{\Theta_{ii}\Theta_{jj}}$. More discussion is given in Sect. 2.4.

2.1.3 Quadratic Form, Mean, and Variance

The covariance matrix often appears in the moments of the quadratic form that is defined below. Let $\mathbf{A} = (a_{ij})$ be a $p \times p$ symmetric matrix with elements a_{ij} for $i = 1, \cdots, p$ and $j = 1, \cdots, p$. In addition, let $Y = (y_1, \cdots, y_p)'$ be a $p \times 1$ random vector. Then the quadratic form in Y is defined as $Y'\mathbf{A}Y = \sum_{i=1}^{n}\sum_{j=1}^{n} a_{ij} y_i y_j$. If $Y'\mathbf{A}Y > 0$ for any $Y \neq 0$, we call $\mathbf{A}$ a positive definite matrix. It can be shown that $\mathbf{A}$ is a positive definite matrix if and only if (iff) the p eigenvalues of $\mathbf{A}$ are all positive.

Suppose that the random vector Y has finite fourth moments with mean $\boldsymbol{\mu}$ and covariance $\boldsymbol{\Sigma}$. Then $E(Y'\mathbf{A}Y) = \boldsymbol{\mu}'\mathbf{A}\boldsymbol{\mu} + tr(\mathbf{A}\boldsymbol{\Sigma})$. In addition, $\text{var}(Y'\mathbf{A}Y) = tr\{(\mathbf{A} \otimes \mathbf{A})E(YY' \otimes YY')\} - \{tr(\mathbf{A}\boldsymbol{\Sigma}) + \boldsymbol{\mu}'\mathbf{A}\boldsymbol{\mu}\}^2$, where $E(YY' \otimes YY')$ is the fourth moment of Y. In addition, let $\mathbf{B}$ be a $p \times p$ symmetric matrix. Then $\text{cov}(Y'\mathbf{A}Y, Y'\mathbf{B}Y) = tr\{(\mathbf{A} \otimes \mathbf{B})E(YY' \otimes YY')\} - \{tr(\mathbf{A}\boldsymbol{\Sigma}) + \boldsymbol{\mu}'\mathbf{A}\boldsymbol{\mu}\}\{tr(\mathbf{B}\boldsymbol{\Sigma}) + \boldsymbol{\mu}'\mathbf{B}\boldsymbol{\mu}\}$; see Seber (2008). Furthermore, assume that Y follows a multivariate normal distribution, $N(\boldsymbol{\mu}, \boldsymbol{\Sigma})$. We then have $\text{var}(Y'\mathbf{A}Y) = 2tr\{(\mathbf{A}\boldsymbol{\Sigma})^2\} + 4\boldsymbol{\mu}'\mathbf{A}\boldsymbol{\Sigma}\mathbf{A}\boldsymbol{\mu}$, and $\text{cov}(Y'\mathbf{A}Y, Y'\mathbf{B}Y) = 2tr\{(\mathbf{A}\boldsymbol{\Sigma})(\mathbf{B}\boldsymbol{\Sigma})\} - 4\boldsymbol{\mu}'\mathbf{A}\boldsymbol{\Sigma}\mathbf{B}\boldsymbol{\mu}$. In addition to the mean and variance of a general quadratic form, its tail inequality is presented in Sect. 1.7.4.8.

2.2 Covariance Matrix Estimation

2.2.1 Moment Estimation

Suppose that $Y_i = (y_{i1}, \cdots, y_{ip})'$ is i.i.d. random vector with mean $\boldsymbol{\mu}$ and covariance $\boldsymbol{\Sigma}$ for $i = 1, \cdots, n$. Then the first moment estimator is $\widehat{\boldsymbol{\mu}}_s = \bar{Y} = n^{-1}\sum_{i=1}^n Y_i$ and the second moment estimator about the mean is $\widehat{\boldsymbol{\Sigma}}_s = n^{-1}\sum_{i=1}^n (Y_i - \bar{Y})(Y_i - \bar{Y})' = n^{-1}(\mathbf{Y} - \mathbf{1}\bar{Y}')'(\mathbf{Y} - \mathbf{1}\bar{Y}')$, where $\mathbf{Y} = (Y_1, \cdots, Y_n)'$ is an $n \times p$ matrix and $\mathbf{1} = (1, \cdots, 1)'$ is an $n \times 1$ vector. We also call $\widehat{\boldsymbol{\mu}}_s$ and $\widehat{\boldsymbol{\Sigma}}_s$ the sample mean and sample covariance (or sample covariance matrix), respectively. It can be shown that $E(\widehat{\boldsymbol{\mu}}_s) = \boldsymbol{\mu}$ and $E(\widehat{\boldsymbol{\Sigma}}_s) = \frac{n-1}{n}\boldsymbol{\Sigma}$. Hence, $\widehat{\boldsymbol{\Sigma}}_u = \frac{n}{n-1}\widehat{\boldsymbol{\Sigma}}_s = (n-1)^{-1}\sum_{i=1}^n (Y_i - \bar{Y})(Y_i - \bar{Y})'$ is an unbiased estimator of $\boldsymbol{\Sigma}$. Suppose that Y_i has been standardized (i.e., $Z_i = (\widehat{\boldsymbol{\Sigma}}_s)^{-1/2}(Y_i - \bar{Y})$). Then the resulting covariance matrix is the sample correlation matrix $\mathbf{R} = n^{-1}\sum_{i=1}^n Z_i Z_i'$. Without causing confusion, we sometimes denote $\widehat{\boldsymbol{\mu}}_s$ and $\widehat{\boldsymbol{\Sigma}}_s$ by $\widehat{\boldsymbol{\mu}}$ and $\widehat{\boldsymbol{\Sigma}}$, respectively.

2.2.2 Maximum Likelihood Estimation

Suppose that a $p \times 1$ vector Y_i is i.i.d. $N(\boldsymbol{\mu}, \boldsymbol{\Sigma})$ for $i = 1, \cdots, n$. Then the log-likelihood function after removing the constant $-(np/2)\log(2\pi)$ is

$$\begin{aligned} L(\boldsymbol{\mu}, \boldsymbol{\Sigma}) = \log f(\mathbf{Y}|\boldsymbol{\mu}, \boldsymbol{\Sigma}) = & -\frac{n}{2}\log(|\boldsymbol{\Sigma}|) - \frac{1}{2}\sum_{i=1}^n (Y_i - \boldsymbol{\mu})'\boldsymbol{\Sigma}^{-1}(Y_i - \boldsymbol{\mu}) \\ = & -\frac{n}{2}\log(|\boldsymbol{\Sigma}|) - \frac{1}{2}tr\{(\mathbf{Y} - \mathbf{1}\boldsymbol{\mu}')\boldsymbol{\Sigma}^{-1}(\mathbf{Y} - \mathbf{1}\boldsymbol{\mu}')'\}. \end{aligned} \tag{2.1}$$

By maximizing $L(\boldsymbol{\mu}, \boldsymbol{\Sigma})$ with respect to $\boldsymbol{\mu}$ and $\boldsymbol{\Sigma}$ in (2.1), we obtain the maximum likelihood estimators (MLEs) $\widehat{\boldsymbol{\mu}}_m = \bar{Y}$ and $\widehat{\boldsymbol{\Sigma}}_m = n^{-1}\sum_{i=1}^n (Y_i - \bar{Y})(Y_i - \bar{Y})' = n^{-1}(\mathbf{Y} - \mathbf{1}\bar{Y}')'(\mathbf{Y} - \mathbf{1}\bar{Y}')$, which are identical to their corresponding moment estimators. When the dataset contains outliers, robust estimation techniques such as trimmed means or weighted covariance estimators may be more appropriate; see Olive (2017) for details.

2.2.3 Penalized and Thresholding Estimation

In linear (i.e., mean) regression models, Tibshirani (1996) proposed a seminal approach, least absolute shrinkage and selection operator (lasso), to shrink unnecessary regression coefficients to zero and improve the accuracy of predictions. Specifically, lasso is an L_1 penalized estimation method, and it enjoys some of the nice features of both subset selection and ridge regression. To better understand lasso, Tibshirani (1996) discussed the differences between lasso, Breiman's (1995) non-negative garotte estimator, and the ridge regression estimator, and he also studied variable selection. In the last two decades, lasso has been widely used in high-dimensional data (i.e., where the number of variables is larger than the number of observations), across various fields. Useful references can be found in Bühlmann and van de Geer (2011), Tibshirani (2011), and Hastie et al. (2015). Note that this subsection and the following sections mainly focus on high-dimensional data.

After lasso was introduced in 1996, various types of lasso have been proposed such as elastic net (Zou and Hastie 2005), fused lasso (Tibshirani et al. 2005), adaptive lasso (Zou 2006), and Dantzig selector (Candes and Tao 2007). In high-dimensional data, lasso may shrink some relevant variables and introduce bias; see, e.g., Hastie et al. (2015) and Freijeiro-González et al. (2022). This over-shrinking problem is induced by lasso's L_1 penalty. Hence, nonconvex penalties have been proposed to overcome such challenge; see, e.g., smoothly clipped absolute deviation (SCAD) (Fan and Li 2001) and minimax concave (MCP) (Zhang 2010). Note that the objective functions of the above methods include penalties. Thus, we call the resulting estimators penalized (or regularized) estimators. A good review of penalty functions as well as variable selection in high-dimensional data can be found in Fan and Lv (2010).

It is known that the covariance matrix estimator, $\widehat{\boldsymbol{\Sigma}}_s = n^{-1}\sum(Y_i - \bar{Y})(Y_i - \bar{Y})'$, discussed in Sect. 2.2.1, is consistent when $n \to \infty$ and p is fixed. However, $\widehat{\boldsymbol{\Sigma}}_s$ may not perform well as $p \to \infty$ (see Bai 1999; Johnstone 2001). The fundamental reason is that the number of unknown parameters, $p(p+1)/2$, is too large to be accurately estimated by a limited sample size. Hence, it is natural to assume most of the off-diagonal elements to be zero or nearly to zero. To this end, Bickel and Levina (2008a) and Rothman et al. (2009) proposed the following uniformity class of sparse matrices:

$$U_q(s_0(p), c) = \left\{\boldsymbol{\Sigma} : \sigma_{ii} \le c \text{ and} \sum_{j=1}^{p} |\sigma_{ij}|^q \le s_0(p) \text{ for all } i\right\}, \tag{2.2}$$

where $0 \le q < 1$, $c > 0$, and $s_0(p) > 0$. In addition, for $q = 0$,

$$U_0(s_0(p), c) = \left\{\boldsymbol{\Sigma} : \sigma_{ii} \le c \text{ and} \sum_{j=1}^{p} \mathbf{1}(\sigma_{ij} \ne 0) \le s_0(p) \text{ for all } i\right\}.$$

Hence, the smaller $s_0(p)$, the sparser $\boldsymbol{\Sigma}$. Later, Fan et al. (2016) adopted the above definition and considered the sparsity given below. That is,

$$m_p = \begin{cases} \max_{i \leq p} \sum_{j=1}^{p} \mathbf{1}(\sigma_{ij} \neq 0), & \text{if } q = 0; \\ \max_{i \leq p} \sum_{j=1}^{p} |\sigma_{ij}|^q, & \text{if } 0 < q < 1; \end{cases} \tag{2.3}$$

where $\mathbf{1}(\cdot)$ is the indicator function. Then the resulting $\boldsymbol{\Sigma}$ is a sparse matrix when m_p is either bounded or grows slowly as $p \to \infty$.

Under the assumption of sparsity, there are various thresholding estimators being proposed, some good references are Bickel and Levina (2008a), Karoui (2008), Rothman et al. (2009), Pourahmadi (2013), Liu et al. (2014), Fan et al. (2016), and Lam (2020). It is worth noting that Rothman et al. (2009) defined a generalized thresholding operator and applied it to each of the elements of $\boldsymbol{\Sigma}$. That is, for $i, j = 1, \cdots, p$, the shrinkage function, $s_\lambda(z) = s_\lambda(\sigma_{ij})$, satisfies the following three conditions: (i) $s_\lambda(z) \leq |z|$; (ii) $s_\lambda(z) = 0$ for $|z| \leq \lambda$, where $\lambda \geq 0$; and (iii) $|s_\lambda(z) - z| \leq \lambda$. These three conditions encompass shrinkage, thresholding, and bound. Accordingly, the generalized thresholding operator includes a number of commonly used shrinkage operators, such as the soft thresholding $s_\lambda(z) = \text{sign}(z)(|z| - \lambda)_+$ and the hard thresholding $s_\lambda(z) = z\mathbf{1}(|z| > \lambda)$. In addition, it consists of the following two known thresholding operators.

(i) The smoothly clipped absolute deviation (SCAD) thresholding operator (Fan and Li 2001; Fan et al. 2016):

$$s_\lambda(z) = \begin{cases} sign(z)(|z| - \lambda)_+, & |z| \leq 2\lambda; \\ \{(a-1)z - sign(z)a\lambda\}/(a-2), & 2\lambda < |z| \leq a\lambda; \\ z, & |z| > a\lambda; \end{cases} \tag{2.4}$$

where $a > 2$.

(ii) The minimax concave penalty (MCP) thresholding operator (Zhang 2010; Fan et al. 2016):

$$s_\lambda(z) = \begin{cases} \{(b/(b-1)\}sign(z)(|z| - \lambda)_+, & |z| \leq b\lambda; \\ z, & |z| > b\lambda; \end{cases} \tag{2.5}$$

where $b > 1$. Moreover, Cai and Liu (2011) employed a general thresholding function analogous to those used by Rothman et al. (2009) and proposed the adaptive thresholding operator, $s_{\lambda_{ij}}(\sigma_{ij})$, where λ_{ij} adapts to the variability of individual entries of the sample covariance matrix $\widehat{\boldsymbol{\Sigma}}_s$.

In addition to thresholding operators, banding (Bickel and Levina 2008b) and tapering (Furrer and Bengtsson 2007; Cai et al. 2010) methods were introduced to estimate $\boldsymbol{\Sigma}$. Specifically, the banding and tapering estimators are defined below:

$B_\gamma(\mathbf{\Sigma}) = [\sigma_{ij}\mathbf{1}(|i-j| \le \gamma)]$ and $T_W(\mathbf{\Sigma}) = \mathbf{\Sigma} * \mathbf{W} = (\sigma_{ij}W_{ij})$, where $0 \le \gamma \le p-1$, and $\mathbf{W} = (W_{ij})$ is a positive definite tapering matrix such as the trapezoidal matrix (see, e.g., Pourahmadi 2013).

In practice, $\mathbf{\Sigma}$ is unknown, and we can obtain the thresholding, banding, and tapering estimators of $\mathbf{\Sigma}$ by replacing the above σ_{ij} with its sample estimator $\widehat{\sigma}_{ij}$, where $\widehat{\sigma}_{ij}$ is the ij-th element of $\widehat{\mathbf{\Sigma}}_s$ for $i, j = 1, \cdots, p$. To assure the covariance estimator is consistent or to achieve some optimal convergence rate under the Gaussian or sub-Gaussian distribution of Y, one often assumes that $\log(p)/n = o(1)$ (or $(\log(p)/n) \cdot s_i \ll 1$, where s_i is the sparsity index of parameters) and the tuning parameter λ_n is proportional to $\sqrt{\log(p)/n}$; see, e.g., Bickel and Levina (2008a), Bühlmann and van de Geer (2011), and Pourahmadi (2013). For example, Bickel and Levina (2008a) has demonstrated the following consistency results.

Theorem 2.1 *Assume that Y follows the Gaussian distribution. In addition, assume that $log(p)/n = o(1)$. If $\lambda_n = m\sqrt{log(p)/n}$, then the hard thresholding estimator $\mathbf{G}_{\lambda_n}(\widehat{\mathbf{\Sigma}}_s) = (\widehat{\sigma}_{ij}\mathbf{1}(|\widehat{\sigma}_{ij}| \ge \lambda_n))$ satisfies*

$$||\mathbf{G}_{\lambda_n}(\widehat{\mathbf{\Sigma}}_s) - \mathbf{\Sigma}|| = O_p(s_0(p)(log(p)/n)^{(1-q)/2}),$$

uniformly on $U_q(s_0(p), c)$, for a sufficiently large number m, where $U_q(s_0(p), c)$ is defined in (2.2).

The above result is also valid under the sub-Gaussian assumption; see Bickel and Levina (2008a).

2.2.4 Robust Estimation

The above covariance matrix estimation methods heavily rely on Gaussian or sub-Gaussian assumptions. When the sub-Gaussian assumption is violated, robust estimation methods are more preferable. To take into account heavy-tailed data under elliptical distributions, Maronna (1976) and Tyler (1987) proposed Huber-type M-estimators and demonstrated their theoretical properties. Later, Chen et al. (2011) extended Tyler's (1987) method to a high-dimensional setting with $p > n$ using shrinkage regularization techniques. Additional useful references include Fang and Zhang (1990) and Hubert et al. (2008). Recently, Wang et al. (2023) integrated robust generalized M-estimators with a modified Cholesky decomposition technique to estimate high-dimensional precision matrices.

To broaden the scope of elliptical distributions by allowing non-symmetric heavy-tailed data, Lu et al. (2021) proposed the pair-elliptical distribution defined as follows: a vector Y is a pair-elliptical distribution if and only if any pair of random variables $(y_j, y_k)'$ of Y is elliptically distributed. It is obvious that the family of pair-elliptical distributions contains elliptical distributions. In addition, Lu et al.

(2021) studied the relationship between the pair-elliptical and some heavy tailed distributions, such as the non-paranormal, elliptical, and trans-elliptical. Moreover, they employed quantile-based statistics to estimate covariance matrices and showed their theoretical properties.

Without imposing the parametric assumption of elliptical distributions, Avella-Medina et al. (2018) proposed robust covariance matrix estimators (i.e., robust pilot estimators), such as a rank based estimator and a Huber-type M-estimator. They showed that, under a bounded fourth-moment assumption, their proposed estimators yield the same minimax convergence rate as those under sub-Gaussian assumptions. In addition, they proved the consistency of their robust estimators under the weaker condition of bounded $2+\epsilon$ moments with $\epsilon \in (0, 2)$. In addition, Ke et al. (2019) introduced the elementwise and spectrumwise truncation operators for the sample covariance matrix. Under the condition that the fourth moments are bounded, they established the convergence rate of their proposed covariance matrix estimators.

In addition to the above covariance estimation methods, factor models have attracted considerable attention for estimating high-dimensional covariance matrices; see, e.g., the review papers of Bartholomew (2005) and Fan et al. (2021). We will provide detailed discussion in Chap. 8. Since the covariance matrix can be decomposed by spectral decomposition, it is natural to impose constraints on eigenvectors and eigenvalues given below.

2.3 Eigenvector Thresholding and Eigenvalue Shrinkage

In multivariate data, principal component analysis (PCA) has played an important role for dimension reduction. This technique creates new variables (components) through linear combinations of original variables that not only successively maximize variance but also are uncorrelated with each other. To this end, one can employ spectral decomposition on the sample covariance matrix or the correlation matrix to obtain eigenvalues and their associated eigenvectors. In high-dimensional data scenarios, however, classical PCA faces significant limitations, primarily because the sample covariance matrix (a key input for classical PCA) becomes highly unreliable. Hence, Jolliffe et al. (2003) adapted Tibshirani's (1996) lasso approach and proposed the Simplified Component Technique–lasso (SCoTLASS) procedure to find sparse eigenvectors sequentially. Specifically, SCoTLASS yields the first thresholding eigenvector of the sample covariance matrix ($\widehat{\boldsymbol{\Sigma}}_s$) or the sample correlation matrix ($\mathbf{R}$) by solving the following optimization problem:

$$\max_{\mathbf{v}} \mathbf{v}' \widehat{\boldsymbol{\Sigma}}_s \mathbf{v} \quad subjects\ to\ ||\mathbf{v}||_2^2 \leq 1\ and\ ||\mathbf{v}||_1 \leq c, \tag{2.6}$$

where $\mathbf{v} \in \mathbb{R}^p$ and $c > 0$. Denote the resulting vector by $\mathbf{v}_1$. Then the subsequent thresholding eigenvectors can be obtained in the same manner except that the i-th eigenvector $\mathbf{v}_i$ is orthogonal to the previous eigenvectors, $\mathbf{v}_1, \cdots, \mathbf{v}_{i-1}$ for $i = 2, \cdots, p$.

Solving the above optimization problem is a challenging task since it requires maximizing a convex objective function. To tackle this issue, Witten et al. (2009) proposed an efficient procedure to find the sparse principal components (SPC) by solving the penalized matrix decomposition (PMD) model in their Algorithm 3 with a special setting given below; see also Pourahmadi (2013).

$$\max_{\mathbf{u},\mathbf{v}} \mathbf{u}'\mathbf{Y}\mathbf{v} \quad \text{subject to} \quad ||\mathbf{v}||_2^2 \le 1, \;\; ||\mathbf{v}||_1 \le c, \;\; \text{and} \;\; ||\mathbf{u}||_2^2 \le 1, \tag{2.7}$$

where $\mathbf{u}$ is an $n \times 1$ vector.

The implementation of Witten et al.'s (2009) procedure is as follows:

1. Initialize $\mathbf{v}$ so that $||\mathbf{v}||_2 = 1$.
2. Conduct the following iterative process until convergence:

   (i) $\mathbf{u} \leftarrow \frac{\mathbf{Y}\mathbf{v}}{||\mathbf{Y}\mathbf{v}||_2}$;

   (ii) $\mathbf{v} \leftarrow \frac{s_\lambda(\mathbf{Y}'\mathbf{u})}{||s_\lambda(\mathbf{Y}'\mathbf{u})||_2}$, where $s_\lambda(\cdot)$ is the soft thresholding.

   In addition, if $\mathbf{v}$ satisfies $||\mathbf{v}||_1 \le c$, then $\lambda = 0$; otherwise, set $||\mathbf{v}||_1 = c$ by choosing a positive constant λ.

   Note that the SPC algorithm does not impose the orthogonal constraint on the regularized eigenvectors.

In addition to studying thresholding eigenvectors, Ledoit and Wolf (2004) proposed a procedure to shrink eigenvalues. Specifically, they considered the following setting:

$$\widehat{\boldsymbol{\Sigma}} = a_1\mathbf{I} + a_2\widehat{\boldsymbol{\Sigma}}_s, \tag{2.8}$$

where a_1 and a_2 are chosen to minimize the risk function $E||\widehat{\boldsymbol{\Sigma}} - \boldsymbol{\Sigma}||_{F*}^2$, $||\cdot||_{F*} = p^{-1/2}||\cdot||_F$, $||\cdot||_F$ is the Frobenius norm, and $\mathbf{I}$ is a $p \times p$ matrix. Let $b_1 = tr(\boldsymbol{\Sigma})/p$, $b_2^2 = ||\boldsymbol{\Sigma} - b_1\mathbf{I}||_{F*}^2$, $b_3^2 = E||\widehat{\boldsymbol{\Sigma}}_s - \boldsymbol{\Sigma}||_{F*}^2$, and $b_4^2 = E||\widehat{\boldsymbol{\Sigma}}_s - b_1\mathbf{I}||_{F*}^2$, where $b_2^2 + b_3^2 = b_4^2$. Then, Ledoit and Wolf (2004) demonstrated that the optimal solution of (2.8) is $\widehat{\boldsymbol{\Sigma}} = (b_3^2/b_4^2 b_1)\mathbf{I} + (b_2^2/b_4^2)\widehat{\boldsymbol{\Sigma}}_s$. In practice, b_1, b_2, b_3, and b_4 are unknown. Hence, they further obtained the consistent estimator of $\boldsymbol{\Sigma}$ given below:

$$\tilde{\boldsymbol{\Sigma}} = (b_{3n}^2/b_{4n}^2 b_{1n})\mathbf{I} + (b_{2n}^2/b_{4n}^2)\widehat{\boldsymbol{\Sigma}}_s, \tag{2.9}$$

where $b_{1n} = tr(\widehat{\boldsymbol{\Sigma}}_s)/p$, $b_{2n}^2 = b_{4n}^2 - b_{3n}^2$, $b_{3n}^2 = \min\{b_{4n}^2, b_{5n}^2\}$, $b_{4n}^2 = ||\widehat{\boldsymbol{\Sigma}}_s - b_{1n}\mathbf{I}||_{F*}^2$, $b_{5n}^2 = \sum_{i=1}^n ||Y_{(i)}'Y_{(i)} - \widehat{\boldsymbol{\Sigma}}_s||_{F*}^2/n^2$, and $Y_{(i)}$ is the i-th row of $\mathbf{Y}$ for $i = 1, \cdots, n$. It is worth noting that (2.8) is designed to shrink eigenvalues rather than eigenvectors. A useful reference is Tsukuma and Kubokawa (2020).

Note that while eigenvalue shrinkage is critical in the context of high-dimensional covariance matrices, it is inherently related to the properties of eigenvalues. For the sake of simplicity, we assume that the elements y_{ij} are i.i.d. with mean 0, variance $\sigma^2 = 1$, and the fourth moment is bounded. Thus, $\boldsymbol{\Sigma} = \mathbf{I}$, and its corresponding sample covariance matrix is $\widehat{\boldsymbol{\Sigma}}_s = n^{-1}\mathbf{Y}'\mathbf{Y}$. Suppose that

$p/n \to \delta$ and $0 < \delta < 1$. Then, Bai and Yin (1993) showed that the smallest and largest eigenvalues of $\widehat{\boldsymbol{\Sigma}}_s$ converge almost surely to $(1 - \sqrt{\delta})^2$ and $(1 + \sqrt{\delta})^2$, respectively, when both n and p tend to infinity. In contrast, assume that $\delta > 1$. Bai and Yin (1993) demonstrated that the $(p - n + 1)$-th smallest eigenvalue of $\widehat{\boldsymbol{\Sigma}}_s$ also converges almost surely to $(1 - \sqrt{\delta})^2$. In addition, Yin et al. (1988) showed that the largest eigenvalues of $\widehat{\boldsymbol{\Sigma}}_s$ converge almost surely to $(1 + \sqrt{\delta})^2$.

In addition to the convergence of extreme eigenvalues, Jonsson (1982) proved the asymptotic distribution of the empirical spectral distribution under the i.i.d. standard normal assumption. Specifically, let $\widehat{\gamma}_1 \geq \cdots \geq \widehat{\gamma}_p \geq 0$ be the eigenvalues of $\widehat{\boldsymbol{\Sigma}}_s$, and, for any given z, define the empirical spectral distribution of $\widehat{\boldsymbol{\Sigma}}_s$ as $\widehat{F}(z) = p^{-1} \sum_{i=1}^{p} \mathbf{1}(\widehat{\gamma}_i < z)$. Then, $\widehat{F} \to F$ almost surely when both n and p tend to infinity, and F is the Marchenko–Pastur's distribution. Specifically, for $0 < \delta < 1$, the density function of F is given by

$$F'(z) = \frac{1}{2\pi z \delta} \Big\{ (\tau_+ - z)(z - \tau_-) \Big\}^{1/2} \quad \text{if } \tau_- < z < \tau_+, \tag{2.10}$$

and 0 otherwise, where $\tau_- = (1 - \sqrt{\delta})^2$ and $\tau_+ = (1 + \sqrt{\delta})^2$. The result for $\delta > 1$ is also provided in Jonsson (1982). The above result shows that the sample eigenvalues depart from their corresponding true value of 1. Specifically, the largest sample eigenvalues tend to have a positive bias, and the smallest sample eigenvalues tend to have a negative bias. In addition, the larger the value of δ, the more spread out the sample eigenvalues are. Useful references on eigenvalues of large covariance matrices can be found in Bai (1999), Bai and Shi (2011), Johnstone and Paul (2018), Hu et al. (2019), and Zagidullina (2021).

Recently, the spiked covariance model proposed by Johnstone (2001) has garnered considerable attention and popularity. Basically, the spikes mean the eigenvalues of the matrix are detached from the bulk of the eigenvalue spectrum. The spiked covariance model is given below:

$$\boldsymbol{\Sigma} = \mathbf{U} \boldsymbol{\Gamma} \mathbf{U}', \tag{2.11}$$

where $\mathbf{U}$ is a $p \times p$ orthonormal matrix, $\boldsymbol{\Gamma} = \mathrm{diag}(\gamma_1, \cdots, \gamma_L, 1, \cdots, 1)$ is a $p \times p$ diagonal matrix, and $\gamma_1 > \cdots > \gamma_L > 1$ for $L \in [1, p)$. That is, all eigenvalues of $\boldsymbol{\Sigma}$ are equal to 1, except for the first L largest eigenvalues. Let $\widehat{\gamma}_1 \geq \cdots \geq \widehat{\gamma}_p \geq 0$ be the eigenvalues of $\widehat{\boldsymbol{\Sigma}}_s$. Based on model (2.11), by applying the results of Baik and Silverstein (2006), we can show that for $0 < \delta < 1$,

$$\widehat{\gamma}_j \to \begin{cases} \gamma_j + \frac{\delta \gamma_j}{\gamma_j - 1}, & j = 1, \cdots, L_0; \\ \tau_+ & j = L_0 + 1; \end{cases},$$

almost surely as $\{n, p\} \to \infty$. Here, L_0 denotes the number of indices j for which $\gamma_j > \tau_+^{1/2}$, where $\tau_+^{1/2} = 1 + \sqrt{\delta}$ is the transition point (Baik et al. 2005). The results for the case with $\delta \geq 1$ are also provided in Baik and Silverstein (2006).

Some references of spiked covariance matrices can be found in Paul (2007), Wang and Fan (2017), Donoho et al. (2018), Ding and Yang (2021), Bao et al. (2022), Bi et al. (2022), Li et al. (2024), and Feldman (2025).

2.4 Precision Matrix Estimation

In multivariate analysis, studying the estimation of a large inverse covariance matrix (i.e., precision matrix or concentration matrix) has attracted a great deal of attention in the last two decades. This is because the precision matrix can be widely used in many areas such as classification and discrimination, factor analysis, neurology, social networks, and visualizations. It is known that the inverse covariance matrix $\Theta = \mathbf{\Sigma}^{-1} = (\Theta_{jk})$ allows us to examine the conditional independence between any two random variables, y_j and y_k, for $(j \neq k) = 1, \cdots, p$. Hence, the precision matrix is closely related to undirected Gaussian graphical model; see some relevant books such as Whittaker (1991), Lauritzen (1996), Edwards (2000), Højsgaard et al. (2012), and Pourahmadi (2013). For the sake of illustration, this section assumes that the $Y_i = (y_{i1}, \cdots, y_{ip})' \in \mathbb{R}^p$ are i.i.d. $N(\boldsymbol{\mu}, \mathbf{\Sigma})$ for $i = 1, \cdots, n$. Consider an undirected graph $G = (V, E)$, where $V = \{1, \cdots, p\}$ is the set of nodes that correspond to the p variables in Y_i, and the edge $(j, k) \in E$ iff $\Theta_{jk} \neq 0$. Accordingly, the graph G describes the conditionally independent relationship among y_{ij} and y_{ik} for $j \neq k$. Specifically, $(j, k) \notin E$ iff $\Theta_{jk} = 0$ iff y_{ij} and y_{ik} are independent given $Y_{i\cdot\backslash\{j,k\}} = \{y_{il} : l \neq j, k\}$. Hence, identifying the components (j, k) for which $\Theta_{jk} = 0$ in the precision matrix becomes the covariance selection problem (Dempster 1972). It can also be viewed as the model selection problem; see, e.g., Jones and West (2005), Yuan and Lin (2007), and Cox and Wermuth (2014).

In the previous section, we discussed several approaches for estimating covariance matrices. In this section, we mainly focus on the use of penalized methods to estimate precision matrices. Based on the normality assumption, the penalized negative log-likelihood function is proportional to

$$L_p(\Theta) = -\log(|\Theta|) + tr(\widehat{\mathbf{\Sigma}}_s\Theta) + \sum_{i\neq j}^{p} p_\lambda(|\Theta_{ij}|), \tag{2.12}$$

where the Θ_{ij} are the ij-th elements of Θ for $i, j = 1, \cdots, p$, and the penalty function $p_\lambda(\cdot)$ is often assumed to be nondecreasing and continuously differentiable on $(0, \infty)$ with the thresholding parameter λ; see, e.g., L_1 penalty, SCAD, and MCP. Note that this equation can be viewed as the negative quasi-log-likelihood function even though the Y_i are not normally distributed; see, e.g., Lee (2004). The penalized estimator $\widehat{\Theta}$ can be obtained by minimizing equation (2.12). For example, Yuan and Lin (2007), Banerjee et al. (2008), and Friedman et al. (2008) employed the L_1-norm and obtained the lasso-type estimator of Θ. Later, Cai et al. (2011) proposed the constrained L_1-minimization for inverse covariance matrix estimation (CLIME), which is computationally attractive. Furthermore, they demonstrated the rates of convergence between $\widehat{\Theta}$ and Θ under the spectral norm, Frobenius norm, and the element-wise L_∞-norm, respectively. More studies on the L_1 penalty include D-trace loss (Zhang and Zou 2014) and the generalized D-trace loss (Avagyan 2022). Alternatively to the L_1 penalty, Lam and Fan (2009) considered the SCAD and hard-thresholding penalty functions and provided the rate of convergence of resulting estimators under the Frobenius norm. It is worth noting that SCAD is a folded-concave function that can reduce the bias of the true parameter with a relatively large magnitude.

We have discussed an element-wise approach to estimate the precision matrix. Supposing that all of the columns of Θ are sparse, Meinshausen and Bühlmann (2006) proposed the column-by-column estimation method. Basically, they exploited the relationship between the conditional distribution of a multivariate normal and the precision matrix to obtain the sparse estimator of each column of Θ separately. Then, they combined these estimators together; see below for detailed discussion.

Denote the random vector Y excluding its j-th component, y_j, as $Y_{\bullet\backslash\{j\}} \in \mathbb{R}^{p-1}$. In addition, denote $\boldsymbol{\Sigma}$ excluding its i-th row and j-th column as $\boldsymbol{\Sigma}_{\bullet\backslash\{ij\}} \in \mathbb{R}^{(p-1)\times(p-1)}$. Moreover, denote by $\boldsymbol{\Sigma}_{i,\backslash\{j\}}$ and $\boldsymbol{\Sigma}_{\backslash\{i\},j}$ the i-th row of $\boldsymbol{\Sigma}$ excluding the j-th element, and the j-th column of $\boldsymbol{\Sigma}$ excluding the i-th element, respectively. Using the fact that $Y \sim N(\boldsymbol{\mu}, \boldsymbol{\Sigma})$, we have

$$y_j | Y_{\bullet\backslash\{j\}} \sim N(\alpha_j + Y'_{\bullet\backslash\{j\}}\boldsymbol{\beta}_j, \sigma_j^2), \tag{2.13}$$

where $\boldsymbol{\mu} = (\mu_1, \cdots, \mu_p)'$, $\alpha_j = \mu_j - \boldsymbol{\Sigma}_{j,\backslash\{j\}}\boldsymbol{\Sigma}^{-1}_{\bullet\backslash\{jj\}}\boldsymbol{\mu}_{\bullet\backslash\{j\}}$, $\boldsymbol{\mu}_{\bullet\backslash\{j\}} = E(Y_{\bullet\backslash\{j\}})$, $\boldsymbol{\beta}_j = \boldsymbol{\Sigma}^{-1}_{\bullet\backslash\{jj\}}\boldsymbol{\Sigma}_{\backslash\{j\},j} \in \mathbb{R}^{p-1}$, and $\sigma_j^2 = \boldsymbol{\Sigma}_{jj} - \boldsymbol{\Sigma}_{j,\backslash\{j\}}\boldsymbol{\Sigma}^{-1}_{\bullet\backslash\{jj\}}\boldsymbol{\Sigma}_{\backslash\{j\},j}$. In fact, (2.13) can be expressed as the linear regression model given below:

$$y_j = \alpha_j + Y'_{\bullet\backslash\{j\}}\boldsymbol{\beta}_j + \varepsilon_j, \tag{2.14}$$

where $\varepsilon_j \sim N(0, \sigma_j^2)$ for $j = 1, \cdots, p$.

Let $\Theta_{\backslash\{j\},j}$ be the j-th column of Θ excluding the j-th element. After algebraic simplification, we can show that $\Theta_{jj} = \text{var}(\varepsilon_j)^{-1}$ and $\Theta_{\backslash\{j\},j} = -\text{var}(\varepsilon_j)^{-1}\boldsymbol{\beta}_j$. In addition, the partial correlation of y_j and y_l given $Y_{\bullet\backslash\{jl\}}$ is $\rho_{jl_\bullet} = -\Theta_{jl}/\sqrt{\Theta_{jj}\Theta_{ll}}$; see also Lauritzen (1996). Using the regression expression, we can collect n i.i.d. samples, $Y_i = (y_{i1}, \cdots, y_{ip})'$ for $i = 1, \cdots, n$, and then employ shrinkage methods to estimate unknown parameters in (2.14). For the sake of simplicity, we can either assume that $\boldsymbol{\mu} = 0$ or center each variable $Y_{(j)} = (y_{1j}, \cdots, y_{nj})'$ by its sample mean (i.e., $Y_{(j)} - \bar{Y}_{(j)}$), for $j = 1, \cdots, p$. Accordingly, we can focus on estimating $\boldsymbol{\beta}_j$ by solving the following penalized objective function with the L_1 penalty proposed by Meinshausen and Bühlmann (2006).

$$\widehat{\boldsymbol{\beta}}_j = \underset{\boldsymbol{\beta}_j}{\text{argmin}}\, \frac{1}{n}||Y_{(j)} - \mathbf{Y}_{\backslash\{j\}}\boldsymbol{\beta}_j||_2^2 + \lambda_j||\boldsymbol{\beta}_j||_1, \tag{2.15}$$

where $\mathbf{Y}_{\setminus\{j\}} \in \mathbb{R}^{n\times(p-1)}$ is the matrix $\mathbf{Y}$ excluding the j-th column of $\mathbf{Y}$ and λ_j is the tuning parameter. In the graphical model, the non-zero elements in $\widehat{\boldsymbol{\beta}}_j$ constitute the neighborhoods of the graph. The graph estimate $\widehat{\mathbf{G}}$ is obtained by merging the neighborhoods from all p nodes. Specifically, any two nodes are linked by an edge if they identify each other as neighbors. As a result, their associated variables are conditionally dependent, and their corresponding entry in the estimator of Θ is non-zero. It is worth noting that Friedman et al. (2008) introduced a graphical lasso algorithm to speed up the computation of sparse precision matrices. Recently, Zhao et al. (2024) extended this algorithm and proposed the Trans-Glasso algorithm for precision matrix estimation.

In addition to introducing column-by-column penalized estimators for high dimensional precision matrix, Meinshausen and Bühlmann (2006) also demonstrated that these estimators enjoy oracle properties under mild conditions. Furthermore, Janková and van de Geer (2017) proposed a refined version of these estimators. In addition to the lasso estimator, Yuan (2010) and Cai et al. (2011), Liu and Luo (2015), and Liu and Wang (2017) proposed the Dantzig selector named by Candes and Tao (2007), the CLIME estimator, the SCIO estimator, and the TIGER estimator, respectively. They also showed the theoretical properties of their proposed estimators. Finally, Meng et al. (2024) proposed a class of debiased estimators for statistical inference in Gaussian graphical regressions with high dimensional data. They also showed that the debiased estimators not only enjoy a fast convergence rate but also are asymptotically normal.

2.5 Time-Dependent Covariance Matrix Estimation

In the previous sections, we mainly assumed that the data were independent and identically distributed. In practice, however, the data may be collected over a period of time. Hence, estimating large covariance and precision matrices that have a time-dependent structure can be much more challenging. This section briefly reviews three types of methods for large covariance estimation when there is time dependence. For the sake of illustrating dependence, we mainly use the data matrix $\mathbf{Y}' \in \mathbb{R}^{p\times n}$ in this section.

The first method is based on a weak temporal dependence structure proposed by Bhattacharjee and Bose (2014). To this end, we adapt the mean-restricted matrix-variate normal distribution from Allen and Tibshirani (2010) and obtain that $\mathbf{Y}' \sim N_{p,n}(\boldsymbol{\nu}, \boldsymbol{\mu}, \boldsymbol{\Delta}, \boldsymbol{\Sigma})$, where $\mathbf{Y}' = (Y_1, \cdots, Y_n)$ is a $p \times n$ matrix, the row mean is $\boldsymbol{\nu} \in \mathbb{R}^p$, the row covariance is $\boldsymbol{\Sigma} \in \mathbb{R}^{p\times p}$, the column mean is $\boldsymbol{\mu} \in \mathbb{R}^n$, and the column covariance is $\boldsymbol{\Delta} \in \mathbb{R}^{n\times n}$. Accordingly, $\mathrm{vec}(\mathbf{Y}') \sim N_{pn}(\mathrm{vec}(\boldsymbol{\xi}), \boldsymbol{\Omega})$, where $\boldsymbol{\xi} = \boldsymbol{\nu}\mathbf{1}_n' + \mathbf{1}_p\boldsymbol{\mu}' \in \mathbb{R}^{p\times n}$, and $\boldsymbol{\Omega} = \boldsymbol{\Delta} \otimes \boldsymbol{\Sigma}$. In order to take into account dependence across the covariance structure, Bhattacharjee and Bose (2014) extended Bickel and Levina's (2008b) approach and considered the following class of bandable matrices for given $\alpha > 0$ and $\varepsilon > 0$:

$$\tilde{U}(\alpha, \varepsilon, c) = \left\{\mathbf{\Sigma} \in \mathbf{A}(\varepsilon) \cap \mathbf{B} : \max_j \sum_{i=1}^{p} \{|\sigma_{ij}|; |i-j| > k\} \leq ck^{-\alpha},\ \forall k > 0\right\}, \tag{2.16}$$

where $\mathbf{A}(\varepsilon) = \{\mathbf{\Sigma} : 0 < \varepsilon \leq \gamma_{\min}(\mathbf{\Sigma}) \leq \gamma_{\max}(\mathbf{\Sigma}) \leq \varepsilon^{-1} < \infty\}$, $\mathbf{B} = \{\mathbf{\Sigma} : \sigma_{ij} \neq 0$ for $i \neq j\}$, and $c > 0$. Then they introduced the column-dependence structure of $\mathbf{Y}'$ as follows. Consider $\text{vec}(\mathbf{Y}') \sim N_{pn}(\text{vec}(\boldsymbol{\xi}), \mathbf{\Psi})$, where $\mathbf{\Psi} = (\mathbf{\Sigma} * \mathbf{\Lambda}_{ij}\mathbf{1}(i \neq j) + \mathbf{\Sigma} * \mathbf{J}\mathbf{1}(i = j)) \in \mathbb{R}^{pn \times pn}$, the matrices $\mathbf{\Lambda}_{ij} = \mathbf{\Lambda}'_{ij} \in \mathbb{R}^{p \times p}$, for $i, j = 1, \cdots, n$, $\mathbf{J}$ is a $p \times p$ matrix of ones, and "$*$" is the Hadamard (Schur) product. A few examples of the above covariance setting can be found in Bhattacharjee and Bose (2014) and Bose and Bhattacharjee (2018). In addition, let $\{a_n\}$ be a sequence of non-negative integers such that $a_n/n < 1$. When $|i - j|$ is large, the weak dependence between the i-th and j-th columns is defined by $\max_{a_n \leq |i-j| \leq n} ||\mathbf{\Lambda}_{ij}||_\infty = O(a_n/n^2)$.

Suppose that $\mathbf{\Sigma} \in \tilde{U}(\alpha, \varepsilon, c)$ and $\boldsymbol{\xi} = \mathbf{0}$. Under the weak dependence condition and its related assumptions, Bhattacharjee and Bose (2014) demonstrated the banded estimator $B_k(\widehat{\mathbf{\Sigma}}_s) = (\widehat{\sigma}_{ij}\mathbf{1}(|i-j| \leq k))$ (Bickel and Levina 2008a) satisfies $||B_k(\widehat{\mathbf{\Sigma}}_s) - \mathbf{\Sigma}||_2 = O_p((\log(p)/n)^{\frac{\alpha}{2(\alpha+1)}})$ when $k = k_n \sim (\log(p)/n)^{-\frac{1}{2(\alpha+1)}}$. Accordingly, $||B_k(\widehat{\mathbf{\Sigma}}_s) - \mathbf{\Sigma}||_2 \to 0$ if $\log(p)/n \to 0$. In sum, they extended Bickel and Levina's (2008b) banded estimator to weakly dependent data.

The second type of estimation method is for high-dimensional stationary time series. Consider a p-dimensional stationary process Y_i that satisfies $Y_i = (y_{i1}, \cdots, y_{ip})' = g(F_i)$, where $g(\mathcal{F}_i) = (g_1(\mathcal{F}_i), \cdots, g_p(\mathcal{F}_i))' \in \mathbb{R}^{p\times 1}$ is a measurable function, $\mathcal{F}_i = (\cdots, \varepsilon_{i-1}, \varepsilon_i)$ is a shift operator of innovations up to the time period i, and ε_i is i.i.d. random vector for $i = 1, \cdots, n$. Then, Chen et al. (2013) defined the functional dependence measure by

$$\eta_{i,j,\zeta} = \{E[|y_{ij} - y^*_{ij}|^\zeta]\}^{1/\zeta}, \tag{2.17}$$

where $y_{ij} = g_j(\mathcal{F}_i)$, $y^*_{ij} = g_j(\mathcal{F}^*_i)$, $\mathcal{F}^*_i = (\cdots, \varepsilon_{-1}, \varepsilon^*_0, \varepsilon_1, \cdots, \varepsilon_i)$, ε^*_0 is an i.i.d. copy of ε_0, and $\zeta > 0$. In addition, they assumed that the following condition holds: $H_{l,\zeta} = \max_{1 \leq j \leq p} \sum_{m=l}^{\infty} \eta_{m,j,\zeta} < \infty$, where $l > 0$. Afterward, Chen et al. (2013) considered the hard thresholding estimator $T_\lambda(\widehat{\mathbf{\Sigma}}_s) = (\widehat{\sigma}_{jk}\mathbf{1}(|\widehat{\sigma}_{jk}| \geq \lambda))_{1 \leq j,k \leq p}$. Assume that there exist $q > 2$, $\alpha > 0$, $c_1 > 0$, and $0 < c_2 < \infty$ such that $\max_{j \leq p}\{E[|y_{ij}|^{2q}]\}^{1/2q} \leq c_1$ and $H_{l,2q} \leq c_2 l^{-\alpha}$ for all $l \geq 1$. By the above assumptions, together with some regularization conditions, they showed the rates of convergence for $T_\lambda(\widehat{\mathbf{\Sigma}}_s)$ under the normalized Frobenius norm. Chen et al. (2013) further indicated that $T_\lambda(\widehat{\mathbf{\Sigma}}_s)$ has the same rate of convergence as that obtained under independence when the weaker dependence condition, $\alpha > 1/2 - 1/q$, holds.

As for estimating a large precision matrix, Chen et al. (2013) adopted Friedman et al.'s (2008) L_1 penalty function given in (2.12) and obtained the penalized estimator of Θ under the same assumptions mentioned above. They also demonstrated the rates of convergence of the L_1 penalized estimator. Moreover, they compared their results with those of Rothman et al. (2008), Cai et al. (2011), and Ravikumar et al. (2011).

Finally, the third type of estimation method is proposed by Shu and Nan (2019), who took a different approach to model dependence. Specifically, they considered the following linear spatiotemporal model:

$$\mathrm{vec}(\mathbf{Y}') = \mathbf{D}\boldsymbol{\varepsilon} + \mathbf{1}_n \otimes \boldsymbol{\mu}_p, \tag{2.18}$$

where $\mathbf{D} \in \mathbb{R}^{pn\times m}$ is a real deterministic matrix, $\boldsymbol{\varepsilon} = (\varepsilon_1, \cdots, \varepsilon_m)'$, and ε_i is independent random variable with mean 0 and variance 1 for $i = 1, \cdots, m$. Accordingly, $\mathrm{Cov}(\mathrm{vec}(\mathbf{Y}')) = \mathbf{DD}'$, and its diagonal sub-matrices are $\boldsymbol{\Sigma}$ and the off-diagonal sub-matrices are associated with temporal correlations.

Recall that $\mathbf{Y} = (Y_1, \cdots, Y_n)' = (Y_{(1)}, \cdots, Y_{(p)})$, where $Y_{(j)} = (y_{1j}, \cdots, y_{nj})'$ is the $n\times 1$ vector for $j = 1, \cdots, p$. Then $Y'_{(j)}$ is the p row vector for $j = 1, \cdots, p$. Denote by $\mathbf{R}_{(j)} \in \mathbb{R}^{n\times n}$ the correlation matrix of $Y'_{(j)}$. To quantify the temporal dependence of $\mathbf{R}_{(j)}$ via the Frobenius norm and the spectral norm, Shu and Nan (2019) defined c_F and c_S such that

$$\max_{1\le j\le p} \frac{1}{n}||\mathbf{R}_{(j)}||_F^2 \le c_F \text{ and } \max_{1\le j\le p} ||\mathbf{R}_{(j)}||_2 \le c_S, \tag{2.19}$$

where $1 \le c_F \le c_S \le n$.

Suppose that $\mathbf{Y}'$ is generated via (2.18), ε_i is sub-Gaussian with the same tail bound for $i = 1, \cdots, n$, and $\mathbf{R}_{(j)}$ satisfies (2.19). Then, for $\boldsymbol{\Sigma}$ being uniform on $U_q(s_0(p), c)$, as defined in (2.2), Shu and Nan (2019) showed that, for a sufficiently large constant $c_G > 0$, if $\lambda = c_G d$ and $d = \max\{c_S \log p/n, (c_F \log p/n)^{1/2}\} = o(1)$, then the generalized thresholding estimator, $s_\lambda(\widehat{\sigma}_{ij}) = s_\lambda(\widehat{\boldsymbol{\Sigma}}_s)$, defined in Sect. 2.2.3, has the following rates of convergence: (i) $||s_\lambda(\widehat{\boldsymbol{\Sigma}}_s) - \boldsymbol{\Sigma}||_2 = O_p(s_0(p)d^{1-q})$ and (ii) $p^{-1}||s_\lambda(\widehat{\boldsymbol{\Sigma}}_s) - \boldsymbol{\Sigma}||_F = O_p(s_0(p)d^{2-q})$. Furthermore, Shu and Nan (2019) extended Cai et al.'s (2011) results of CLIME and obtained the rates of convergence for the thresholding estimators of the precision matrix under temporal dependence. More references on covariance matrix estimation for high-dimensional time series can be found in Mcmurry and Politis (2015), Basu and Michailidis (2015), Wu and Wu (2016), Zhang and Wu (2017), and Zhang (2021), among others.

2.6 Positive Definiteness

In sparse estimation of high-dimensional covariance and precision matrices, one can impose the constraint that the covariance or precision matrix is positive definite. In practice, however, the computation of a positive definite matrix is still a challenging task due to complicating factors such as convexity, uniqueness, and sparsity. For example, Rothman (2012) proposed an algorithm to compute the positive definite minimizer of the following objective function h:

$$h(\mathbf{\Sigma}) = \frac{1}{2}||\mathbf{\Sigma} - \mathbf{M}||_F^2 - \xi \log(\mathbf{\Sigma}) + \lambda||\mathbf{\Sigma}||_1^-, \tag{2.20}$$

where $\mathbf{M} \in \mathbb{R}^{p\times p}$ is a symmetric matrix with positive diagonal entries (e.g., $\widehat{\mathbf{\Sigma}}_s$), $\xi > 0$, $\lambda \geq 0$, and $||\cdot||_F$ and $||\cdot||_1^-$ denote the Frobenius norm and the L_1-norm with all off-diagonal elements, respectively. However, they could not guarantee that their algorithm yields a global minimizer. It is of interest to note that Xue et al. (2012) studied the positive definite L_1-penalized covariance estimator by solving the following objective function:

$$\widehat{\mathbf{\Sigma}}^+ = \arg\min_{\mathbf{\Sigma}\geq\delta\mathbf{I}} \frac{1}{2}||\widehat{\mathbf{\Sigma}}_s - \mathbf{\Sigma}||_F^2 + \lambda||\mathbf{\Sigma}||_1^-, \tag{2.21}$$

where δ is some arbitrarily small positive number, $\lambda > 0$ is a tuning (thresholding) parameter, and $\mathbf{I}$ is a $p \times p$ identity matrix. In addition, they not only developed an efficient algorithm to solve the optimization problem but also established its convergence properties.

Later, Fan et al. (2016) discussed two approaches to address positive definiteness. The first approach is the selection of a thresholding constant. Let $\lambda_n = m\sqrt{\log(p)/n}$ be given in Theorem 2.1, and denote by $\widehat{\mathbf{\Sigma}}(m) = \mathbf{G}_{\lambda_n}(\widehat{\mathbf{\Sigma}}_s)$ the penalized estimator of $\mathbf{\Sigma}$. As m gets sufficiently large, the estimator becomes a diagonal matrix. Accordingly, the minimum eigenvalue is strictly positive. Then, they proposed decreasing the value of m until it reaches

$$m_{\min} = \inf\{m > 0 : \gamma_{\min}(\widehat{\mathbf{\Sigma}}(m^*)) > 0, \text{ for all } m^* > m\}.$$

In practice, Fan et al. (2016) recommended choosing m in the range of $(m_{\min} + \delta, m^*)$ via cross-validation, where δ is a small number and m^* is a large number.

The second approach is solving the constrained optimization problem, and the resulting positive definite covariance matrix can be obtained below.

$$\tilde{\mathbf{\Sigma}}^+ = \underset{\gamma_{\min}(\mathbf{\Sigma})\geq\tau}{\operatorname{argmin}} \left\{\frac{1}{2}||\widehat{\mathbf{\Sigma}}_s - \mathbf{\Sigma}||_F^2 + \sum_{i\neq j}^{p} p_\lambda(\sigma_{ij})\right\}, \tag{2.22}$$

where $p_\lambda(\cdot)$ is the penalty function defined in (2.12) and $\tau > 0$ is given a priori and controls the smallest eigenvalue of $\tilde{\mathbf{\Sigma}}^+$. When $p_\lambda(\cdot)$ is the L_1 penalty function, Liu et al. (2014) developed an efficient algorithm to compute $\tilde{\mathbf{\Sigma}}^+$.

It is worth noting that the cross-validation and its variants have been commonly used to select the tuning parameter λ; see, e.g., Fang et al. (2016). Other approaches have also been developed; see, e.g., Lian's (2011) BIC, Gao et al.'s (2012) BIC, and Liu and Wang's (2017) TIGER. Moreover, Qiu and Liyanage (2019) proposed an optimal threshold level for the consistent estimator of the covariance matrix. In sum, the task of choosing the tuning parameters is critical and needs further study.

2.7 Tests of Covariance Matrices

In the estimation of high-dimensional covariance matrices discussed above, we have used the assumption that there exists a true covariance structure. In practice, however, this assumption needs to be examined. Hence, testing the covariance structure can play an important role in real applications of high-dimensional data analysis (see, e.g., detecting Schizophrenia risk genes in Zhu et al. 2017), and testing the cross-sectional uncorrelatedness of idiosyncratic stock returns (Yang et al. 2021). This section discusses two types of covariance tests: single-population covariance tests and multiple-population covariance tests. Note that two-population tests are a special case of multiple-population tests. A valuable review article on high-dimensional covariance tests is given by Cai (2017).

The single-population covariance tests for testing sphericity ($H_0 : \boldsymbol{\Sigma} = \sigma^2\mathbf{I}_p$ vs $H_1 : \boldsymbol{\Sigma} \neq \sigma^2\mathbf{I}_p$) and identity ($\boldsymbol{\Sigma} = \mathbf{I}_p$ vs $H_1 : \boldsymbol{\Sigma} \neq \mathbf{I}_p$) have been studied when $p < n$, where $\sigma > 0$; see, e.g., Mardia et al. (1979) and Anderson (2003). In high-dimensional data with $p/n \to c \in (0, \infty)$, Ledoit and Wolf (2002) considered two test statistics for testing sphericity and identity, separately. When $c > 1$, they referred to their asymptotic results as (n, p) asymptotics.

To test sphericity, Ledoit and Wolf (2002) revisited the following test statistic proposed by John (1971),

$$U = \frac{(1/p)tr(\widehat{\boldsymbol{\Sigma}}_s^2)}{[(1/p)tr(\widehat{\boldsymbol{\Sigma}}_s)]^2} - 1. \tag{2.23}$$

Suppose that the Y_i are normally distributed random vectors with mean $\boldsymbol{\mu}$ and covariance $\boldsymbol{\Sigma}$ for $i = 1, \cdots, n$. Under the sphericity hypothesis and appropriate assumptions, Ledoit and Wolf (2002) showed that $nU - p \xrightarrow{d} N(1, 4)$. Subsequently, they proposed the following test statistic:

$$W = \frac{1}{p}tr\{(\widehat{\boldsymbol{\Sigma}}_s - \mathbf{I}_p)^2\} - \frac{p}{n}\{\frac{1}{p}tr(\widehat{\boldsymbol{\Sigma}}_s)\}^2 + \frac{p}{n}. \tag{2.24}$$

Then, under the identity hypothesis and appropriate assumptions, Ledoit and Wolf (2002) demonstrated that $nW - p \xrightarrow{d} N(1, 4)$. To relax the normality assumption and accommodate the "large p and small n" scenario, Chen et al. (2010) introduced two statistics to test sphericity and identity, separately. Basically, their novel test statistics are, respectively, modified from U and W in (2.23) and (2.24). Note that Bai et al. (2009) proposed the corrected likelihood ratio test (CLRT) to assess identity under Gaussian data with known means, and Wang et al. (2013) extended CLRT to non-Gaussian with unknown means. In addition, Zhong et al. (2017) proposed an adjusted goodness-of-fit test to assess a broad range of covariance structures. Recently, Yang et al. (2021) developed tests for testing sphericity and identity for any non-negative definite matrix under generalized elliptical distributions.

It is of interest to note that Lan et al. (2015) proposed a bias-corrected test to assess the significance of the off-diagonal elements of the covariance matrix of the random error vectors in multivariate regression models. However, their test assumed that random error vectors are i.i.d. with multivariate normal distribution. Recently, Guo and Tang (2021) relaxed the Gaussian assumption by considering the exponential tails of random error vectors and assumed that the explanatory variables can be unobservable. Then, they developed a maximum discrepancy statistic to test the off-diagonal elements as well as to test the correlation matrix.

We next study multiple-population covariance tests. Let $Y_{im} = (y_{i1m}, \cdots, y_{ipm})' \in \mathbb{R}^{p\times 1}$ be the i.i.d. random vectors with the mean vectors $\boldsymbol{\mu}_m \in \mathbb{R}^{p\times 1}$ and the covariance matrices $\boldsymbol{\Sigma}_m \in \mathbb{R}^{p\times p}$, for $m = 1, \cdots, M$ and $i = 1, \cdots, n_m$. The sample estimators of the mean and covariance matrix are

$$\widehat{\boldsymbol{\mu}}_m = \bar{Y}_m = \frac{1}{n_m}\sum_{i=1}^{n_m} Y_{im} \text{ and } \widehat{\boldsymbol{\Sigma}}_m = \frac{1}{n_m}\sum_{i=1}^{n_m}(Y_{im} - \bar{Y}_m)(Y_{im} - \bar{Y}_m)'. \tag{2.25}$$

In addition, define $\mathbf{V_m} = n_m\widehat{\boldsymbol{\Sigma}}_m$, $\mathbf{V} = \sum_{m=1}^{M}\mathbf{V_m}$, and $N = \sum_{m=1}^{M}(n_m - 1)$. Since there are M different covariance matrices, it is sensible to conduct a test based on the following null and alternative hypotheses:

$$H_0 : \boldsymbol{\Sigma}_1 = \cdots = \boldsymbol{\Sigma}_M \text{ vs } H_1 : \boldsymbol{\Sigma}_m \neq \boldsymbol{\Sigma}_l \text{ for some } m \neq l \text{ and } m, l = 1, \cdots, M. \tag{2.26}$$

Under the Gaussian setting, Schott (2007) introduced the test statistic based on the estimators of the squared Frobenius norm of the differences of the sample covariance matrices. Then, Srivastava and Yanagihara (2010) proposed the test given below:

$$T_M{}^2 = \widehat{\mathbf{a}}_2'\left(\widehat{\mathbf{G}} - \frac{\widehat{\mathbf{G}}\mathbf{1}\mathbf{1}'\widehat{\mathbf{G}}}{\mathbf{1}'\widehat{G}\mathbf{1}}\right)\widehat{\mathbf{a}}_2, \tag{2.27}$$

where $\widehat{\mathbf{a}}_2 = (\widehat{a}_{21}, \cdots, \widehat{a}_{2M})'$, $\widehat{a}_{2m} = \frac{1}{p(n_m-2)(n_m+1)}\{tr(\mathbf{V}_m^2) - \frac{1}{(n_m-1)}[tr(\mathbf{V_m})]^2\}$, $\widehat{\mathbf{G}} = \text{diag}(\frac{1}{\widehat{\eta}_1^2}, \cdots, \frac{1}{\widehat{\eta}_M^2})$, $\widehat{\eta}_m^2 = \frac{4}{(n_m-1)^2}\widehat{a}_2^2(1+\frac{2(n_m-1)\widehat{a}_4}{p\widehat{a}_2^2})$, $\widehat{a}_2 = \frac{1}{p(N-1)(N+2)}\{tr(\mathbf{V}^2) - \frac{1}{N}[tr(\mathbf{V})]^2\}$, $\widehat{a}_4 = \frac{1}{d_0}[\frac{1}{p}tr(\mathbf{V}^4) - pd_1\widehat{a}_1 - p^2d_2\widehat{a}_1^2\widehat{a}_2 - pd_3\widehat{a}_2^2 - Np^3\widehat{a}_1^4]$, $\widehat{a}_1 = \frac{1}{Np}tr(\mathbf{V})$, $d_0 = N(N^3 + 6N^2 + 21N + 18)$, $d_1 = 2N(2N^2 + 6N + 9)$, $d_2 = 2N(3N+2)$, $d_3 = N(2N^2+5N+7)$, and $m = 1, \cdots, M$. Under H_0 in (2.26), together with appropriate assumptions, they then showed that T_M^2 is distributed as χ^2_{M-1} when $(p, N) \to \infty$. Recently, Zhong et al. (2019) developed a homogeneity test to assess the validity of H_0 in high-dimensional longitudinal data. To take into account the temporal dependence, we assume that $Y_{im} = (y_{i1m}, \cdots, y_{ipm})'$ are random vectors and Y_{im} and Y_{il} are dependent, where the $(m \neq l) = 1, \cdots, M$ are temporal indices. Under some conditions, Zhong et al. (2019) showed the asymptotic property of their proposed test statistic.

For $M = 2$, several researchers proposed various tests to assess the equality of two-population covariance matrices. For example, without the normality assumption, Li and Chen (2012) introduced a test statistic by using a combination of U-statistics. Their test also focused on the squared Frobenius norm of the differences of the two covariance matrices' estimators. Since the Frobenius norm testing procedure may not be powerful against sparse alternatives, Cai et al. (2013) proposed the following test.

Denote $\widehat{\boldsymbol{\Sigma}}_1 = (\widehat{\sigma}_{rs1}) \in \mathbb{R}^{p \times p}$ and $\widehat{\boldsymbol{\Sigma}}_2 = (\widehat{\sigma}_{rs2}) \in \mathbb{R}^{p \times p}$, where $r, s = 1, \cdots, p$. Let $\widehat{\theta}_{rs1} = \frac{1}{n_1}\sum_{i=1}^{n_1}[(Y_{ir1} - \bar{Y}_{r1})(Y_{is1} - \bar{Y}_{s1}) - \widehat{\sigma}_{rs1}]^2$, $\bar{Y}_{r1} = \frac{1}{n_1}\sum_{i=1}^{n_1} Y_{ir1}$, $\bar{Y}_{s1} = \frac{1}{n_1}\sum_{i=1}^{n_1} Y_{is1}$, $\widehat{\theta}_{rs2} = \frac{1}{n_2}\sum_{i=1}^{n_2}[(Y_{ir2} - \bar{Y}_{r2})(Y_{is2} - \bar{Y}_{s2}) - \widehat{\sigma}_{rs2}]^2$, $\bar{Y}_{r2} = \frac{1}{n_2}\sum_{i=1}^{n_2} Y_{ir2}$, and $\bar{Y}_{s2} = \frac{1}{n_2}\sum_{i=1}^{n_2} Y_{is2}$, where $r, s = 1, \cdots, p$. In addition, they defined standardized statistics,

$$M_{rs} = \frac{(\widehat{\sigma}_{rs1} - \widehat{\sigma}_{rs2})^2}{\widehat{\theta}_{rs1}/n_1 + \widehat{\theta}_{rs2}/n_2}, \tag{2.28}$$

where $1 \le r \le s \le p$. Subsequently, Cai et al. (2013) proposed the test statistic for testing $H_0 : \boldsymbol{\Sigma}_1 = \boldsymbol{\Sigma}_2$ given below:

$$M_n = \max_{1 \le r \le s \le p} M_{rs}. \tag{2.29}$$

Moreover, they showed that M_n enjoys certain optimality, and it is powerful against sparse alternatives without assuming normality. Based on M_{rs} and the multi-level thresholding method, Chen et al. (2023) recently constructed a test statistic that can detect sparse and faint signal differences between any two covariance matrices. It would be of interest to extend their results to test the differences among multiple-population covariance matrices.

2.8 Empirical Examples

To illustrate applications of covariance matrices, this section briefly discusses some examples that were studied in literature. After reviewing those examples, we classify them into three categories: (i) biological and medical examples; (ii) assets, stock prices, and stock returns examples; and (iii) machine learning examples. In category (i), please refer to the articles of Banerjee et al. (2008), Rothman et al. (2009), Chen et al. (2010), Cai et al. (2011), Cai et al. (2013), Janková and van de Geer (2017), Liu and Wang (2017), and Shu and Nan (2019). For example, Cai et al. (2013) used two classes of gene expression data (Singh et al. 2002) to compare the covariance matrices of 5000 genes in samples of prostate tissue with a tumor and without. Their unequal covariance findings can play an important role for follow-up studies such as identifying genes with changes in covariances between the two classes. In addition, Liu and Wang (2017) employed their TIGER method to estimate precision matrices based on the gene expression arrays from *Arabidopsis thaliana* data that appeared in Wille et al. (2004). They then used them to analyze gene network structures. As a result, they not only identified the within-pathway connections of the two isoprenoid metabolic pathways (i.e., the mevalonate and plastidial pathways) but also determined cross-pathway links. This finding indicates that cross-talks exist, which warrants further investigation.

As for category (ii), the following five articles used real examples to illustrate their associated methods: Yuan and Lin (2007), Lan et al. (2015), Lam (2020), Yang et al. (2021), and Chen et al. (2023). For example, Lan et al. (2015) collected weekly stock returns from the Chinese stock market between 2008 and 2009 and then applied their proposed test to assess the off-diagonal elements of residual covariance matrices after separately fitting the capital asset pricing model (Sharpe 1964) and the Fama–French three-factor model (Fama and French 1993). The results indicate that the Fama–French three-factor model is more appropriate in high-dimensional data. In addition, Chen et al. (2023) used the daily returns of the S&P 500 stocks to test if there were any changes in the covariance matrices before and after the outbreak of the COVID-19 pandemic. They found that some stock pairs of covariances changed significantly. In particular, they found that the covariances of stock pairs within the sectors such as information technology and health care were either negatively enhanced or decreased after COVID-19 pandemic.

In category (iii), Rothman (2012) applied the quadratic discriminant analysis technique, together with his proposed positive definite covariance matrix method, to classify healthy individuals and Parkinson's patients based on the speech signal data introduced by Little et al. (2009). The results showed that the average percentage of misclassification is around 22%. In addition, Zhong et al. (2017) employed their proposed test to analyze Canadian weather data (see Ramsay and Silverman 2005), which are daily temperature and precipitation measures collected from 35 weather stations between 1960 and 1994. They found that the compound symmetry matrix is an appropriate covariance matrix for predictions in that case.

The above examples illustrate some applications of covariance estimations and hypothesis tests. Due to the availability of high-dimensional data and technological advancements, more wide-ranging applications are expected to emerge over time.

Chapter 3
Structured Covariance Matrices and Unconstrained Parameterization

This chapter presents four structured covariance matrices that indicate specific relationships among the multivariate responses. Then, it introduces linear covariance models, which posit a linear relationship between the covariance matrix and a set of known symmetric matrices. To assure the covariance matrix is positive definite, two unconstrained parameterizations are considered. In addition, four empirical examples are presented to illustrate the usefulness of structured covariance matrices.

3.1 Structured Covariance Matrices

In Chap. 2, we basically do not impose any specific structure in covariance estimation. However, those estimation methods usually require that the number of observations tends toward infinity to ensure reliable estimators. In practice, this requirement is not necessarily satisfied. Hence, we introduce structured covariance matrices, which can be estimated with a limited sample size. As described in Chap. 2, the covariance matrix, $\mathbf{\Sigma}$, characterizes the pairwise relationships between the components of a multivariate response vector $Y = (y_1, \cdots, y_p)'$. In practice, there can be a specific structure that describes these relationships and provides a meaningful interpretation. Four known structured covariance matrices and their associated parameter estimators are given below, and more structured covariance matrices can be found in Janiszewska (2022).

(i) The covariance matrix of an autoregressive process with order 1, also called AR(1), is

W. Lan, C.-L. Tsai, *Covariance Analysis and Beyond*,
https://doi.org/10.1007/978-3-032-08796-6_3

$$\boldsymbol{\Sigma}_A = \frac{\sigma^2}{1-\phi^2}\begin{bmatrix} 1 & \phi & \phi^2 & \cdots & \phi^{p-1} \\ \phi & 1 & \phi & \cdots & \phi^{p-2} \\ \phi^2 & \phi & 1 & \cdots & \phi^{p-3} \\ \vdots & \vdots & \vdots & \ddots & \vdots \\ \phi^{p-1} & \phi^{p-2} & \phi^{p-3} & \cdots & 1 \end{bmatrix}, \tag{3.1}$$

where $\sigma > 0$ and $-1 < \phi < 1$. This covariance matrix structure shows that the relationship between any two components in Y depends on their index (or time) lag. In fact, (3.1) is a special case of Toeplitz matrices (see Gray 2006), and its inverse matrix is

$$\boldsymbol{\Sigma}_A^{-1} = \frac{1}{\sigma^2}S(\phi) = \frac{1}{\sigma^2}\begin{bmatrix} 1 & -\phi & 0 & \cdots & 0 \\ -\phi & 1+\phi^2 & -\phi & \cdots & 0 \\ \vdots & \vdots & \ddots & \vdots & \vdots \\ 0 & \cdots & -\phi & 1+\phi^2 & -\phi \\ 0 & \cdots & 0 & -\phi & 1 \end{bmatrix}. \tag{3.2}$$

In addition, the eigenvalue structure of the AR(1) covariance matrix can be found in Grenander and Szego (1958) and Sherman (2023). Moreover, the determinant of $\boldsymbol{\Sigma}_A$ is $|\boldsymbol{\Sigma}_A| = \sigma^{2p}(1-\phi^2)^{-1}$.

To estimate the unknown parameters ϕ and σ^2, we consider the AR(1) model given below:

$$y_t = \phi y_{t-1} + \varepsilon_t, \tag{3.3}$$

where $y_1 = \varepsilon_1$ follows a normal distribution $N(0, \sigma^2/(1-\phi^2))$, and ε_t is i.i.d. random variable with a normal distribution $N(0, \sigma^2)$ for $t = 2, \cdots, p$, and independent of ε_1. The resulting log-likelihood function is

$$L(\phi, \sigma^2) = -\frac{p}{2}\log(2\pi) - \frac{1}{2}\log|\boldsymbol{\Sigma}_A| - \frac{1}{2}Y'\boldsymbol{\Sigma}_A^{-1}Y. \tag{3.4}$$

For a given value of σ^2, we obtain the maximum likelihood estimator $\tilde{\phi}$ by maximizing $L(\phi, \sigma^2)$ with respect to ϕ. Subsequently, the MLE of σ^2 is $\tilde{\sigma}^2 = Y'S(\widehat{\phi})Y/p$. We then repeat this iterative process until convergence, which yields the MLEs $\widehat{\phi}_{mle}$ and $\widehat{\sigma}^2_{mle}$. It can be shown that $\sqrt{p}(\widehat{\phi}_{mle} - \phi) \to N(0, (1-\phi^2))$, as $p \to \infty$; see, e.g., Shumway and Stoffer (2017).

Although $\widehat{\phi}_{mle}$ does not have a closed form, the least squares estimator obtained via the conditional distribution of $y_t|y_{t-1}$ with y_1 fixed has a closed form. Specifically, the resulting estimator is $\widehat{\phi}_{ols} = (Y'_{(-p)}Y_{(-p)})^{-1}Y'_{(-p)}Y_{(-1)}$, where $Y_{(-p)} = (y_1, \cdots, y_{p-1})'$ and $Y_{(-1)} = (y_2, \cdots, y_p)'$. Then the OLS estimator of σ^2 is $\widehat{\sigma}^2_{ols} = (Y_{(-1)} - Y_{(-p)}\widehat{\phi}_{ols})'(Y_{(-1)} - Y_{(-p)}\widehat{\phi}_{ols})/(p-1)$. It can also be shown that $\sqrt{p}(\widehat{\phi}_{ols} - \phi) \to N(0, (1-\phi^2))$, as $p \to \infty$.

(ii) The covariance matrix of a moving average process with order 1, also called MA(1), is

$$\boldsymbol{\Sigma}_M = \sigma^2 \begin{bmatrix} 1+\theta^2 & \theta & 0 & \cdots & 0 \\ \theta & 1+\theta^2 & \theta & \cdots & 0 \\ 0 & \theta & 1+\theta^2 & \cdots & 0 \\ \vdots & \vdots & \vdots & \ddots & \vdots \\ 0 & \cdots & \theta & 1+\theta^2 & \theta \\ 0 & 0 & \cdots & \theta & 1+\theta^2 \end{bmatrix}, \tag{3.5}$$

where $\sigma > 0$ and $|\theta| < 1$. This covariance matrix structure shows that the main diagonal elements are $(1+\theta^2)\sigma^2$, the first off-diagonal elements are $\theta\sigma^2$, and the rest of the elements are 0. It is the tridiagonal covariance structure and can also be viewed as a banded covariance structure. The approximation of the precision matrix of (3.5) is

$$\boldsymbol{\Sigma}_M^{-1} \approx \frac{1}{\sigma^2(1-\theta^2)} \begin{bmatrix} 1 & -\theta & \theta^2 & \cdots & (-\theta)^{p-1} \\ -\theta & 1 & -\theta & \cdots & (-\theta)^{p-2} \\ \theta^2 & -\theta & 1 & \cdots & (-\theta)^{p-3} \\ \vdots & \vdots & \vdots & \ddots & \vdots \\ (-\theta)^{p-1} & (-\theta)^{p-2} & (-\theta)^{p-3} & \cdots & 1 \end{bmatrix}; \tag{3.6}$$

see Durbin and Koopman (2012). The exact precision matrix can be found in Uppuluri and Carpenter (1969).

When $\sigma^2 = 1$, we can apply the result from Noschese et al. (2013) to obtain the eigenvalues of the MA(1) covariance matrix as $\gamma_j = 1 + \theta^2 + 2\theta cos(j\pi/(p+1))$ for $j = 1, \cdots, p$. In addition, applying the result of Qi et al. (2019), we can show that the determinant of $\boldsymbol{\Sigma}_M$ is $|\boldsymbol{\Sigma}_M| = \sigma^{2p}(1 - \theta^{2(p+1)})/(1-\theta^2)$.

To estimate unknown parameters θ and σ^2, we consider the MA(1) model given below

$$y_t = \varepsilon_t + \theta\varepsilon_{t-1}, \tag{3.7}$$

where ε_t is i.i.d. $N(0, \sigma^2)$ for $t = 0, \cdots, p$. The resulting log-likelihood function is

$$L(\theta, \sigma^2) = -\frac{p}{2}\log(2\pi) - \frac{1}{2}\log|\boldsymbol{\Sigma}_M| - \frac{1}{2}Y'\boldsymbol{\Sigma}_M^{-1}Y. \tag{3.8}$$

Then the maximum likelihood estimators of $(\theta, \sigma^2)'$ are $(\widehat{\theta}_{mle}, \widehat{\sigma}^2_{mle})' = \arg\max_{(\theta,\sigma^2)} L(\theta, \sigma^2)$. It can be shown that $\sqrt{p}(\widehat{\theta}_{mle} - \theta) \to N(0, (1-\theta^2))$, as $p \to \infty$; see, e.g., Shumway and Stoffer (2017). One can also assume that $\varepsilon_0 = 0$ and obtain the conditional least squares estimator by minimizing $\sum_{t=1}^{p} \varepsilon_t^2$. Since it is a nonlinear optimization problem, the resulting estimator θ_{ols} does not have a closed form. Note that the covariance matrices and precision matrices of AR(1) and MA(1) can be extended to autoregressive moving average (ARMA) process; see, e.g., Haddad (2004). In addition, the estimation of ARMA can be found in Hamilton (1994), Box et al. (2015), and Shumway and Stoffer (2017).

(iii) The covariance matrix with compound symmetry or equicorrelation (also called CS) is

$$\mathbf{\Sigma}_C = \sigma^2 \begin{bmatrix} 1 & \rho & \rho & \cdots & \rho \\ \rho & 1 & \rho & \cdots & \rho \\ \rho & \rho & 1 & \cdots & \rho \\ \vdots & \vdots & \vdots & \ddots & \vdots \\ \rho & \rho & \rho & \cdots & 1 \end{bmatrix}, \tag{3.9}$$

where $\sigma > 0$ and $-(p-1)^{-1} < \rho < 1$. This covariance matrix structure shows that the correlation between any two components in Y is the same. Note that the above compound symmetric matrix is a special case of the expression of $a\mathbf{I} + b(\mathbf{1}\mathbf{1}')$, where a and b are scalars, $\mathbf{I}$ is a p-dimensional identity matrix, and $\mathbf{1}$ represents a p-dimensional vector of ones.

Let $\mathbf{J} = \mathbf{1}\mathbf{1}'$ be the $p \times p$ matrix of ones. Then, the precision matrix and the determinant of $\mathbf{\Sigma}_C$ are $\mathbf{\Sigma}_C^{-1} = \frac{1}{\sigma^2(1-\rho)}(\mathbf{I} - \frac{\rho}{1+(p-1)\rho}\mathbf{J})$ and $|\mathbf{\Sigma}_C| = \sigma^{2p}(1-\rho)^{p-1}[1 + (p-1)\rho]$, respectively; see, e.g., Tong (1990). In addition, the eigenvalues of $\mathbf{\Sigma}_C$ are $\gamma_1 = \sigma^2\{1 + (p-1)\rho\}$ and $\gamma_2 = \cdots = \gamma_p = \sigma^2(1-\rho)$ when $\rho > 0$; see, e.g, Mardia et al. (1979).

To estimate unknown parameters $\boldsymbol{\mu}$, ρ, and σ^2, we assume that Y is distributed with $N(\boldsymbol{\mu}, \mathbf{\Sigma}_C)$, where $\boldsymbol{\mu} = (\mu_1, \cdots, \mu_p)'$. The resulting log-likelihood function is

$$L(\boldsymbol{\mu}, \rho, \sigma^2) = -\frac{p}{2}\log(2\pi) - \frac{1}{2}\log|\mathbf{\Sigma}_C| - \frac{1}{2}(Y - \boldsymbol{\mu})'\mathbf{\Sigma}_C^{-1}(Y - \boldsymbol{\mu}). \tag{3.10}$$

Then, the maximum likelihood estimator of $(\boldsymbol{\mu}', \rho, \sigma^2)'$ is $(\widehat{\boldsymbol{\mu}}', \widehat{\rho}, \widehat{\sigma}^2)' = \arg\max_{\boldsymbol{\mu},\rho,\sigma^2} L(\boldsymbol{\mu}, \rho, \sigma^2)$.

(iv) The symmetric Toeplitz covariance matrix is

$$\mathbf{\Sigma}_T = \sigma^2 \text{Toep}(\theta_0, \theta_1, \cdots, \theta_{p-1}) = \sigma^2 \begin{bmatrix} \theta_0 & \theta_1 & \theta_2 & \cdots & \theta_{p-1} \\ \theta_1 & \theta_0 & \theta_1 & \cdots & \theta_{p-2} \\ \theta_2 & \theta_1 & \theta_0 & \cdots & \theta_{p-3} \\ \vdots & \vdots & \vdots & \ddots & \vdots \\ \theta_{p-1} & \theta_{p-2} & \theta_{p-3} & \cdots & \theta_0 \end{bmatrix},$$

where $\sigma > 0$, $\theta_0 > 0$, and $\theta_1, \cdots, \theta_{p-1}$ are selected such that $\boldsymbol{\Sigma}_T$ is a positive definite matrix. This covariance matrix structure indicates that $(\boldsymbol{\Sigma}_T)_{ij} = \sigma^2\theta_{|i-j|}$. Note that the covariance matrices of AR(1), MA(1), and CS are all special cases of the Toeplitz covariance matrix. In addition, under some conditions, Rodman and Shalom (1992) derived the precision matrix of $\boldsymbol{\Sigma}_T$. A good review of Toeplitz matrices is given by Gray (2006). Note that four covariance matrices in this section (matrices (i)–(iv)) can be used to construct a block covariance structure; see, e.g., Szatrowski (1982) and Liang et al. (2022). In addition, they can be applied to the composition of Kronecker product covariance matrices or separable covariance matrices; see, e.g., Leng and Pan (2018), Drton et al. (2021), Guggenberger et al. (2023), Hoff et al. (2023), Masak et al. (2023), and Puchkin and Rakhuba (2024).

It is of interest to note that the structured covariance matrices (i)–(iii) can be estimated based on the p components of Y with only one observation. In the areas of longitudinal data analysis (Diggle et al. 2002), multivariate analysis (Anderson 2003), social science (Frees 2004), econometric analysis of panel data (Wooldridge 2010), and medicine (Brown and Prescott 2015), among others, one can collect the n random sample observations, $Y_i = (y_{i1}, \cdots, y_{ip})'$ $(i = 1, \cdots, n)$, so as to improve the consistency and efficiency of parameter estimators. Based on n independent samples and under the normality assumption, Mardia et al. (1979) indicated that the maximum likelihood estimators of the compound symmetry are as follows: $\widehat{\boldsymbol{\mu}} = \bar{Y}$, $\widehat{\rho} = \frac{1}{\widehat{\sigma}^2}\sum_{j=1}^{p}\sum_{k=1,k\neq j}^{p}\frac{\widehat{\sigma}_{jk}}{p(p-1)}$, and $\widehat{\sigma}^2 = p^{-1}\sum_{j=1}^{p}\widehat{\sigma}_{jj}$, where $\widehat{\sigma}_{jk}$ is the jk-th element of the sample covariance matrix $\widehat{\boldsymbol{\Sigma}}_s$; similar results are given in Seber (2004).

In the symmetric Toeplitz covariance matrix, which has $p + 1$ unknown parameters, one also needs to collect the n random sample observations in order to estimate those parameters. Based on n samples and large p, Cai et al. (2013) obtained the rate-optimal estimators and established the optimal rate of convergence under the spectral norm. In addition, they studied the properties of tapering and banding estimators. Recently, Yang et al. (2021) proposed a method to simultaneously select the sparse or nearly sparse Toeplitz covariance structure and estimate the covariance matrix under the entropy loss function with non-convex penalty. Furthermore, Wang and Zhang (2024) introduced an adaptive approach for selecting the optimal Toeplitz target matrix in the context of shrinkage estimation. Note that the above studies employed various thresholding or penalized methods to estimate large dimensional parameters of Toeplitz covariance matrices. In sum, there are two types of structured covariance matrices. One type is naturally induced by the structure of the random vector Y; see, e.g., the covariance matrices in (i)–(iv). The other type is a direct result of applying constraints such as bandable, tapering, spiking, and thresholding discussed in Chap. 2.

We next introduce the linear combination of covariance matrices, which includes moving average, compound symmetry, and banded covariance as its special cases.

3.2 Linear Covariance Models

This section presents a general approach to forming structured covariance matrices. Specifically, we adopt Anderson's (1973) approach and consider the following family of linear covariance models:

$$\mathbf{\Sigma} = \sum_{k=0}^{K} \beta_k \mathbf{W}_k, \tag{3.11}$$

where the $\mathbf{W}_k$s are $p \times p$ symmetric and known matrices and the β_ks are the associated unknown parameters for $k = 0, \cdots, K$. Accordingly, various known examples of a $p \times p$ covariance matrix $\mathbf{\Sigma}$ can be formed in (3.11) by choosing the appropriate $\mathbf{W}_k$s. For example, the spherical matrix is formed by assuming that $\mathbf{W}_0 = \mathbf{I}$, $\beta_0 = \sigma^2$, and $\beta_k = 0$ for $k = 1, \cdots, K$. In addition, the compound symmetry matrix can be expressed as a linear covariance model by letting $\mathbf{W}_0 = \mathbf{I}$, $\mathbf{W}_1 = \sum_{|j_1 - j_2| \geq 1} \mathbf{G}_{j_1 j_2}$, $\beta_0 = \sigma^2$ and $\beta_1 = \rho\sigma^2$, where $\mathbf{G}_{j_1 j_2}$ is the $p \times p$ matrix with 1 on the (j_1, j_2)-th entry and 0 otherwise. Moreover, the symmetric Toeplitz covariance matrix can exhibit the linear covariance structure by letting $\beta_0 = \sigma^2\theta_0$, $\beta_k = \sigma^2\theta_k$, $\mathbf{W}_0 = \mathbf{I}$, $\mathbf{W}_k = \text{Toep}(0, \cdots, 0, 1, 0, \cdots, 0)$, where 1 appears in the $(k+1)$-th position and $k = 1, \cdots, p-1$. That is, the (i, j)-th component of $\mathbf{W}_k$ equals to 1 when $|i - j| = k$ and 0 otherwise; see Marin and Dhorne (2002).

Since linear covariance modeling was proposed more than 50 years ago, many studies have appeared in the literature. For example, Pourahmadi (2011, 2013) introduced precision models by replacing $\mathbf{\Sigma}$ in (3.11) with $\mathbf{\Sigma}^{-1}$. In addition, Qu et al. (2000) employed a precision model to generalize the estimating equations in longitudinal data analysis. To assure positive definiteness, Leonard and Hsu (1992), Chiu et al. (1996) and Battey (2017) considered log-covariance models by replacing $\mathbf{\Sigma}$ in (3.11) with $\log(\mathbf{\Sigma})$. Note that $\log(\mathbf{\Sigma})$ means $\mathbf{U}\log(\Gamma)\mathbf{U}'$, where $\mathbf{U}\Gamma\mathbf{U}'$ is a spectral decomposition of $\mathbf{\Sigma}$. Suppose there are n random samples, and the covariance matrix of the i-th sample is $\mathbf{\Sigma}^{(i)}$ for $i = 1, \cdots, n$. Then, we can construct linear covariance models for each $\mathbf{\Sigma}^{(i)}$ as follows:

$$\mathbf{\Sigma}^{(i)} = \sum_{k=0}^{K} \beta_k^{(i)} \mathbf{W}_k^{(i)}, \tag{3.12}$$

where the $\mathbf{W}_k^{(i)}$s are $p \times p$ symmetric and known matrices and the $\beta_k^{(i)}$s are their associated unknown parameters for $k = 0, \cdots, K$ and $i = 1, \cdots, n$. The above model can be applied in various areas such as mixed models, longitudinal data analysis, and social network analysis.

Under the normality assumption, Anderson (1973) obtained the maximum likelihood estimators of the β_ks in (3.11) and showed their asymptotic properties. Later, Szatrowski (1980) provided the necessary and sufficient conditions for explicit maximum likelihood estimators of the β_ks. Then Zwiernik et al. (2017) found that the maximum likelihood estimation for linear covariance models is a non-convex optimization problem, which usually yields many local optima. They went on to provide sufficient conditions for any hill-climbing method to converge to the global optimum. All of these works have broadened the usefulness of linear covariance models.

Recently, Zheng et al. (2019) developed a unified framework for testing linear covariance models. They established the asymptotic properties of their proposed tests in high-dimensional data. Since we have discussed several tests for high-dimensional covariance matrices in Sect. 2.7, we next briefly present their estimation and testing results. Suppose that the random vector $Y = (y_1, \cdots, y_p)'$ has a finite fourth moment with mean $\boldsymbol{\mu}$ and covariance $\boldsymbol{\Sigma}$. In addition, there are n i.i.d. samples, Y_i, for $i = 1, \cdots, n$. Assume that $p/(n-1) \to c \in (0, \infty)$. Thus, both the sample size n and the response dimension p are allowed to go to infinity. Without assuming a specific distribution, Zheng et al. (2019) proposed a parameter estimator of $\boldsymbol{\beta} = (\beta_0, \beta_1, \cdots, \beta_K)'$ in (3.11) via the squared loss function given below:

$$\min_{\boldsymbol{\beta}} tr[(\widehat{\boldsymbol{\Sigma}}_u - \sum_{k=0}^{K} \beta_k \mathbf{W}_k)^2], \tag{3.13}$$

where $\widehat{\boldsymbol{\Sigma}}_u$ is the unbiased estimator of $\boldsymbol{\Sigma}$ defined in Sect. 2.2. After algebraic simplification, the resulting estimator is

$$\widehat{\boldsymbol{\beta}} = \mathbf{G}^{-1}\mathbf{h}, \tag{3.14}$$

where $\mathbf{G} = (tr(\mathbf{W}_i\mathbf{W}_j)) \in \mathbb{R}^{(K+1)\times(K+1)}$ and $\mathbf{h} = (tr(\widehat{\boldsymbol{\Sigma}}_u\mathbf{W}_0), tr(\widehat{\boldsymbol{\Sigma}}_u\mathbf{W}_1), \cdots, tr(\widehat{\boldsymbol{\Sigma}}_u\mathbf{W}_K))'$. They also demonstrated that $\widehat{\beta_k} = \beta_k + O_p(n^{-1})$, for $k = 0, 1, \cdots, K$. It is worth noting that $\widehat{\boldsymbol{\beta}}$ is similar to the unconstrained OLS estimator of Zou et al. (2017) with $n = 1$, and we will discuss its properties and related estimators extensively in Chap. 4.

Based on $\widehat{\boldsymbol{\beta}}$, we can obtain the linear covariance estimator $\widehat{\boldsymbol{\Sigma}} = \sum_{k=0}^{K} \widehat{\beta}_k \mathbf{W}_k$. Then, it is useful to assess whether model (3.11) is adequate. To this end, Zheng et al. (2019) considered the test $H_0 : \boldsymbol{\Sigma} = \sum_{k=0}^{K} \beta_k \mathbf{W}_k$, and introduced the following two test statistics:

$$T_1 = tr(\widehat{\boldsymbol{\Sigma}}_u\widehat{\boldsymbol{\Sigma}}^{-1}) - \log|\widehat{\boldsymbol{\Sigma}}_u\widehat{\boldsymbol{\Sigma}}^{-1}| - p, \text{ when } p < n-1; \tag{3.15}$$

$$T_2 = tr[(\widehat{\boldsymbol{\Sigma}}_u\widehat{\boldsymbol{\Sigma}}^{-1} - \mathbf{I})^2]. \tag{3.16}$$

Denote by $\widehat{\gamma}_1 \geq \cdots \geq \widehat{\gamma}_p$ the eigenvalues of $\widehat{\boldsymbol{\Sigma}}_u^{1/2}\widehat{\boldsymbol{\Sigma}}^{-1}\widehat{\boldsymbol{\Sigma}}_u^{1/2}$, and define $q = \min\{p, n-1\}$. Then T_1 can be extended to the case where $p \geq n-1$, and its unified form is given as follows:

$$T_1 = q(p^{-1}\sum_{j=1}^{q}\widehat{\gamma}_j - q^{-1}\sum_{j=1}^{q}\log\widehat{\gamma}_j) - q. \tag{3.17}$$

Under appropriate assumptions, Zheng et al. (2019) showed that the standardized T_1 and T_2 test statistics are asymptotically standard normal. In addition, they demonstrated that T_2 is a consistent test. Their simulation results indicated that both T_1 and T_2 control size well, but T_2 yields larger power.

3.3 Unconstrained Parameterization

In theory, the covariance matrix $\boldsymbol{\Sigma}$ is positive definite. In practice, however, it can be a challenging task to find an optimal estimator of $\boldsymbol{\Sigma}$ that is positive definite, particularly in high-dimensional data. Hence, Pinheiro and Bates (1996) proposed five unconstrained parameterizations to guarantee the resulting covariance matrix is positive definite. These five parameterizations include the Cholesky decomposition, the log-Cholesky decomposition, the spherical parameterization, the log-spectral decomposition, and the "Givens" parameterization. Since these methods may not lead to a simple statistical interpretation, this section focuses on the modified Cholesky decomposition (MCD) approach considered by Pourahmadi (1999) and the logarithmic-based parameterization studied by Leonard and Hsu (1992), Chiu et al. (1996), Pinheiro and Bates (1996), and Archakov and Hansen (2021).

3.3.1 Modified Cholesky Decomposition

The covariance matrix $\boldsymbol{\Sigma}$ is positive definite so that it can be factored by the Cholesky decomposition as

$$\boldsymbol{\Sigma} = \mathbf{G}\mathbf{G}', \tag{3.18}$$

where $\mathbf{G} = (g_{ij}) \in \mathbb{R}^{p\times p}$ is a lower-triangular matrix with positive diagonal elements. To interpret such a decomposition, (3.18) can be re-expressed in the following form:

$$\boldsymbol{\Sigma} = \mathbf{L}\mathbf{D}\mathbf{L}', \tag{3.19}$$

where $\mathbf{L} = \mathbf{G}\mathbf{Q}^{-1}$, $\mathbf{D} = \mathbf{Q}^2$, and $\mathbf{Q} = \mathrm{diag}(g_{11}, \cdots, g_{pp})$. The (3.19) is usually called the modified Cholesky decomposition (MCD) of $\boldsymbol{\Sigma}$, where $\mathbf{L}$ is a unit lower-triangular matrix and $\mathbf{D}$ is a positive diagonal matrix (see Pourahmadi 1999). Note that this decomposition employs the fact that a symmetric matrix is positive definite if and only if there exists a unit lower-triangular matrix and a unique diagonal matrix with positive diagonal elements (see, e,g., Newton 1988; Lyche 2020).

To illustrate the interpretation of (3.19), let $Y = (y_1, \cdots, y_p)'$ be a normal random vector with mean $\mathbf{0}$ and positive definite covariance matrix $\boldsymbol{\Sigma}$. In addition, for $j = 1, \cdots, p$, let $\widehat{y}_j$ be the least squares predictor of y_j conditional on its predecessors, $y_1, \cdots, y_{j-1}$, and let $\varepsilon_j = y_j - \widehat{y}_j$ be the prediction error with variance $\sigma_j^2 = \mathrm{var}(\varepsilon_j)$. Accordingly, there are unique scalars θ_{jk} satisfying

$$y_j = \sum_{k=1}^{j-1} \theta_{jk} y_k + \varepsilon_j, \tag{3.20}$$

such that $\widehat{y}_j = \sum_{k=1}^{j-1} \theta_{jk} y_k$, for $j = 2, \cdots, p$. In addition, assume that $\widehat{y}_1 = E(y_1) = 0$. Denote $\boldsymbol{\theta}_j = (\theta_{j1} \cdots, \theta_{j,j-1})'$, $\boldsymbol{\Sigma}_j$ as the $(j-1) \times (j-1)$ leading principal minor of $\boldsymbol{\Sigma}$, σ_{jj} as the j-th diagonal element of $\boldsymbol{\Sigma}$, and σ_j^* as the column vector of the first $j-1$ elements of the j-th column of $\boldsymbol{\Sigma}$. Then employing standard regression estimation theory (see, e.g., Anderson 2003), we obtain that $\boldsymbol{\theta}_j = \boldsymbol{\Sigma}_j^{-1}\sigma_j^*$ and $\sigma_j^2 = \sigma_{jj} - \sigma_j^{*'}\boldsymbol{\Sigma}_j^{-1}\sigma_j^*$.

Let $\boldsymbol{\varepsilon} = (\varepsilon_1, \cdots, \varepsilon_p)'$, $\mathrm{Cov}(\boldsymbol{\varepsilon}) = \mathbf{D} = \mathrm{diag}(\sigma_1^2, \cdots, \sigma_p^2)$, and

$$\mathbf{L}^{-1} = \begin{bmatrix} 1 & 0 & 0 & \cdots & 0 \\ -\theta_{21} & 1 & 0 & \cdots & 0 \\ -\theta_{31} & -\theta_{32} & 1 & \cdots & 0 \\ \vdots & \vdots & \vdots & \ddots & \vdots \\ -\theta_{p1} & -\theta_{p2} & -\theta_{p3} & \cdots & 1 \end{bmatrix},$$

be a unit-lower-triangular matrix, where θ_{jk} is the scalar in (3.20) for $j = 2, \cdots, p$ and $k = 1, \cdots, j-1$. Then we have $\boldsymbol{\varepsilon} = \mathbf{L}^{-1}Y$, $\mathrm{Cov}(\boldsymbol{\varepsilon}) = \mathbf{L}^{-1}\boldsymbol{\Sigma}\mathbf{L}^{-1'} = \mathbf{D} = \mathrm{diag}(\sigma_1^2, \cdots, \sigma_p^2)$, and $\boldsymbol{\Sigma}^{-1} = \mathbf{L}^{-1'}\mathbf{D}^{-1}\mathbf{L}^{-1}$. According to Theorem 4.2 of Lyche (2020), $\boldsymbol{\Sigma}^{-1}$ is positive definite. In sum, the θ_{jk}s are not only unconstrained for composing a positive definite matrix $\boldsymbol{\Sigma}$, but they are also the regression coefficients of model (3.20). More detailed illustrations of MCD can be found in Pourahmadi (1999).

As mentioned above, we assume that Y is normally distributed with mean $\mathbf{0}$ and covariance $\boldsymbol{\Sigma}$. Since model (3.20) contains $p(p+1)/2$ unknown parameters, we collect n random samples, $Y_1, \cdots, Y_n$, from $N(\mathbf{0}, \boldsymbol{\Sigma})$ with $n > p(p+1)/2$. Define $\mathbf{Y} = (Y_1, \cdots, Y_n)' \in \mathbb{R}^{n \times p}$ and $\mathbf{E} = (\boldsymbol{\varepsilon}_1, \cdots, \boldsymbol{\varepsilon}_n)' = (\varepsilon_{ij}) \in \mathbb{R}^{n \times p}$, for $i = 1, \cdots, n$ and $j = 1, \cdots, p$. Then, we have $\mathbf{Y} = \mathbf{E}\mathbf{L}'$ and $\mathbf{Y}\boldsymbol{\Sigma}^{-1}\mathbf{Y}' = \mathbf{E}\mathbf{D}^{-1}\mathbf{E}'$. In addition, the log-likelihood function after removing the constant $-(np/2)\log(2\pi)$ is

$$
\begin{aligned}
L(\boldsymbol{\Sigma}) = \log f(\mathbf{Y}|\boldsymbol{\Sigma}|) &= -\frac{n}{2}\log(|\boldsymbol{\Sigma}|) - \frac{1}{2}\sum_{i=1}^{n} Y_i'\boldsymbol{\Sigma}^{-1}Y_i \\
&= -\frac{n}{2}\log(|\mathbf{D}|) - \frac{1}{2}tr(\mathbf{Y}\boldsymbol{\Sigma}^{-1}\mathbf{Y}') \\
&= -\frac{n}{2}\sum_{j=1}^{p}\log(\sigma_j^2) - \frac{1}{2}\sum_{i=1}^{n}\sum_{j=1}^{p}\frac{\varepsilon_{ij}^2}{\sigma_j^2}. \qquad (3.21)
\end{aligned}
$$

After algebraic simplification, we obtain the following maximum likelihood estimators of $\boldsymbol{\Sigma}$, $\boldsymbol{\theta}_j$, σ_j^2, $\boldsymbol{\Sigma}_j$, σ_{jj}, and σ_j^*. They are $\widehat{\boldsymbol{\Sigma}} = \widehat{\boldsymbol{\Sigma}}_m = (\widehat{\sigma}_{ij})$, $\widehat{\boldsymbol{\theta}}_j = \widehat{\boldsymbol{\Sigma}}_j^{-1}\widehat{\sigma}_j^*$, $\widehat{\sigma}_j^2 = \widehat{\sigma}_{jj} - \widehat{\sigma}_j^{*\prime}\widehat{\boldsymbol{\Sigma}}_j^{-1}\widehat{\sigma}_j^*$, $\widehat{\boldsymbol{\Sigma}}_j$ is the $(j-1)\times(j-1)$ leading principal minor of $\widehat{\boldsymbol{\Sigma}}$, and $\widehat{\sigma}_j^*$ is the column vector of the first $j-1$ elements of the j-th column of $\widehat{\boldsymbol{\Sigma}}$, where $\widehat{\boldsymbol{\Sigma}}_m$ is the maximum likelihood estimator of $\boldsymbol{\Sigma}$ defined in Chap. 2.

In practice, it is useful to assess the significance of $\boldsymbol{\theta}_j$. Let $Y_{(j)}$ be the j-th column of $\mathbf{Y}$ for $j = 1, \cdots, p$. Then, Pourahmadi (1999) regressed $Y_{(j)}$ on $Y_{(1)}, \cdots, Y_{(j-1)}$, and obtained the multiple correlation coefficient R_j^2. Subsequently, he proposed a test statistic, $F_j = (p-j+1)/(j-2)[R_j^2/(1-R_j^2)]$, to test the null hypothesis $H_0 : \boldsymbol{\theta}_j = \mathbf{0}$. He also proposed a t-test statistic to examine the significance of each individual coefficient, θ_{jk}, for $j = 2, \cdots, p$ and $k = 1, \cdots, j-1$.

In addition to hypothesis testing, Huang et al. (2006) proposed a penalized covariance estimator to shrink irrelevant parameters in model (3.20). This approach is critical in high-dimensional data with $p(p+1)/2 \gg n$. Specifically, they considered the following penalized likelihood function with the L_m penalty:

$$
-2L(\boldsymbol{\Sigma}) + \lambda\sum_{j=2}^{p}\sum_{k=1}^{j-1}|\theta_{kj}|^m, \qquad (3.22)
$$

where λ is a tuning (thresholding) parameter and $m > 0$. They also developed an algorithm to compute the maximum penalized likelihood estimators and employed the cross-validation and generalized cross-validation methods to select the tuning parameter λ.

It is worth noting that the above Cholesky parameterization considered by Pourahmadi (1999) and Huang et al. (2006) relies on an important assumption, i.e., that Y_i is an ordered vector. As a result, changing the order of entries in Y_i can lead to a different covariance matrix. This motivates us to study an alternative approach in the next section.

3.3.2 Logarithmic-Based Parameterization

In practice, the observed data may not follow a natural order such as the chronology of time. Hence, it is essential to find a parameterization that can not only be easily interpreted without imposing constraints and ensure the covariance matrix is positive definite, but can also be invariant to the ordering of variables. To assure positive definiteness with unconstrained parameters, one possible approach is taking a logarithm transformation; see, e.g., Leonard and Hsu (1992), Chiu et al. (1996), Dieci (1996), Pinheiro and Bates (1996), and Deng and Tsui (2013).

Since $\mathbf{\Sigma}$ is positive definite, its spectral decomposition is $\mathbf{\Sigma} = \mathbf{V}\mathbf{\Omega}\mathbf{V}'$, where $\mathbf{V}$ is an orthogonal matrix such that $\mathbf{V}'\mathbf{V} = \mathbf{I}$, $\mathbf{\Omega} = \text{diag}(\omega_1, \cdots, \omega_p)$, and ω_j is the eigenvalue of $\mathbf{\Sigma}$, which is associated to its eigenvector V_j (i.e., the j-th column of $\mathbf{V}$), for $j = 1, \cdots, p$. Then the logarithm transformation of $\mathbf{\Sigma}$ is defined as $\mathbf{A} = \log(\mathbf{\Sigma}) = \mathbf{V}\log(\mathbf{\Omega})\mathbf{V}'$, where $\log(\mathbf{\Omega}) = \text{diag}(\log(\omega_1), \cdots, \log(\omega_p))$ and the ω_js are unconstrained parameters for $j = 1, \cdots, p$. Based on this definition, the logarithm transformation is invariant regardless of the order of variables.

As considered in the previous section, we collect n random samples, $Y_1, \cdots, Y_n$, from $N(\mathbf{0}, \mathbf{\Sigma})$, and define $\mathbf{Y} = (Y_1, \cdots, Y_n)' \in \mathbb{R}^{n \times p}$. Then, the log-likelihood function after removing the constant $-(np/2)\log(2\pi)$ is

$$L(\mathbf{A}) = \log f(\mathbf{Y}|\mathbf{A}) = -\frac{n}{2}tr(\mathbf{A}) - \frac{n}{2}tr[\widehat{\mathbf{\Sigma}}_s exp(-\mathbf{A})], \tag{3.23}$$

where $\widehat{\mathbf{\Sigma}}_s$ is the sample covariance defined in Chap. 2. This equation allows us to estimate a symmetric matrix $\mathbf{A}$ rather than estimating a positive definite matrix $\mathbf{\Sigma}$. However, it is a challenging task to obtain the MLE via (3.23) directly since this log-likelihood function includes the exponential term of $\mathbf{A}$. To this end, Leonard and Hsu (1992) employed the Volterra integral equation (Bellman 1997) iteratively to approximate the second term on the right hand side of (3.23) and obtained a likelihood-approximation function. The initial value used in their iteration process is the sample covariance $\widehat{\mathbf{\Sigma}}_s$, which can be singular when $p > n$. Thus, Deng and Tsui (2013) introduced a different initial estimator (e.g., $\widehat{\mathbf{\Sigma}}_s + \delta\mathbf{I}$, where δ is a small and known positive value) and obtained an alternative likelihood-approximation function. Subsequently, they proposed the logarithm-transformed matrix estimator (Log-ME) by taking into account the penalty function of the matrix $\mathbf{A}$ with the squared Frobenius norm. They also demonstrated that their likelihood-approximation function performs well.

Although the logarithm transformation of $\mathbf{\Sigma}$ has several merits, it lacks clear interpretation. Hence, Archakov and Hansen (2021) proposed a parameterization with some interpretations. Their logarithm parameterization consists of two main steps: (i) take a logarithmic transformation of the correlation matrix, $\log(\mathbf{C})$, and (ii) stack the lower off-diagonal elements into a vector $\boldsymbol{\gamma} = \boldsymbol{\gamma}(\mathbf{C}) := \text{vecl}(\log(\mathbf{C}))$ with dimension $p(p-1)/2$, where $\mathbf{C}$ is a correlation matrix and vecl represents a column vector formed by arranging the lower off-diagonal elements of a given matrix. For the sake of convenience, we name this procedure logarithmic-based parameterization (LBP). Define a vector $\boldsymbol{\theta} = (\frac{1}{2}\log(\sigma_{11}), \cdots, \frac{1}{2}\log(\sigma_{pp}), \boldsymbol{\gamma}')'$ with dimension $p(p+1)/2$, that is unconstrained, where σ_{jj} is the j-th diagonal element of $\boldsymbol{\Sigma}$, for $j = 1, \cdots, p$. Then, Archakov and Hansen (2021) showed that any positive definite $p \times p$ covariance matrix, $\boldsymbol{\Sigma}$, can be expressed as a unique vector $\boldsymbol{\theta}$ in $\mathbb{R}^{p(p+1)/2}$. Note that the relationship between $\mathbf{C}$ and $\boldsymbol{\Sigma}$ is given below:

$$\mathbf{C} = [\text{diag}(\boldsymbol{\Sigma})]^{-1/2}\boldsymbol{\Sigma}[\text{diag}(\boldsymbol{\Sigma})]^{-1/2}. \tag{3.24}$$

We next use $p = 2$ to illustrate LBP in detail. The 2×2 correlation matrix is

$$\mathbf{C} = \begin{bmatrix} 1 & \rho \\ \rho & 1 \end{bmatrix}, \tag{3.25}$$

where $-1 < \rho < 1$. Applying the spectral decomposition to $\mathbf{C}$, we obtain that $\mathbf{C} = \mathbf{U}\Gamma\mathbf{U}'$, where

$$\mathbf{U} = \frac{1}{\sqrt{2}}\begin{bmatrix} 1 & 1 \\ 1 & -1 \end{bmatrix} \text{ and } \Gamma = \begin{bmatrix} 1+\rho & 0 \\ 0 & 1-\rho \end{bmatrix}. \tag{3.26}$$

Accordingly,

$$\log(\mathbf{C}) = \mathbf{U}\log(\Gamma)\mathbf{U}' = \frac{1}{2}\begin{bmatrix} \log(1-\rho^2) & \log(\frac{1+\rho}{1-\rho}) \\ \log(\frac{1+\rho}{1-\rho}) & \log(1-\rho^2) \end{bmatrix}, \tag{3.27}$$

which leads to $\gamma(\mathbf{C}) = \frac{1}{2}\log(\frac{1+\rho}{1-\rho})$ and $\boldsymbol{\theta} = \frac{1}{2}(\log(\sigma_{11}), \log(\sigma_{22}), \log(\frac{1+\rho}{1-\rho}))'$. It is worth noting that $\gamma(\mathbf{C})$ is Fisher's transformation (see Fisher 1915), and its estimator, $\frac{1}{2}\log(\frac{1+\widehat{\rho}}{1-\widehat{\rho}})$, has been widely used in testing the correlation between any two continuous random variables, where $\widehat{\rho}$ is the sample correlation coefficient.

When $p > 2$, $\mathbf{C}$ is the compound symmetry matrix in (3.9) with $\sigma^2 = 1$. Then Archakov and Hansen (2021) showed that the off-diagonal elements of $\log(\mathbf{C})$ are identical and equal to $\frac{1}{p}\log(1 + \frac{p\rho}{1-\rho})$. In addition, Archakov and Hansen (2021) proposed the element-wise Fisher transformed correlations so that one can employ their corresponding sample correlations to test the relationship between any paired variables.

To enhance the applicability of logarithmic-based parameterization, Archakov and Hansen (2021) provided theoretical foundations in establishing the one-to-one correspondence between the correlation matrix $\mathbf{C}$ and $\boldsymbol{\gamma}(\mathbf{C})$. As a result, the inverse mapping, $\mathbf{C}(\boldsymbol{\gamma})$, is well-defined. Under LBP, they not only developed an algorithm for computing $\mathbf{C}(\boldsymbol{\gamma})$ but also demonstrated invariance to the reordering of variables. To show the asymptotic properties of the estimator of $\boldsymbol{\gamma}$ ($\widehat{\boldsymbol{\gamma}}$), we calculate the sample estimator $\widehat{\mathbf{C}}$ via the n random samples, $Y_1, \cdots, Y_n$, from $N(\boldsymbol{\mu}, \boldsymbol{\Sigma})$ (or from a probability distribution whose first four moments exist). Employing the result of Neudecker and Wesselman (1990), we then have that $\sqrt{n}[\text{vec}(\widehat{\mathbf{C}}) - \text{vec}(\mathbf{C})] \xrightarrow{d}$

$N(\mathbf{0}, \mathbf{\Lambda})$, as $n \to \infty$, where $\mathbf{\Lambda} \in \mathbb{R}^{p^2 \times p^2}$ is the asymptotic covariance of $\sqrt{n}\text{vec}(\widehat{\mathbf{C}})$. Using this result, Archakov and Hansen (2021) demonstrated that

$$\sqrt{n}(\widehat{\boldsymbol{\gamma}} - \boldsymbol{\gamma}) \xrightarrow{d} N(\mathbf{0}, \mathbf{\Lambda}_{\gamma}), \tag{3.28}$$

as $n \to \infty$, where $\mathbf{\Lambda}_{\gamma} = \tilde{\mathbf{E}}\tilde{\mathbf{J}}\mathbf{\Lambda}\tilde{\mathbf{J}}'\tilde{\mathbf{E}}'$, $\tilde{\mathbf{J}} = \partial\text{vec}(\log \mathbf{C})/\partial\text{vec}(\mathbf{C}) \in \mathbb{R}^{p^2 \times p^2}$ is a Jacobian matrix, $\tilde{\mathbf{E}} \in \mathbb{R}^{\frac{1}{2}p(p-1)\times p^2}$ is an elimination matrix so that $\tilde{\mathbf{E}}(\text{vec}(\mathbf{M})) = \text{vecl}(\mathbf{M})$ for any $p \times p$ matrix $\mathbf{M}$. The detailed expression of $\tilde{\mathbf{J}}$ can be found in Linton and McCrorie (1995) and Archakov and Hansen (2021).

In sum, LBP has some nice and important features such as being unconstrained, invariant, and easily interpretable. Hence, it has the potential to be used broadly across many subjects and studies, such as in financial time-series models and social network models.

3.4 Empirical Examples

To illustrate the application of structured covariance matrices and unconstrained param-etrizations, this section briefly discusses four empirical examples that were studied in the literature.

The first example considers the covariance matrix of an AR(1) process. The dataset is taken from Hatch (1980), who investigated the effect of different amounts of human heart rate biofeedback on subjects' heart rate. Later, Hearne et al. (1983) only used the data from Session 4 (speeding performance), which consists of three different groups (i.e., amounts of feedback: 100%, 75% and control) with six subjects per group. In addition, heart rate was monitored for 3 minutes across five periods, and each period is 2 minutes apart. Then, Hearne et al. (1983) employed the likelihood ratio test to assess if the covariance matrix is spherical. After this hypothesis was rejected, they subsequently tested whether the AR(1) covariance structure is adequate. The result did not reject the first-order autocorrelation process.

The second example is related to the compound symmetry structure. The dataset can be found in Potthoff and Roy (1964). It includes 11 girls and 16 boys, and the response of each subject is the distance (mm) from the center of the pituitary gland to the pterygomaxillary fissure, measured at the ages of 8, 10, 12, and 14. Based on this dataset, Jennrich and Schluchter (1986) studied eight different models, such as unstructured, sphericity, AR(1), banded, and compound symmetry. They found that the linear longitudinal model with the compound symmetry covariance matrix performs well. In practice, one may not have knowledge about the true model and its covariance structure. This example indicates that we can consider several candidate models and then find one or two adequate models based on model fitting, hypothesis testing, computational efficiency, forecasting accuracy, and/or model selection.

The third example is the logarithm transformation studied by Deng and Tsui (2013). They used the ionosphere data from the UC Irvine Machine Learning

Repository (see Sigillito et al. 1989). These data were collected by a system that consists of a phased array of high-frequency antennas. As a result, the data contain 351 observations with 34 variables. In addition, the response is coded as 1 for "Good" or -1 for "Bad" with respect to the radar returns from the ionosphere. Based on these data, Deng and Tsui (2013) conducted a classification analysis for two categories. Specifically, they employed their proposed Log-ME method to compute the covariance matrix via the training data with 40 observations and then use the testing data to calculate misclassification errors based on linear discriminate analysis. They also compared their results with those using different covariance matrix estimates. In sum, Log-ME performs well.

The fourth example is associated with the modified Cholesky decomposition of a large covariance matrix. Specifically, Huang et al. (2006) employed their penalized covariance estimate to forecast the arrival pattern of telephone calls. The dataset consists of 239 days, and each day collects phone calls over 102 intervals with the interval length of 10 minutes. The goal of their study is to use the arrival patterns at earlier times of the day (i.e., the first 51 intervals) to forecast the number of arrivals at later times in the day (i.e., the last 51 intervals). We denote them as Y_1 and Y_2, respectively. Then, they split the data into the training dataset with the first 205 days and the testing dataset with the last 34 days. Subsequently, Huang et al. (2006) used the training dataset with the square root transformation to estimate the mean and covariance matrix. Afterward, they conducted forecasts based on the conditional mean $E(Y_2|Y_1)$ of the multivariate normal distribution of $Y = (Y_1, Y_2)$ from the testing dataset. The average absolute forecast error indicates that the penalized covariance estimate of Huang et al. (2006) is superior to the sample covariance estimate.

Chapter 4
Covariance Regression Models

We provide our main motivation for introducing covariance regression models. Then, we present covariance regression models and study theoretical properties of regression parameter estimators. We subsequently analyze non-Gaussian covariance regression models and explore the covariance model with general linear structure. Finally, three empirical examples are briefly discussed.

4.1 Motivation

In regression analysis, one usually links the response variable $y \in \mathbb{R}^1$ to the explanatory vector $X = (x_1, \cdots, x_q)' \in \mathbb{R}^q$ via the following mean response model:

$$E(y|X) = \mu = f(X'\boldsymbol{\alpha}), \tag{4.1}$$

where f is a known or unknown function and $\boldsymbol{\alpha} \in \mathbb{R}^q$ is a known or unknown vector. By varying data characteristics, parameter constraints, and variable structures, this mean regression setting can encompass many known models, such as linear and nonlinear regressions, generalized parametric regression, robust regression, quantile regression, random effects, mixed, error in variables, functional data, varying coefficient, nonparametric, semi-parametric and single-index regressions, regression trees, and autoregressive (AR) processes, among others. When the dimension of the response variable is greater than 1, the specific settings of f, X and $\boldsymbol{\alpha}$ can lead to multivariate regression, seemingly unrelated regression, longitudinal (or panel) data analysis, functional data analysis, and vector autoregressive (VAR) processes, among others.

It is worth noting that some of the above models implicitly link the variance components of X and $\boldsymbol{\alpha}$ to the variation in the response variables, such as the

W. Lan, C.-L. Tsai, *Covariance Analysis and Beyond*,
https://doi.org/10.1007/978-3-032-08796-6_4

random effects model, functional data, longitudinal data, and AR processes. In addition, some models not mentioned above can directly link the variance of responses to covariates in regression settings. For example, Carroll and Ruppert (1988) considered the following variance function:

$$\text{var}(y|X, Z) = \sigma^2 m^2(\mu, Z, \boldsymbol{\theta}), \tag{4.2}$$

where y, X, and μ are defined in (4.1), m is some function, $\sigma > 0$ is a scaled parameter, $Z \in \mathbb{R}^r$ is a known vector of covariates, and $\boldsymbol{\theta} \in \mathbb{R}^s$ is an unknown parameter vector. This motivates us to link the covariance matrix of multivariate responses to explanatory matrices in a regression setting such that each element of the covariance matrix is a combination of its associated explanatory elements. The detailed discussion is given below.

4.2 Covariance Regression Models

In multivariate regression, it is common to relate the mean vector of the response, $Y = (y_1, \cdots, y_p)'$, to the K components of the explanatory vector. As an alternative to the mean vector, however, Anderson (1973) proposed the linear covariance model, defined in Sect. 3.2, to link the covariance $\boldsymbol{\Sigma}$ of Y to a set of known $p \times p$ symmetric matrices. He also discussed several applications of this model such as mixed models (see, e.g., Rao 1972; McCulloch et al. 2008; Demidenko 2013) and moving average processes (see, e.g., Shumway and Stoffer 2017).

To broaden the application of Anderson's (1973) linear covariance model, Zou et al. (2017) introduced the covariance regression model. Let $Y_i = (y_{i1}, \cdots, y_{ip})'$ be the response vector collected from the i-th subject (or observation or sample) for $i = 1, \cdots, n$. For each subject, we suppress the subscript i and let y_j be the j-th response and denote its associated auxiliary information vector by $Z_j = (z_{j1}, \cdots, z_{jK})' \in \mathbb{R}^K$, which can be continuous or discrete. For example, in spatial data, y_j can be the j's housing price associated with the geographical coordinates of the location whose auxiliary information, such as household income, is captured in Z_j. In addition, in social network analysis, y_j can be a measure of actor j's activity level where Z_j captures the actor's demographic information and social behaviors. Moreover, in finance, y_j can be the j-th firm's stock return and Z_j its financial fundamentals, such as book-to-market ratio, market value, earnings per share, return on assets, current ratio, and cash flow, etc.

To measure the distance between any two auxiliary information variables Z_{j_1} and Z_{j_2} in K-dimensional space, we adapt the concept of Johnson and Wichern's (2007) pairwise comparisons and consider the similarity measure $w(Z_{j_1}, Z_{j_2})$. For continuous variables, the similarity can be defined as a decreasing function of distance, e.g., $w(Z_{j_1}, Z_{j_2}) = \exp\left\{-d^2(Z_{j_1}, Z_{j_2})\right\}$, where $d(\cdot, \cdot)$ can be Euclidean distance or Mahalanobis distance. For discrete variables, one can employ similarity measures as given in Johnson and Wichern (2007). For example, the similarity in social network studies (Wasserman and Faust 1994) can be defined as follows:

$$w\left(Z_{j_1}, Z_{j_2}\right) = \begin{cases} 1 & \text{if two individuals } j_1 \text{ and } j_2 \text{ know each other;} \\ 0 & \text{otherwise.} \end{cases}$$

With a slight abuse of notation, the above Z_{j_1} and Z_{j_2} are scalars.

The above definition indicates that responses with higher similarity to each other are more likely to be correlated. Therefore, it is natural to model the covariance matrix $\boldsymbol{\Sigma}$ as a function of the similarity measure $w\left(Z_{j_1}, Z_{j_2}\right)$. For the sake of simplicity, Zou et al. (2017) considered the case with $n = 1$, i.e., $Y = Y_1$. They then proposed the following covariance regression model:

$$YY' = \beta_0 \mathbf{W}_0 + \beta_1 \mathbf{W}(Z^{(1)}) + \cdots + \beta_K \mathbf{W}(Z^{(K)}) + \mathscr{E}, \tag{4.3}$$

where the response Y is standardized to have mean zero, $Z^{(k)} := (z_{1k}, \cdots, z_{pk})' \in \mathbb{R}^p$ for $k = 1, \cdots, K$, $\mathbf{W}\left(Z^{(k)}\right) = \mathbf{W}_k = \left(w\left(z_{j_1k}, z_{j_2k}\right)\right)_{p\times p} \in \mathbb{R}^{p\times p}$ is the similarity matrix with $j_1, j_2 = 1, \cdots, p$, $\mathbf{W}_0 = \mathbf{I}_p$ is the $p \times p$ identity matrix, $\boldsymbol{\beta} = (\beta_0, \beta_1, \cdots, \beta_K)'$ is the vector of regression coefficients, and $\mathscr{E}$ is a random matrix that satisfies $E(\mathscr{E}) = 0$. As a result,

$$\boldsymbol{\Sigma}(\boldsymbol{\beta}) = \beta_0 \mathbf{I}_p + \sum_{k=1}^{K} \beta_k \mathbf{W}_k, \tag{4.4}$$

where $\boldsymbol{\Sigma}(\boldsymbol{\beta}) = E(YY')$. For the sake of convenience, we also refer to (4.4) as the covariance regression model. This model relates the covariance of responses to the auxiliary information, and β_k describes the similarity of the comovement of $\boldsymbol{\Sigma}$ and $\mathbf{W}_k$ for $k = 1, \cdots, K$. Specifically, the β_k represents the change in the (j_1, j_2)-th component of $\boldsymbol{\Sigma}$ when the (j_1, j_2)-th component of $\mathbf{W}_k$ increases by one unit, for all $j_1 \neq j_2$ and $k = 1, \cdots, K$.

Hereafter, we assume that the $\mathbf{W}_k$ are known, for $k = 1, \cdots, K$, and K is fixed. In addition, we assume that the true covariance matrix of Y is $\mathbb{E}(YY') = \boldsymbol{\Sigma}(\boldsymbol{\beta}^0)$, where $\boldsymbol{\beta}^0 = (\beta_0^0, \cdots, \beta_K^0) \in \mathbb{R}^{K+1}$ is the true regression vector of $\boldsymbol{\beta}$ in (4.4). For the sake of simplicity, we denote $\boldsymbol{\Sigma}_0 = \boldsymbol{\Sigma}(\boldsymbol{\beta}^0)$. Model (4.4) incorporates the auxiliary information, which broadens the application of the linear covariance model proposed by Anderson (1973) and discussed in Sect. 3.2. Note that one can also replace $\boldsymbol{\Sigma}(\boldsymbol{\beta})$ in (4.4) with the correlation matrix that links to similarity matrices. In addition, extending the covariance regression model above to the case with unknown $\mathbf{W}_k$ would expand its applications; see, e.g., covariance matrix with missing observations (Lounici 2012).

Model (4.4) contains several known structured covariance matrices such as a banded covariance matrix (Bickel and Levina 2008a), i.e., $\boldsymbol{\Sigma} = (\theta_{j_1 j_2} \mathbf{1}\{|j_1 - j_2| \leq b\})_{p\times p}$, where b is any integer satisfying $0 \leq b \leq p - 1$. When the banded covariance matrix has equal diagonal elements σ^2, Zou et al. (2017) expressed it by $\beta_0 \mathbf{I}_p + \sum_{k=1}^{K} \beta_k \mathbf{W}_k$, where $K = b(2p - b - 1)/2$, $\beta_0 = \sigma^2$, $\beta_1 = \theta_{21}$,

$\beta_2 = \theta_{32}, \beta_{p-1} = \theta_{p,p-1}, \beta_p = \theta_{31}, \cdots, \beta_{b(2p-b-1)/2} = \theta_{p,p-b}, \mathbf{W}_1 = \mathbf{G}_{21}+\mathbf{G}'_{21}$, $\mathbf{W}_2 = \mathbf{G}_{32} + \mathbf{G}'_{32}, \cdots, \mathbf{W}_{p-1} = \mathbf{G}_{p,p-1} + \mathbf{G}'_{p,p-1}, \mathbf{W}_p = \mathbf{G}_{31} + \mathbf{G}'_{31}, \cdots$, and $\mathbf{W}_{b(2p-b-1)/2} = \mathbf{G}_{p,p-b} + \mathbf{G}'_{p,p-b}$, where $\mathbf{G}_{j_1 j_2}$ is the $p \times p$ matrix with 1 on the (j_1, j_2)-th entry and 0 otherwise, as defined in Sect. 3.2. In addition, the covariance matrix of MA(1) ($\mathbf{\Sigma}_M$) in Sect. 3.1 is a special case of (4.4) when $\beta_0 = (1+\theta^2)\sigma^2$, $\beta_1 = \theta\sigma^2$, $\mathbf{W}_0 = \mathbf{I}_p$, and $\mathbf{W}_1$ has 1 on the first off-diagonal elements and 0 elsewhere. Finally, Lan et al.'s (2018) polynomial network covariance model is a special application of (4.4).

In practice, $\mathbf{\Sigma}(\beta)$ in (4.4) is unknown. Thus, Zou et al. (2017) proposed three parameter estimators: the maximum likelihood estimator (MLE), the ordinary least squares (OLS) estimator, and the feasible generalized least squares (FGLS) estimator. To demonstrate the theoretical properties of these three estimators, Zou et al. (2017) introduced the following three conditions:

(C4.1) For any symmetric matrices in $\{\mathbf{W}_k \in \mathbb{R}^{p\times p} : k = 0, \cdots, K, K < \infty\}$, there exists $b_1 > 0$ such that $\sup_{p\geq 1} \|\mathbf{W}_k\|_2 \leq b_1 < \infty$, where $\|\mathbf{W}_k\|_2$ is the spectral norm of $\mathbf{W}_k$.

(C4.2) There exists $b_2 > 0$ such that $\lambda_p(\mathbf{\Sigma}_0) > b_2$ for any $p \geq 1$, where $\lambda_p(\mathbf{\Sigma}_0)$ is the smallest eigenvalue of $\mathbf{\Sigma}_0$.

(C4.3) The matrix $p^{-1}(\mathrm{tr}\,(\mathbf{\Sigma}_0^d \mathbf{W}_k \mathbf{\Sigma}_0^d \mathbf{W}_l))_{(K+1)\times(K+1)}$ converges to a positive definite matrix $\mathcal{Q}_d \in \mathbb{R}^{(K+1)\times(K+1)}$, respectively, for $d = -2, -1, 0, 1$, where $\mathbf{\Sigma}_0^0 := \mathbf{I}_p$.

Suppose that Y is multivariate normal with mean 0 and covariance $\mathbf{\Sigma}_0$. Then the maximum likelihood estimator $\widehat{\boldsymbol{\beta}}_{p,MLE}$ can be obtained by maximizing the log-likelihood function given below.

$$\ell_p(\boldsymbol{\beta}) = -\frac{p}{2}\log(2\pi) - \frac{1}{2}\sum_{j=1}^{p}\log\{\lambda_j(\mathbf{\Sigma}(\boldsymbol{\beta}))\} - \frac{1}{2}Y'\mathbf{\Sigma}^{-1}(\boldsymbol{\beta})Y, \tag{4.5}$$

where λ_j is the j-th largest eigenvalue of $\mathbf{\Sigma}(\boldsymbol{\beta})$. Define the parameter space as follows:

$$\mathcal{B} := \{\boldsymbol{\beta} : \mathbf{\Sigma}(\boldsymbol{\beta}) > 0\}, \tag{4.6}$$

where $\mathbf{\Sigma}(\boldsymbol{\beta}) > 0$ indicates that $\mathbf{\Sigma}(\boldsymbol{\beta})$ is a positive definite matrix. Note that the function in (4.5) includes the log-eigenvalues of matrix $\mathbf{\Sigma}(\boldsymbol{\beta})$, which implies that $\widehat{\boldsymbol{\beta}}_{p,MLE} \in \mathcal{B}$. However, finding the maximal value of (4.5) can be computationally infeasible when p is large. Hence, Zou et al. (2017) applied the conjugate gradient algorithm of Fletcher (1987) to find the MLE that involves only the first-order derivative of the log-likelihood function (4.5).

We next present the asymptotic property of $\widehat{\boldsymbol{\beta}}_{p,MLE}$ from Zou et al. (2017).

Theorem 4.1 *Under Conditions **(C4.1)–(C4.3)**,*

$$p^{1/2}\left(\widehat{\boldsymbol{\beta}}_{p,MLE}-\boldsymbol{\beta}^{0}\right)\xrightarrow{d}N\left(0,2\mathcal{Q}_{-1}^{-1}\right),$$

as $p\to\infty$.

The proofs of this theorem and three subsequent theorems can be found in Zou et al. (2017).

As mentioned above, finding the maximal value of (4.5) becomes computationally challenging when p is large. Hence, Zou et al. (2017) proposed the OLS estimator, which is obtained by minimizing

$$D_p(\boldsymbol{\beta}):=\left\|YY'-\boldsymbol{\Sigma}(\boldsymbol{\beta})\right\|_F^2, \tag{4.7}$$

where $\|\mathbf{A}\|_F=\{\mathrm{tr}\left(\mathbf{A}'\mathbf{A}\right)\}^{1/2}$ denotes the Frobenius norm for any generic matrix $\mathbf{A}$, and it is also defined in Sect. 1.6.2. The resulting estimator is

$$\widehat{\boldsymbol{\beta}}_{p,OLS}=\left(\mathrm{tr}\left(\mathbf{W}_k\mathbf{W}_l\right)\right)^{-1}_{(K+1)\times(K+1)}\left(Y'\mathbf{W}_kY\right)_{(K+1)\times 1}, \tag{4.8}$$

The calculation of the OLS estimator is simpler and faster than that of the MLE since it does not require an iterative process. The asymptotic property of $\widehat{\boldsymbol{\beta}}_{p,OLS}$ is given below.

Theorem 4.2 *Under Conditions (C4.1)–(C4.3),*

$$p^{1/2}\left(\widehat{\boldsymbol{\beta}}_{p,OLS}-\boldsymbol{\beta}^{0}\right)\xrightarrow{d}N\left(0,2\mathcal{Q}_{0}^{-1}\mathcal{Q}_1\mathcal{Q}_{0}^{-1}\right),$$

as $p\to\infty$.

Although $\widehat{\boldsymbol{\beta}}_{p,OLS}$ can be calculated effectively, the associated covariance matrix estimator, $\boldsymbol{\Sigma}(\widehat{\boldsymbol{\beta}}_{p,OLS})$, may not be positive definite. To fix this problem, we can impose the constraint $\boldsymbol{\beta}\in\mathcal{B}$. This yields the constrained OLS (OLS+) estimator, $\widehat{\boldsymbol{\beta}}_{p,OLS+}=\arg\min_{\boldsymbol{\beta}\in\mathcal{B}}D_p(\boldsymbol{\beta})$. Zou et al. (2017) demonstrated that $\widehat{\boldsymbol{\beta}}_{p,OLS+}$ has the same asymptotic distribution as that of $\widehat{\boldsymbol{\beta}}_{p,OLS}$. By Theorems 4.1 and 4.2 and the above discussion, we have that $\widehat{\boldsymbol{\beta}}_{p,MLE}$, $\widehat{\boldsymbol{\beta}}_{p,OLS}$, and $\widehat{\boldsymbol{\beta}}_{p,OLS+}$ converge to multivariate normal and yield a $p^{1/2}$-estimator of $\boldsymbol{\beta}^0$ as $p\to\infty$. Based on these results, Zou et al. (2017) showed the stochastic convergence rates of $\boldsymbol{\Sigma}(\widehat{\boldsymbol{\beta}})$ and $\boldsymbol{\Sigma}^{-1}(\widehat{\boldsymbol{\beta}})$ given below, where $\widehat{\boldsymbol{\beta}}$ can be $\widehat{\boldsymbol{\beta}}_{p,MLE}$, $\widehat{\boldsymbol{\beta}}_{p,OLS}$, or $\widehat{\boldsymbol{\beta}}_{p,OLS+}$.

Theorem 4.3 *Under Conditions (C4.1)–(C4.3),*

$$\left\|\boldsymbol{\Sigma}(\widehat{\boldsymbol{\beta}})-\boldsymbol{\Sigma}_0\right\|_2=O_p\left(p^{-1/2}\right)\ \text{and}\ \left\|\boldsymbol{\Sigma}^{-1}(\widehat{\boldsymbol{\beta}})-\boldsymbol{\Sigma}_0^{-1}\right\|_2=O_p\left(p^{-1/2}\right);$$

$$\left\|\boldsymbol{\Sigma}(\widehat{\boldsymbol{\beta}})-\boldsymbol{\Sigma}_0\right\|_F^2\xrightarrow{d}\mathcal{Z}'\mathcal{Q}_0\mathcal{Z}\ \text{and}\ \left\|\boldsymbol{\Sigma}^{-1}(\widehat{\boldsymbol{\beta}})-\boldsymbol{\Sigma}_0^{-1}\right\|_F^2\xrightarrow{d}\mathcal{Z}'\mathcal{Q}_{-2}\mathcal{Z},$$

as $p\to\infty$*, where* $\mathcal{Z}$ *is the convergence in multivariate normal distribution of* $p^{1/2}(\widehat{\boldsymbol{\beta}}-\boldsymbol{\beta}^0)$.

Accordingly, the orders of $p^{-1/2}\|\mathbf{\Sigma}(\widehat{\boldsymbol{\beta}}) - \mathbf{\Sigma}_0\|_F$ and $p^{-1/2}\|\mathbf{\Sigma}^{-1}(\widehat{\boldsymbol{\beta}}) - \mathbf{\Sigma}_0^{-1}\|_F$ are exactly equal to $p^{-1/2}$. Furthermore, the above theorem, in conjunction with the following inequality:

$$p^{-1/2}\left\|\mathbf{\Sigma}(\widehat{\boldsymbol{\beta}}) - \mathbf{\Sigma}_0\right\|_F \leq \left\|\mathbf{\Sigma}(\widehat{\boldsymbol{\beta}}) - \mathbf{\Sigma}_0\right\|_2, \tag{4.9}$$

implies that the order of $\|\mathbf{\Sigma}(\widehat{\boldsymbol{\beta}}) - \mathbf{\Sigma}_0\|_2$ and $\|\mathbf{\Sigma}^{-1}(\widehat{\boldsymbol{\beta}}) - \mathbf{\Sigma}_0^{-1}\|_2$ is also exactly equal to $p^{-1/2}$. Consequently, $\mathbf{\Sigma}(\widehat{\boldsymbol{\beta}})$ and $\mathbf{\Sigma}^{-1}(\widehat{\boldsymbol{\beta}})$ are consistent estimators of $\mathbf{\Sigma}_0$ and $\mathbf{\Sigma}_0^{-1}$, respectively, in both Frobenius norm and spectral norm.

Based on Theorems 4.1 and 4.2, we find that $\widehat{\boldsymbol{\beta}}_{p,OLS}$ is less efficient than $\widehat{\boldsymbol{\beta}}_{p,MLE}$ since its asymptotic variance, $2\mathcal{Q}_0^{-1}\mathcal{Q}_1\mathcal{Q}_0^{-1}$, is larger than that of $\widehat{\boldsymbol{\beta}}_{p,MLE}$, $2\mathcal{Q}_{-1}^{-1}$. By Theorem 4.3, Zou et al. (2017) proposed the FGLS estimator, which improves the efficiency of the OLS estimator. Specifically, the FGLS estimator can be obtained by minimizing

$$\widetilde{D}_p(\boldsymbol{\beta}) = \mathrm{vec}'\left(YY' - \mathbf{\Sigma}(\boldsymbol{\beta})\right)\left(\widehat{\mathbf{\Sigma}}^{-1} \otimes \widehat{\mathbf{\Sigma}}^{-1}\right)\mathrm{vec}\left(YY' - \mathbf{\Sigma}(\boldsymbol{\beta})\right), \tag{4.10}$$

where $\widehat{\mathbf{\Sigma}}^{-1} = \mathbf{\Sigma}^{-1}(\widehat{\boldsymbol{\beta}}_{p,OLS+})$ is the consistent estimator of $\mathbf{\Sigma}_0^{-1}$ and $\otimes$ represents the Kronecker product of two matrices, as defined in Sect. 1.6. After algebraic simplification, the FGLS estimator is given as follows:

$$\widehat{\boldsymbol{\beta}}_{p,FGLS} = \left(\mathrm{tr}\left(\widehat{\mathbf{\Sigma}}^{-1}\mathbf{W}_k\widehat{\mathbf{\Sigma}}^{-1}\mathbf{W}_l\right)\right)^{-1}_{(K+1)\times(K+1)}\left(Y'\widehat{\mathbf{\Sigma}}^{-1}\mathbf{W}_k\widehat{\mathbf{\Sigma}}^{-1}Y\right)_{(K+1)\times 1}. \tag{4.11}$$

In the following theorem, Zou et al. (2017) showed that $\widehat{\boldsymbol{\beta}}_{p,FGLS}$ is asymptotically as efficient as $\widehat{\boldsymbol{\beta}}_{p,MLE}$.

Theorem 4.4 *Under Conditions(C4.1) – (C4.3),*

$$p^{1/2}\left(\widehat{\boldsymbol{\beta}}_{p,FGLS} - \boldsymbol{\beta}^0\right) \xrightarrow{d} N\left(0, 2\mathcal{Q}_{-1}^{-1}\right),$$

as $p \to \infty$.

Note that $\widehat{\boldsymbol{\beta}}_{p,FGLS}$ may not lie in the open set $\mathcal{B}$. Hence, Zou et al. (2017) considered FGLS estimator $\widehat{\boldsymbol{\beta}}_{p,FGLS+} = \arg\min_{\boldsymbol{\beta}\in\mathcal{B}} \widetilde{D}_p(\boldsymbol{\beta})$. It can be shown that $\widehat{\boldsymbol{\beta}}_{p,FGLS+}$ and $\widehat{\boldsymbol{\beta}}_{p,FGLS}$ have the same asymptotic distribution as $p \to \infty$.

In sum, we have obtained parameter estimators of $\boldsymbol{\beta}$ in covariance regression models and studied their theoretical properties. It is worth noting that the estimator $\widehat{\boldsymbol{\beta}}_{p,FGLS+}$ is not only efficient but also computationally convenient. To assure OLS and FGLS estimators lie in the parameter space $\mathcal{B}$, Zou et al. (2017) provided a detailed algorithm to calculate $\widehat{\boldsymbol{\beta}}_{p,OLS+}$ and $\widehat{\boldsymbol{\beta}}_{p,FGLS+}$. In addition, they presented an empirical example using stock returns to illustrate the usefulness of covariance regression models and these estimators. In addition to the covariance regression model, it is worth noting that Xu and Li (2025) employed the Frèchet regression model on Bures–Wasserstein manifolds (Petersen and Müller 2019) to obtain the estimators of covariance matrices and to introduce a test statistic.

4.3 Non-Gaussian Covariance Regression Models

In the last section, we presented linear covariance regression models and introduced five parameter estimators: the maximum likelihood, ordinary least squares, constrained ordinary least squares, feasible generalized least squares, and constrained feasible generalized least squares estimators. Under the normality assumption, Zou et al. (2017) demonstrated the theoretical property of the maximum likelihood estimator. In practice, however, the collected data may not follow a normal distribution. Hence, to broaden the usefulness of the covariance regression model, Zou et al. (2020) relaxed the normality assumption and employed Lee's (2004) approach to obtain the inference for all five estimators mentioned above.

Suppose that the responses are not truly normally distributed. Based on the quasi-maximum likelihood estimation (QMLE) method of Lee (2004) and Tsay (2013), Zou et al. (2020) obtained the estimator $\widehat{\boldsymbol{\beta}}_{p,QMLE}$ by maximizing the log-likelihood function (4.5) evaluated under the normal assumption. It can be shown that $\widehat{\boldsymbol{\beta}}_{p,QMLE}$ is a consistent estimator. Note that the unconstrained/constrained OLS and FGLS estimators are still applicable since they do not rely on the normality assumption.

Applying similar techniques to those used in Lee (2004) and Zou et al. (2017), Zou et al. (2020) indicated that they can demonstrate the asymptotic distributions of the unconstrained/constrained OLS and FGLS estimators and QMLE, under the non-normal responses. Assume that, for $d = -1, 0, 1$,

$$\frac{1}{p}\left(\mathrm{tr}\left(\boldsymbol{\Sigma}^d \mathbf{W}_{k_1} \boldsymbol{\Sigma}^d \mathbf{W}_{k_2}\right)\right)_{(K+1)\times(K+1)} \longrightarrow \mathcal{Q}_d \in \mathbb{R}^{(K+1)\times(K+1)} \text{ and} \tag{4.12}$$

$$\frac{1}{p}\left(\mathrm{tr}\left\{\left(\boldsymbol{\Sigma}^{d/2} \mathbf{W}_{k_1} \boldsymbol{\Sigma}^{d/2}\right) * \left(\boldsymbol{\Sigma}^{d/2} \mathbf{W}_{k_2} \boldsymbol{\Sigma}^{d/2}\right)\right\}\right)_{(K+1)\times(K+1)} \longrightarrow \mathcal{P}_d \in \mathbb{R}^{(K+1)\times(K+1)}, \tag{4.13}$$

where $*$ represents the Hadamard product of two matrices defined in Sect. 1.6 and $\mathcal{Q}_d$ and $\mathcal{P}_d$ are $(K+1) \times (K+1)$ positive definite matrices for $d = -1, 0, 1$. In addition, let $Z = \boldsymbol{\Sigma}^{-1/2} Y = (z_j)_{p\times 1} \in \mathbb{R}^p$ and assume the fourth-order moment $\mu^{(4)} = E(z_j{}^4)$ for $j = 1, \cdots, p$. Then, under mild conditions, Zou et al. (2020) could show that the unconstrained/constrained OLS estimator is asymptotically normal with mean $\boldsymbol{\beta}^0$ and covariance matrix $2p^{-1}\mathcal{Q}_0^{-1}\mathcal{Q}_1\mathcal{Q}_0^{-1} + p^{-1}(\mu^{(4)} - 3)\mathcal{Q}_0^{-1}\mathcal{P}_1\mathcal{Q}_0^{-1}$ as $p \to \infty$. They could further demonstrate that the unconstrained/constrained FGLS estimator is asymptotically normal with mean $\boldsymbol{\beta}^0$ and covariance matrix $2p^{-1}\mathcal{Q}_{-1}^{-1} + p^{-1}(\mu^{(4)} - 3)\mathcal{Q}_{-1}^{-1}\mathcal{P}_{-1}\mathcal{Q}_{-1}^{-1}$ as $p \to \infty$. Note that the QMLE $\widehat{\boldsymbol{\beta}}_{p,QMLE}$ is also asymptotically normal with mean $\boldsymbol{\beta}^0$ and covariance matrix $2p^{-1}\mathcal{Q}_{-1}^{-1}+p^{-1}(\mu^{(4)}-3)\mathcal{Q}_{-1}^{-1}\mathcal{P}_{-1}\mathcal{Q}_{-1}^{-1}$ as $p\to\infty$. Therefore, the constrained FGLS estimator and the QMLE have the same asymptotic efficiency for non-normal responses. Since the constrained FGLS estimator is more computationally feasible, Zou et al. (2020) recommended using the constrained FGLS estimator in real practice.

In order to employ the five covariance regression parameter estimators to make inferences, one needs to estimate their corresponding asymptotic covariances mentioned above. In practice, we can estimate them via (4.12) and (4.13) evaluated at $\widehat{\boldsymbol{\Sigma}}=\boldsymbol{\Sigma}(\widehat{\boldsymbol{\beta}})$, where $\widehat{\boldsymbol{\beta}}$ can be one of the five parameter estimators. In addition, $\mu^{(4)}$ can be estimated by its empirical fourth-order moment calculated from all of the components in the vector $\widehat{Z}=\widehat{\boldsymbol{\Sigma}}^{-1/2}Y$. As a result, one can test the significance of regression parameters and assess the effect of the similarities induced by covariates on the covariance structure.

4.4 Covariance Model with General Linear Structure

In the last two sections, we mainly focused on linear covariance regression models. As mentioned in Sect. 3.2, some nonlinear covariance models, such as the logarithm covariance model and the inverse covariance model, have been proposed to construct a nonlinear relationship between the covariance matrix and a linear combination of known matrices. However, there is no unified framework to link the covariance matrix of response variables to the linear combination of similarity (weight) matrices. Hence, Fan et al. (2024a) proposed the covariance model with general linear structure (CMGL) as follows:

$$\mathrm{Cov}(Y)=\boldsymbol{\Sigma}(\boldsymbol{\beta})=\mathbf{G}(\mathbf{I}_p\beta_0+\mathbf{W}_1\beta_1+\cdots+\mathbf{W}_K\beta_K)=:\mathbf{G}(B), \tag{4.14}$$

where $\mathbf{I}_p$, $\mathbf{W}_k$ and β_k, for $k=1,\cdots,K$, are defined in (4.4), and $\mathbf{G}(\cdot)$ is a known function mapping from $\mathbb{R}^{p\times p}$ to $\mathbb{R}^{p\times p}$ such that the output is a positive definite matrix. As a result, the linear, logarithm, and inverse covariance models can be viewed as special cases of CMGL. As denoted in Sect. 4.2, let $\boldsymbol{\beta}^0$ be the true vector of parameters and denote $\boldsymbol{\Sigma}(\boldsymbol{\beta}^0)$ by $\boldsymbol{\Sigma}_0$. Note that Y is standardized to have mean zero. Hence, the true covariance matrix of Y is $E(YY')=\boldsymbol{\Sigma}_0$.

In linear and nonlinear covariance models, one usually assumes that the responses follow some distribution (e.g., the normal distribution) or that the number of known matrices (K) is fixed. In practice, however, the data may not satisfy the normality assumption and the number of known matrices can be large. For example, the stock returns are unlikely to be normally distributed (see, e.g., Karoglou 2010; Kan and Zhou 2017). In addition, the number of similarity matrices induced by the firms' fundamentals can be large (see, e.g., 447 covariates were considered for modeling stock returns in Hou et al. 2020). Hence, Fan et al. (2024a) explored the following four topics to broaden the usefulness of the covariance model. First, as discussed in Sect. 4.2, relax the distribution assumption imposed on the continuous response variable and employ the quasi-likelihood function to obtain parameter estimators; second, allow the number of known matrices to diverge; third, choose the relevant similarity matrices from a large number of candidate matrices; fourth, test the appropriateness of link functions. In this section, we only consider the first three subjects, and the last one can be found in Fan et al. (2024a).

4.4.1 Parameter Estimators

Without making any distribution assumption for responses, Fan et al. (2024a) employed the quasi-maximum likelihood approach (see, e.g., Lee 2004; Tsay 2013) as discussed in Sect. 4.3 to estimate unknown parameters. Based on (4.14), the quasi-log-likelihood function of $\boldsymbol{\beta}$ is

$$\ell_Q(\boldsymbol{\beta}) = -\frac{p}{2}\log(2\pi) - \frac{1}{2}\sum_{j=1}^{p}\log\left\{\lambda_j(\boldsymbol{\Sigma}(\boldsymbol{\beta}))\right\} - \frac{1}{2}Y'\boldsymbol{\Sigma}^{-1}(\boldsymbol{\beta})Y, \tag{4.15}$$

where $\lambda_j(\boldsymbol{\Sigma}(\boldsymbol{\beta}))$ represents the j-th largest eigenvalue of $\boldsymbol{\Sigma}(\boldsymbol{\beta})$. The above function is the same as that of (4.5) except its $\boldsymbol{\Sigma}(\boldsymbol{\beta}) = \mathbf{G}(B)$.

Let $\mathbb{B} = \{\boldsymbol{\beta} : \boldsymbol{\Sigma}(\boldsymbol{\beta}) = \mathbf{G}(B) > 0\}$ be the parameter space such that $\boldsymbol{\Sigma}(\boldsymbol{\beta})$ is a positive definite matrix. Then the quasi-maximum likelihood estimator (QMLE), $\widehat{\boldsymbol{\beta}}_Q$, can be obtained by maximizing (4.15). That is $\widehat{\boldsymbol{\beta}}_Q = \text{argmax}_{\boldsymbol{\beta}\in\mathbb{B}}\,\ell_Q(\boldsymbol{\beta})$. Fan et al. (2024a) proposed calculating QMLE via the Newton-type algorithm; see, e.g., Dennis and Schnabel (1996). The Newton-type algorithm can be implemented via the function "nlm" in R, and the initial parameter $\boldsymbol{\beta}^* = (\beta_0^*, \beta_1^*, \cdots, \beta_K^*)'$ is set to $\beta_j^* = 0$ for $j = 1, \cdots, K$ and $\beta_0^* = \text{argmin}_{\beta_0}(\|YY' - \mathbf{G}(\mathbf{I}_p\beta_0)\|_F^2)$. We find that $\beta_0^* = p^{-1}\sum_{i=1}^{p} Y_i^2$ for the linear link function and $\beta_0^* = \log(p^{-1}\sum_{i=1}^{p} Y_i^2)$ for the exponential link function.

To show theoretical properties of QMLE, Fan et al. (2024a) introduced the following five technical conditions:

(C4.4) Assume $K = O(p^h)$ for some $h < 1/4$ as $p \to \infty$.

(C4.5) Assume $Y = \boldsymbol{\Sigma}_0^{1/2} Z$, where $Z = (z_1, \cdots, z_p)'$ satisfies:

(i) $E(z_i^m) = \mu^{(m)}$ with $\mu^{(1)} = 0$ and $\mu^{(2)} = 1$;

(ii) $E(z_i^{\upsilon}|\mathcal{F}_{i-1}) = E(z_i^{\upsilon})$ for $\upsilon = 1, \cdots, 4$, where $\mathcal{F}_i$ is the σ-field generated by $\{z_j, j = 1, \cdots, i\}$;

(iii) $E(z_{i_1}^{\upsilon_1} z_{i_2}^{\upsilon_2} z_{i_3}^{\upsilon_3} z_{i_4}^{\upsilon_4}) = E(z_{i_1}^{\upsilon_1})E(z_{i_2}^{\upsilon_2})E(z_{i_3}^{\upsilon_3})E(z_{i_4}^{\upsilon_4})$ for any $i_1 \neq i_2 \neq i_3 \neq i_4$ and $\upsilon_1 + \upsilon_2 + \upsilon_3 + \upsilon_4 \leq 8$, where υ_i are integers;

(iv) there exists a finite positive constant η, such that $P(|g(Z) - E(g(Z))| > t) \leq 2\exp(-t^2/\eta^2)$ for any 1-Lipschitz convex function $g : \mathbb{R}^p \to \mathbb{R}$ and $t > 0$.

(C4.6) For any symmetric matrices in $\left\{\frac{\partial \boldsymbol{\Sigma}(\boldsymbol{\beta})}{\partial \beta_k},\ k=0,\cdots,K\right\}$, $\left\{\frac{\partial^2 \boldsymbol{\Sigma}(\boldsymbol{\beta})}{\partial \beta_j \partial \beta_k},\ j,k=0,\cdots,K\right\}$ and $\left\{\frac{\partial^3 \boldsymbol{\Sigma}(\boldsymbol{\beta})}{\partial \beta_j \partial \beta_k \partial \beta_l},\ j,k,l=0,\cdots,K\right\}$, there exists $\tau>0$ and an open ball $B_\delta=\{\boldsymbol{\beta}:\|\boldsymbol{\beta}-\boldsymbol{\beta}^0\|<\delta\}$ with $\delta>0$, such that

$$\max\left(\sup_{k;\boldsymbol{\beta}\in B_\delta}\left\|\frac{\partial \boldsymbol{\Sigma}(\boldsymbol{\beta})}{\partial \beta_k}\right\|_1,\ \sup_{j,k;\boldsymbol{\beta}\in B_\delta}\left\|\frac{\partial^2 \boldsymbol{\Sigma}(\boldsymbol{\beta})}{\partial \beta_j \partial \beta_k}\right\|_1,\right.$$
$$\left.\sup_{j,k,l;\boldsymbol{\beta}\in B_\delta}\left\|\frac{\partial^3 \boldsymbol{\Sigma}(\boldsymbol{\beta})}{\partial \beta_j \partial \beta_k \partial \beta_l}\right\|_1\right)\le\tau<\infty.$$

(C4.7) Assume that there exist three finite positive constants d_1, d_2 and d_3 such that, for any $p\ge 1$, $0<d_1<\inf_{\boldsymbol{\beta}\in B_\delta}\lambda_p(\boldsymbol{\Sigma}(\boldsymbol{\beta}))\le\sup_{\boldsymbol{\beta}\in B_\delta}\lambda_1(\boldsymbol{\Sigma}(\boldsymbol{\beta}))<d_2<\infty$ and $\sup_{\boldsymbol{\beta}\in B_\delta}\left(\|\boldsymbol{\Sigma}^{-1/2}(\boldsymbol{\beta})\|_1+\|\boldsymbol{\Sigma}^{1/2}(\boldsymbol{\beta})\|_1\right)<d_3$, where B_δ is defined in Condition (C4.6).

(C4.8) Assume that

(i) $p^{-1}\left(\mathrm{tr}\left(\boldsymbol{\Sigma}_0^{-1}\frac{\partial \boldsymbol{\Sigma}_0(\boldsymbol{\beta}^0)}{\partial \beta_j}\boldsymbol{\Sigma}_0^{-1}\frac{\partial \boldsymbol{\Sigma}_0(\boldsymbol{\beta}^0)}{\partial \beta_k}\right)\right)_{(K+1)\times(K+1)}$ converges to a finite positive definite matrix $\mathbf{Q}$ in the Frobenius norm, i.e.,

$$\|p^{-1}\left(\mathrm{tr}\left(\boldsymbol{\Sigma}_0^{-1}\frac{\partial \boldsymbol{\Sigma}_0(\boldsymbol{\beta}^0)}{\partial \beta_j}\boldsymbol{\Sigma}_0^{-1}\frac{\partial \boldsymbol{\Sigma}_0(\boldsymbol{\beta}^0)}{\partial \beta_k}\right)\right)_{(K+1)\times(K+1)}-\mathbf{Q}\|_F^2=o(1),$$

where $\lambda_{K+1}(\mathbf{Q})>\nu>0$ for a positive constant ν;

(ii) $p^{-1}\left(\mathrm{tr}\left\{\left(\boldsymbol{\Sigma}_0^{-1/2}\frac{\partial \boldsymbol{\Sigma}_0(\boldsymbol{\beta}^0)}{\partial \beta_j}\boldsymbol{\Sigma}_0^{-1/2}\right)*\left(\boldsymbol{\Sigma}_0^{-1/2}\frac{\partial \boldsymbol{\Sigma}_0(\boldsymbol{\beta}^0)}{\partial \beta_k}\boldsymbol{\Sigma}_0^{-1/2}\right)\right\}\right)_{(K+1)\times(K+1)}$ converges to a matrix $\boldsymbol{\Delta}$ in the Frobenius norm, and $0<\kappa<\lambda_{K+1}\{2\mathbf{Q}+(\mu^{(4)}-3)\boldsymbol{\Delta}\}\le\lambda_1\{2\mathbf{Q}+(\mu^{(4)}-3)\boldsymbol{\Delta}\}=O(K)$, where κ is a finite constant, and $\mu^{(4)}=E(z_1^4)$ is as defined in Sect. 4.3.

Based on the above conditions, Fan et al. (2024a) obtained the asymptotic distribution of $\widehat{\boldsymbol{\beta}}_Q$ given below.

Theorem 4.5 *Under Conditions (C4.4)–(C4.8),*

$$\sqrt{p/K}\mathbf{A}\mathbf{Q}(\widehat{\boldsymbol{\beta}}_Q-\boldsymbol{\beta}^0)\xrightarrow{d}N(0,\mathbf{V}_Q),$$

as $p\to\infty$, where $\mathbf{A}$ is an arbitrary $M\times(K+1)$ matrix satisfying $M<\infty$, $\|\mathbf{A}\|_2<\infty$ and $K^{-1}\mathbf{A}\{2\mathbf{Q}+(\mu^{(4)}-3)\boldsymbol{\Delta}\}\mathbf{A}'\to\mathbf{V}_Q$, and $\mathbf{V}_Q$ is an $M\times M$ nonnegative symmetric matrix.

The detailed proofs of the above theorem and the next four theorems can be found in Fan et al. (2024a). This theorem indicates that $\widehat{\boldsymbol{\beta}}_Q$ is the $(p/K)^{1/2}$-consistent estimator and is asymptotically normal. Set $\mathbf{A}$ in Theorem 4.5 to be an identity matrix with the dimension $K+1$. The resulting asymptotic distribution is the same as that in Theorem 4.1 when the responses are multivariate normal, $\mathbf{G}$ is an identity mapping function $\mathbf{I}$, and K is fixed. In addition, we can make various inferences according to the matrix $\mathbf{A}$. For example, making inference for $(\boldsymbol{\beta}_Q)_k$ $(k = 0, \cdots, K)$, we can choose $\mathbf{A} = (0, \cdots, 1, \cdots, 0)$ with "1" being on the k-th position to obtain the asymptotic variance of $(\widehat{\boldsymbol{\beta}}_Q)_k$.

Based on Theorem 4.5, Fan et al. (2024a) further demonstrated the asymptotic properties of the covariance matrix estimator, $\boldsymbol{\Sigma}(\widehat{\boldsymbol{\beta}}_Q)$, and the precision matrix estimator, $\boldsymbol{\Sigma}^{-1}(\widehat{\boldsymbol{\beta}}_Q)$, given below.

Theorem 4.6 *Under Conditions (C4.4)–(C4.8),*

$$\|\boldsymbol{\Sigma}(\widehat{\boldsymbol{\beta}}_Q) - \boldsymbol{\Sigma}_0\|_2 = O_p(Kp^{-1/2}) \text{ and } \|\boldsymbol{\Sigma}^{-1}(\widehat{\boldsymbol{\beta}}_Q) - \boldsymbol{\Sigma}_0^{-1}\|_2 = O_p(Kp^{-1/2}),$$

as $p \to \infty$.

The above theorem demonstrates that $\boldsymbol{\Sigma}(\widehat{\boldsymbol{\beta}}_Q)$ and $\boldsymbol{\Sigma}^{-1}(\widehat{\boldsymbol{\beta}}_Q)$ are consistent estimators of $\boldsymbol{\Sigma}_0$ and $\boldsymbol{\Sigma}_0^{-1}$, respectively. Note that $p^{-1/2}\|\boldsymbol{\Sigma}(\widehat{\boldsymbol{\beta}}_Q) - \boldsymbol{\Sigma}_0\|_F$ and $p^{-1/2}\|\boldsymbol{\Sigma}^{-1}(\widehat{\boldsymbol{\beta}}_Q) - \boldsymbol{\Sigma}_0^{-1}\|_F$ are correspondingly less than $\|\boldsymbol{\Sigma}(\widehat{\boldsymbol{\beta}}_Q) - \boldsymbol{\Sigma}_0\|_2$ and $\|\boldsymbol{\Sigma}^{-1}(\widehat{\boldsymbol{\beta}}_Q) - \boldsymbol{\Sigma}_0^{-1}\|_2$, and their orders are not larger than $O_p(Kp^{-1/2})$. Accordingly, $\boldsymbol{\Sigma}(\widehat{\boldsymbol{\beta}}_Q)$ and $\boldsymbol{\Sigma}^{-1}(\widehat{\boldsymbol{\beta}}_Q)$ are consistent in the Frobenius norm.

In practice, the asymptotic covariance of $\widehat{\boldsymbol{\beta}}_Q$ is unknown since it is related to $\mu^{(4)}$, $\mathbf{Q}$ and $\boldsymbol{\Delta}$. However, these three unknown quantities can be consistently estimated by $\widehat{\mu}^{(4)} = p^{-1}\sum_{i=1}^{p}\widehat{z}_i^4$, $\widehat{\mathbf{Q}} = p^{-1}\big(\mathrm{tr}(\boldsymbol{\Sigma}^{-1}(\widehat{\boldsymbol{\beta}}_Q)\frac{\partial\boldsymbol{\Sigma}(\widehat{\boldsymbol{\beta}}_Q)}{\partial\beta_k}\boldsymbol{\Sigma}^{-1}(\widehat{\boldsymbol{\beta}}_Q)\frac{\partial\boldsymbol{\Sigma}(\widehat{\boldsymbol{\beta}}_Q)}{\partial\beta_l})\big)_{(K+1)\times(K+1)}$, and $\widehat{\boldsymbol{\Delta}} = p^{-1}\big(\mathrm{tr}\{(\boldsymbol{\Sigma}^{-1/2}(\widehat{\boldsymbol{\beta}}_Q)\frac{\partial\boldsymbol{\Sigma}(\widehat{\boldsymbol{\beta}}_Q)}{\partial\beta_k}\boldsymbol{\Sigma}^{-1/2}(\widehat{\boldsymbol{\beta}}_Q))*(\boldsymbol{\Sigma}^{-1/2}(\widehat{\boldsymbol{\beta}}_Q)\frac{\partial\boldsymbol{\Sigma}(\widehat{\boldsymbol{\beta}}_Q)}{\partial\beta_l}\boldsymbol{\Sigma}^{-1/2}(\widehat{\boldsymbol{\beta}}_Q))\}\big)_{(K+1)\times(K+1)}$, where $\widehat{Z} = (\widehat{z}_1, \cdots, \widehat{z}_p)' = \boldsymbol{\Sigma}^{-1/2}(\widehat{\boldsymbol{\beta}}_Q)Y$.

It is worth noting that the theoretical results of Theorems 4.5 and 4.6 are obtained by letting p go to infinity and n be fixed, but these results can be extended to allow $np \to \infty$. Accordingly, this extension includes the following two scenarios as special cases: (i) $n \to \infty$ and $p \to \infty$; and (ii) $n \to \infty$ and p is fixed. Detailed discussions can be found in Fan et al. (2024a).

In practice, the computation of QMLE can be a challenging task for large p, while the OLS estimator has a closed form under the linear covariance structure with finite number of similarity matrices. Hence, Fan et al. (2024a) studied the OLS estimator and its properties when the number of similarity matrices K is divergent. To this end, let $\mathbf{G}$ in (4.14) be an identity mapping function $\mathbf{I}$. This results in the following linear covariance model:

$$\mathrm{Cov}(Y) = \mathbf{I}_p\beta_0 + \mathbf{W}_1\beta_1 + \cdots + \mathbf{W}_K\beta_K \tag{4.16}$$

as given in (4.4), except that K was not fixed.

By minimizing $\|YY' - \boldsymbol{\Sigma}(\boldsymbol{\beta})\|_F^2$, the OLS estimator can be obtained and it has the following closed form:

$$\widehat{\boldsymbol{\beta}}_{OLS} = (\mathrm{tr}(\mathbf{W}_j\mathbf{W}_k))_{(K+1)\times(K+1)}^{-1}(Y'\mathbf{W}_jY)_{(K+1)\times 1}; \tag{4.17}$$

see also (4.8). It is worth noting that the calculation of $\widehat{\boldsymbol{\beta}}_{OLS}$ is much simpler than that of $\widehat{\boldsymbol{\beta}}_{Q}$. In addition, it can be shown that, as $p \to \infty$, the unconstrained OLS estimator is identical to the constrained OLS estimator with probability tending to 1 even when K is divergent. Hereafter, we only focus on the unconstrained OLS estimator.

To show the theoretical property of the OLS estimator, Fan et al. (2024a) introduced an additional condition given below.

(C4.9) (i) Assume that $p^{-1}(\mathrm{tr}(\boldsymbol{\Sigma}_0^d \mathbf{W}_j \boldsymbol{\Sigma}_0^d \mathbf{W}_k))_{(K+1)\times(K+1)}$ converges to a positive definite matrix $\mathbf{Q}_d$ in the Frobenius norm, i.e., $\| p^{-1}(\mathrm{tr}(\boldsymbol{\Sigma}_0^d \mathbf{W}_j \boldsymbol{\Sigma}_0^d \mathbf{W}_k))_{(K+1)\times(K+1)} - \mathbf{Q}_d\|_F^2 = o(1)$, and $\lambda_p(\mathbf{Q}_d) > \varphi_0 > 0$ for $d = 0, 1$, where φ_0 is a finite constant and $\boldsymbol{\Sigma}_0^0 := \mathbf{I}_p$. (ii) Assume that $p^{-1}(\mathrm{tr}[(\boldsymbol{\Sigma}_0^{1/2}\mathbf{W}_j\boldsymbol{\Sigma}_0^{1/2}) * (\boldsymbol{\Sigma}_0^{1/2}\mathbf{W}_k\boldsymbol{\Sigma}_0^{1/2})])_{(K+1)\times(K+1)}$ converges to a matrix $\boldsymbol{\Delta}_1$ in the Frobenius norm and $0 < \bar{\eta}_1 < \lambda_p(2\mathbf{Q}_1 + (\mu^{(4)} - 3)\boldsymbol{\Delta}_1) \le \lambda_1\{2\mathbf{Q}_1 + (\mu^{(4)} - 3)\boldsymbol{\Delta}_1\} = O(K)$, where $\bar{\eta}_1$ is a finite constant and $\mu^{(4)}$ is defined in Condition (C4.8).

The above condition, together with Conditions (C4.4)–(C4.7), yields the asymptotic result of $\widehat{\boldsymbol{\beta}}_{OLS}$.

Theorem 4.7 *Under Conditions (C4.4)–(C4.7) and (C4.9),*

$$\sqrt{p/K}\mathbf{A}\mathbf{Q}_0(\widehat{\boldsymbol{\beta}}_{OLS} - \boldsymbol{\beta}^0) \xrightarrow{d} N(\mathbf{0}, \mathbf{V}_O),$$

as $p \to \infty$, where $\mathbf{A}$ is an arbitrary $M\times(K+1)$ matrix for any finite M with $\|\mathbf{A}\|_2 < \infty$, $K^{-1}\mathbf{A}\{2\mathbf{Q}_1 + (\mu^{(4)} - 3)\boldsymbol{\Delta}_1\}\mathbf{A}' \to \mathbf{V}_O$, and $\mathbf{V}_O$ is a $M \times M$ nonnegative symmetric matrix.

Theorem 4.7 shows that $\widehat{\boldsymbol{\beta}}_{OLS}$ is $(p/K)^{1/2}$ consistent and asymptotically normal. Set $\mathbf{A}$ to be an identity matrix with dimension $K + 1$. Fan et al. (2024a) then obtained the same asymptotic distribution as given in Sect. 4.2 when K is fixed and $\mu^{(4)} = 3$. Note that the above theorem implies the covariance matrix can be estimated by employing $\boldsymbol{\Sigma}(\widehat{\boldsymbol{\beta}}_{OLS})$. Analogous to Theorem 4.6, they further showed that, under Conditions (C4.4)–(C4.7) and (C4.9), $\|\boldsymbol{\Sigma}(\widehat{\boldsymbol{\beta}}_{OLS}) - \boldsymbol{\Sigma}_0\|_2 = O_p(Kp^{-1/2})$ and $\|\boldsymbol{\Sigma}^{-1}(\widehat{\boldsymbol{\beta}}_{OLS}) - \boldsymbol{\Sigma}_0^{-1}\|_2 = O_p(Kp^{-1/2})$, as $p \to \infty$.

4.4.2 The Selection of Similarity Matrices

This section allows the number of similarity matrices K to tend to infinity. Thus, selecting the relevant similarity matrices from a family of candidate models becomes critical. Fan et al. (2024a) followed the approach of Chen and Chen (2008, 2012) to develop an extended Bayesian information criterion for selecting similarity matrices. To this end, let $s_F = \{0, \cdots, K\}$ be the full model that contains all similarity matrices. Then, let $s_0 \subset s_F$ be the set of the model that includes similarity matrices associated with non-zero elements in $\boldsymbol{\beta}^0$, which is the true model. In addition, for any candidate models $s \subset s_F$, let $\boldsymbol{\beta}(s)$ be the vector of the components in $\boldsymbol{\beta}$ that are associated with the similarity matrices in s, and denote the QMLE of $\boldsymbol{\beta}^0(s)$ by $\widehat{\boldsymbol{\beta}}_Q(s)$. Then, Fan et al. (2024a) proposed the similarity matrix selection given below.

$$\text{EBIC}_Q(s) = -2\ell_Q\{\widehat{\boldsymbol{\beta}}_Q(s)\} + v(s)\log p + 2v(s)\zeta \log K, \tag{4.18}$$

where $\ell_Q(\cdot)$ is defined in (4.15), $v(s)$ is the number of similarity matrices in s, and $\zeta > 0$ is a constant.

Before showing the theoretical properties of EBIC, Fan et al. (2024a) first defined

$$A_0 = A_0(c_0) = \{s : s_0 \subset s;\ v(s) \le c_0\} \text{ and } A_1 = A_1(c_0) = \{s : s_0 \not\subset s;\ v(s) \le c_0\},$$

and then introduced the following two conditions.

(C4.10) Denote $\mathbf{H}_Q(\boldsymbol{\beta}(s)) = -\partial^2\ell_Q(\boldsymbol{\beta}(s))/\partial\boldsymbol{\beta}(s)\partial\boldsymbol{\beta}'(s)$, and assume that there exists a finite positive constant c_0 that is larger than $v(s_0)$. Then, for any s such that $v(s) \le c_0$, assume that $p^{-1}\mathbf{H}_Q(\boldsymbol{\beta}(s)) - p^{-1}\mathbf{H}_Q^0(\boldsymbol{\beta}(s))$ converges to 0 uniformly for some positive matrix $\mathbf{H}_Q^0(\boldsymbol{\beta}(s))$. Furthermore, assume that $0 < \varphi_1 < \inf_{\boldsymbol{\beta}\in B_\delta} \lambda_p(p^{-1}\mathbf{H}_Q^0(\boldsymbol{\beta}(s))) \le \sup_{\boldsymbol{\beta}\in B_\delta} \lambda_1(p^{-1}\mathbf{H}_Q^0(\boldsymbol{\beta}(s))) < \varphi_2 < \infty$ with probability approaching 1, where φ_1 and φ_2 are two positive constants, and B_δ is defined in Condition (C4.6).

(C4.11) Assume that $\sqrt{p/\log p}\,\min_{k\in s_0} |\beta_k^0| \to \infty$, for $k = 1, \cdots, K$, as $p \to \infty$.

Condition (C4.10) is similar to Condition (C4.9), and it bounds the eigenvalues of the Hessian matrix around B_δ. Condition (C4.11) assumes that the order of nonzero coefficients is larger than $\sqrt{\log p/p}$, and a similar condition can be found in linear regression model selections (see, e.g., Wang et al. 2009). The above two conditions, in conjunction with the prior five conditions, yield the following result.

Theorem 4.8 *Under Conditions (C4.4)–(C4.8) and (C4.10)–(C4.11),*

$$(i)\quad \Pr\Big\{\min_{s\in A_0, s\neq s_0} \text{EBIC}_Q(s) \le \text{EBIC}_Q(s_0)\Big\} \to 0 \ \textit{and}$$

$$(ii)\quad \Pr\Big\{\min_{s\in A_1} \text{EBIC}_Q(s) \le \text{EBIC}_Q(s_0)\Big\} \to 0,$$

as $p \to \infty$, for $\zeta > c_0\varphi_4/(8c_1\varphi_1) - 1/(2h)$, where ζ is defined in (4.18), h is defined in Condition (C4.4), c_0 and φ_1 are defined in Condition (C4.10), and c_1 and φ_4 are defined in Lemmas 3 and 5, respectively, in the supplementary material of Fan et al. (2024a).

The above theorem indicates that EBIC_Q can identify the true model consistently as long as p tends to infinity. To implement EBIC_Q, Fan et al. (2024a) employed the backward elimination method (see, e.g., Zhang and Wang 2011; Schelldorfer et al. 2014), which reduces the computational complexity from 2^K to $O(K^2)$. Accordingly, EBIC_Q is computable even when K is divergent.

The selection criterion EBIC_Q is based on the QMLE. For OLS, Fan et al. (2024a) also proposed EBIC for linear covariance model selections given below.

$$\mathrm{EBIC}_{OLS}(s) = \log(\widehat{\sigma}_s^2) + v(s)\log p/p^2 + 2v(s)\zeta\log K/p^2,$$

where $\widehat{\sigma}_s^2 = \|YY' - \boldsymbol{\Sigma}(\widehat{\boldsymbol{\beta}}_{OLS}(s))\|_F^2/p^2$, and $\widehat{\boldsymbol{\beta}}_{OLS}(s)$ is the OLS estimator of $\boldsymbol{\beta}^0(s)$.
Subsequently, they obtained the theoretical properties of $\mathrm{EBIC}_{OLS}(s)$ given below.

Theorem 4.9 *Under Conditions (C4.4)–(C4.7), (C4.9) and (C4.11),*

$$(i)\quad \Pr\Big\{\min_{s\in A_0, s\neq s_0} \mathrm{EBIC}_{OLS}(s) \leq \mathrm{EBIC}_{OLS}(s_0)\Big\} \to 0 \ \textit{and}$$

$$(ii)\quad \Pr\Big\{\min_{s\in A_1} \mathrm{EBIC}_{OLS}(s) \leq \mathrm{EBIC}_{OLS}(s_0)\Big\} \to 0,$$

as $p \to \infty$, for $\zeta > d_2^2 a_0^2 c_0/(2c_1 a_1 \xi_1) - 1/(2h)$, where h and d_2 are defined in Conditions (C4.4) and (C4.7), respectively, c_0 is a finite constant defined in Condition (C4.10), c_1 is defined in Lemma 3, a_0 and a_1 are defined in Lemma 7, ξ_1 is defined in Lemma 10, and Lemmas 3, 7, and 10 are presented in the supplementary material of Fan et al. (2024a).

Theorem 4.9 shows that EBIC_{OLS} is a consistent selection criterion.

The above estimation and selection results are based on a known link function $\mathbf{G}$. In practice, however, $\mathbf{G}$ is unknown. Hence, Fan et al. (2024a) proposed a test to assess the adequacy of link functions, and they also demonstrated the asymptotic property of this test. They further presented simulation studies to show the finite sample performance of theoretical results. Suppose that the test suggests that $\mathbf{G}$ is a linear covariance function. Then we can consider using the oracle estimator of $\boldsymbol{\beta}$ in high dimensional similarity matrices. This estimator was proposed by Gao et al. (2025), who employed a lasso estimator of covariance, together with the local linear approximation (LLA) algorithm (Zou and Li 2008), to develop it.

4.5 Empirical Examples

This section briefly discusses three examples, which illustrate the application of covariance regression models. The first example is from Zou et al. (2017), who employed the covariance regression model to analyze the performance of a portfolio by solving the Markowitz optimization problem (Markowitz 1952). To this end, they used the quarterly returns of $p = 660$ stocks in the Shanghai Stock Exchange and Shenzhen Stock Exchange of China from 2007 to 2011, and this data set was collected from the China Stock Market and Accounting Research (CSMAR) database. The response vector Y contains the returns (in percentages) of the $p = 660$

stocks with a total of 20 quarters. There are $K = 5$ covariates representing firms' fundamentals: **industry** with 14 categories; **leverage** computed by liability-to-asset ratio; **cash flow** of the firm; **size** measured by the logarithm of market value; and **book-to-market ratio.** Based on these five covariates, the corresponding similarity matrices are constructed. Applying (4.4), Zou et al. (2017) estimated the covariance matrix of Y for each quarter and then obtained the optimal portfolio weights by minimizing the variance of the portfolio. Afterward, they used the current estimated weights to calculate the next period's portfolio return. The results indicate that their proposed covariance matrix estimates could yield a better portfolio return than that from the market.

The second example is from Fan et al. (2024a), who also analyzed portfolio returns. However, they considered two different link functions, identity, and exponential link functions, in applying model (4.14). In this study, they collected the monthly stock returns of $p = 400$ stocks in the US stock market from January 2016 to July 2018 with a total of 31 months. For each month, the response vector Y consists of 400 stocks monthly returns. In addition, Fan et al. (2024a) constructed 22 similarity matrices based on their corresponding covariates. Following a similar procedure to that used in the first example, they found that the covariance matrix with the exponential link function is superior to that with the identity link function, and the subset model selected by EBIC performs better than the full model.

The third example is from Zou et al. (2020), who studied herding behaviors of mutual funds in the Chinese stock market. They collected the annual returns of 491 actively managed open-ended mutual funds Y in 2016 from WIND, which is one of the most authoritative financial databases in China. For the sake of simplicity, the response vector Y is centered by subtracting the sample mean. In addition, Zou et al. (2020) constructed five similarity matrices induced by their corresponding covariates: the investment style (STY); the total net asset under management with logarithm transformation (TNA); the concentration of fund portfolio (CON); the fund portfolio's deviation from the market portfolio (DEV); and the return volatility calculated as the standard deviation of the daily return (STD). Note that STY leads to the off-diagonal element of its associated similarity matrix being 1 if two funds have the same investment style and 0 otherwise. The other four continuous covariates were standardized to have zero mean and unit variance.

Regardless of whether Y is multivariate normal or not, Zou et al.'s (2020) method can be applied. Specifically, they employed the covariance model (4.3) with $K = 5$ to explore the herding behavior of mutual funds. Based on the constrained FGLS parameter estimates and their associated standard errors, t-statistics and p-values, Zou et al. (2020) had three insightful findings. (i) STY and STD are positively and significantly related to the covariance between annual returns; (ii) TNA and CON are positively related to the covariance between annual returns, although they are not significant; (iii) DEV is negatively related, but it is not significant. The first two findings indicate that, if two funds have the same investment style (STY) with similar characteristics, such as fund size (TNA), portfolio concentration (CON), and historical return volatility (STD), their performance tends to be similar and have larger correlation. This is in agreement with the empirical evidence in Grinblatt et al. (1995) and Daniel et al. (1997). The third finding shows that, after controlling the impact of the other four covariates, the active level of a mutual fund does not significantly affect the covariance structure. This result is different from that found in the US financial markets, and it is worthy of further study.

Acknowledgments Section 4.2 and the first example of Sect. 4.5 in this chapter are derived in part from *Covariance Regression Analysis* published in the Journal of the American Statistical Association, 2017, American Statistical Association; Sect. 4.4 and the second example of Sect. 4.5 in this chapter are derived in part from *Covariance Model with General Linear Structure and Divergent Parameters* published in the Journal of Business & Economic Statistics, 2024, American Statistical Association; Sect. 4.3 and the third example of Sect. 4.5 in this chapter are derived in part from *Covariance Regression Model for Non-normal Data* published in Handbook of Financial Econometrics, Statistics, Technology, and Risk Management, Eds. Cheng-Few Lee and John C. Lee, 2020, World Scientific.

Chapter 5
Covariance-Mean Regression Models

This chapter first reviews the four known mean regression models with non-spherical (non-identity) covariance matrices: weighted regression, generalized least squares, longitudinal data, and multivariate regression. These models lead us to then introduce covariance-mean regression models. The theoretical properties of regression parameter estimators are established. In addition, two test statistics are presented: one analyzes the necessity of the auxiliary information, and the other assesses the adequacy of the covariance-mean regression model. Two examples are presented to briefly illustrate empirical applications.

5.1 Mean Regression Models with Covariance Structures

This section considers four known mean regression models together with their covariance structures. They are weighted regression, generalized least squares, regression with longitudinal data, and multivariate regression. Without loss of generality, in mean regression we assume that the elements of the first column of covariates are all 1.

5.1.1 Weighted Regression Analysis

In regression analysis, one often assumes that the response variable exhibits homogeneity. In practice, however, each of the responses can have different weights in model fittings; see, e.g., Carroll and Ruppert (1988) and Fotheringham et al. (2002). To this end, we consider the weighted linear regression model. As stated in Chap. 4, let $Y_i = (y_{ij})$ be the response vector collected from the j-th component in the i-th subject (i.e., observation), for $i = 1, \cdots, n$ and $j = 1, \cdots, p$. In weighted

W. Lan, C.-L. Tsai, *Covariance Analysis and Beyond*,
https://doi.org/10.1007/978-3-032-08796-6_5

regression analysis, we only focus on $p = 1$ so that $Y_i = y_{i1} = y_i$ for $i = 1, \cdots, n$. With a slight abuse of notation, we denote $Y = (y_1, \cdots, y_n)' \in \mathbb{R}^n$.

To take into account the heterogeneity of a response variable, we consider the following weighted regression model:

$$Y = \mathbf{X}\boldsymbol{\alpha} + \boldsymbol{\varepsilon}, \tag{5.1}$$

where $\mathbf{X}$ is an $n \times q$ explanatory matrix, $\boldsymbol{\alpha}$ is a $q \times 1$ unknown vector, and $\boldsymbol{\varepsilon} = (\varepsilon_1, \cdots, \varepsilon_n)' \in \mathbb{R}^n$ is a vector of random errors with mean 0 and covariance matrix $\sigma^2\boldsymbol{\Sigma}$, $\sigma > 0$ is a unknown scalar, and $\boldsymbol{\Sigma} = \text{diag}(\sigma_i^2)$ is an $n \times n$ diagonal matrix. To assess the weighting of each observation in model fitting, let $w_i = 1/\sigma_i^2$ for $i = 1, \cdots, n$, which forms a weighted matrix $\mathbf{W} = \text{diag}(w_i) = \boldsymbol{\Sigma}^{-1}$. In general, w_i is an unknown weight without any specific structure. However, it can depend on some covariates z_i via an unknown function g, i.e., $w_i = g(z_i)$, where z_i is an $m \times 1$ vector. Based on practitioners' experience and knowledge, it is also common to assume that w_i is a known function of z_i, i.e., $w_i = h(z_i, \boldsymbol{\beta})$, where h is a known function and $\boldsymbol{\beta}$ is an $m \times 1$ unknown vector.

Based on model (5.1) with a known weight matrix $\mathbf{W}$, one can estimate $\boldsymbol{\alpha}$ by minimizing the following objective function:

$$(Y - \mathbf{X}\boldsymbol{\alpha})'\mathbf{W}(Y - \mathbf{X}\boldsymbol{\alpha}). \tag{5.2}$$

The resulting weighted least squares estimator is

$$\widehat{\boldsymbol{\alpha}} = (\mathbf{X}'\mathbf{W}\mathbf{X})^{-1}\mathbf{X}'\mathbf{W}Y. \tag{5.3}$$

When w_i is neither known nor specified for $i = 1, \cdots, n$, one can employ the absolute residuals or the squared residuals (or the log-scaled of those residuals) to estimate $\mathbf{W}$, where the residuals are obtained from the ordinary least squares fitting; see, e.g., Chew (1970), Rao (1970), and Neter and Wasserman (1996). However, if $w_i = g(z_i)$ as described above, one can adapt Fan and Yao's (1998) nonparametric approach to estimate the unknown function g and obtain the estimator of $\mathbf{W}$. In contrast, if $w_i = h(z_i, \boldsymbol{\beta})$ and h is a known function as mentioned above, one can employ the generalized least squares (Carroll and Ruppert 1988) or pseudo-likelihood or iterative weighted least squares approach to obtain the estimator of $\mathbf{W}$; see, e.g., Davidian and Carroll (1987), Carroll and Ruppert (1988), and Chen and Shao (1993). More detailed discussion about weighted variance estimation can be found in Carroll and Ruppert (1988).

In addition to weighted parameter estimation, there are various tests available in the literature to assess the homogeneity of error variance; see, e.g., Breusch and Pagan (1979), White (1980a), and Cook and Weisberg (1983). It is of interest to note that the diagonal covariance matrix $\boldsymbol{\Sigma}$ can be extended so that its off-diagonal elements are correlated. One example is the first-order autoregressive model with heteroscedasticity, which assumes that the random errors are not only heterogeneous but also have AR(1) structures; see, e.g., Harrison and McCabe (1975) and Tsai (1986).

5.1.2 Generalized Least Squares Estimator

In linear regression analysis, the covariance matrix $\mathbf{\Sigma}$ defined in (5.1) can be modified with a structured form such as AR(1), MA(1), compound symmetry, or Toeplitz as discussed in Sect. 3.1. If $\mathbf{\Sigma}$ is known, then we can follow the same approach as in Sect. 5.1.1 and obtain the generalized least squares (GLS, Aitken 1936; Amemiya 1985) estimator given below.

$$\widehat{\boldsymbol{\alpha}}_{GLS} = (\mathbf{X}'\mathbf{\Sigma}^{-1}\mathbf{X})^{-1}\mathbf{X}'\mathbf{\Sigma}^{-1}Y. \tag{5.4}$$

Accordingly, the weighted least squares estimator in (5.3) can be viewed as an application of GLS.

In practice, $\mathbf{\Sigma}$ is often unknown. In that case, one can replace $\mathbf{\Sigma}$ with its estimator $\widehat{\mathbf{\Sigma}}$ that leads to the feasible generalized least squares (FGLS) estimator:

$$\widehat{\boldsymbol{\alpha}}_{FGLS} = (\mathbf{X}'\widehat{\mathbf{\Sigma}}^{-1}\mathbf{X})^{-1}\mathbf{X}'\widehat{\mathbf{\Sigma}}^{-1}Y. \tag{5.5}$$

For example, assume that $\mathbf{\Sigma}$ is a structured covariance matrix with the unknown parameter vector $\boldsymbol{\beta}$. For given $\boldsymbol{\beta}$, we obtain the estimator $\widehat{\boldsymbol{\alpha}}_{GLS}$ in (5.4). For given $\widehat{\boldsymbol{\alpha}}_{GLS}$, we then find $\widehat{\boldsymbol{\beta}}_{GLS}$ by minimizing the following weighted residual:

$$(Y - \mathbf{X}\widehat{\boldsymbol{\alpha}}_{GLS})'\mathbf{\Sigma}^{-1}(\boldsymbol{\beta})(Y - \mathbf{X}\widehat{\boldsymbol{\alpha}}_{GLS}). \tag{5.6}$$

We iteratively conduct the process between (5.5) and (5.6) starting from the OLS estimator, $\widehat{\boldsymbol{\alpha}}_{OLS} = (\mathbf{X}'\mathbf{X})^{-1}\mathbf{X}'Y$, and stopping only when both estimators converge. We denote them by $\widehat{\boldsymbol{\alpha}}_{FGLS}$ and $\widehat{\boldsymbol{\beta}}_{FGLS}$, respectively. A good illustration of a linear regression model with an AR(1) covariance matrix is given in Amemiya (1985). Some discussion and theoretical properties of $\widehat{\boldsymbol{\alpha}}_{FGLS}$ and $\widehat{\boldsymbol{\beta}}_{FGLS}$ can be found in Amemiya (1985) and Greene (2018).

5.1.3 Longitudinal Data Analysis

In the last three decades, longitudinal data analysis has been widely used across many fields such as biology, business, economics, engineering, social sciences, and science. Useful reference books include Diggle et al. (2002), Singer and Willett (2003), Hedeker and Gibbons (2006), Weiss (2005), Fitzmaurice et al. (2008, 2011), and Twisk (2023). In addition, three review articles can be found in Zeger and

Liang (1992), Locascio and Atri (2011), and Hu and Szymczak (2023). Although we focus on longitudinal data analysis, interested readers could study the closely related subject of "mixed models," which play an important role in analyzing correlated data in both fixed and random effects models. Some good reference books are given as follows: Verbeke and Molenberghs (2001), Davis (2002), McCulloch et al. (2008), Demidenko (2013), Galecki and Burzykowski (2013), West et al. (2014), Brown and Prescott (2015), Jiang (2019), and Jiang and Nguyen (2021). It is worth noting that the term "longitudinal data" is also called "panel data" in the field of econometrics. Some useful reference books about panel data can be found in Frees (2004), Baltagi (2008), Wooldridge (2010), and Hsiao (2022).

We next introduce the linear regression model for longitudinal data. Let $Y_i = (y_{i1}, \cdots, y_{ij}, \cdots, y_{ip_i})'$ be the $p_i \times 1$ response vector collected from the i-th subject in the j-th element of repeated measurements over time or space for $i = 1, \cdots, n$ and $j = 1, \cdots, p_i$. Note that y_{ij} is measured at time t_{ij} or location s_{ij}. In addition, let $X_{im} = (x_{i1m}, \cdots, x_{ijm}, \cdots, x_{ip_im})'$ be a $p_i \times 1$ explanatory vector collected from the i-th subject for the m-th covariate in the j-th measurement for $i = 1, \cdots, n$, $j = 1, \cdots, p_i$, and $m = 1, \cdots, q$. Then the linear model of longitudinal data is given as follows:

$$Y_i = \sum_{m=1}^{q} X_{im}\alpha_m + \boldsymbol{\varepsilon}_i, \tag{5.7}$$

where $\alpha_m \in \mathbb{R}^1$ is an unknown parameter for $m = 1, \cdots, q$, $\boldsymbol{\varepsilon}_i = (\varepsilon_{i1}, \cdots, \varepsilon_{ip_i})'$ is a $p_i \times 1$ vector of independent random errors with mean vector $\mathbf{0} \in \mathbb{R}^{p_i}$ and covariance matrix $\sigma^2\boldsymbol{\Sigma}^i \in \mathbb{R}^{p_i \times p_i}$, and $\sigma > 0$. Let $\mathbb{Y} = (Y_1', \cdots, Y_n')'$ be an $N \times 1$ vector with $N = \sum_{i=1}^{n} p_i$. Then, model (5.7) can be expressed in the following matrix form:

$$\mathbb{Y} = \mathbf{X}\boldsymbol{\alpha} + \mathcal{E}, \tag{5.8}$$

where $\mathbf{X}$ is an $N \times q$ matrix with the $(\sum_{h=0}^{i-1} p_h + j)$-th row $(x_{ij1}, \cdots, x_{ijq})$ for $i = 1, \cdots, n$, $j = 1, \cdots, p_i$ and $p_0 = 0$, $\boldsymbol{\alpha} = (\alpha_1 \cdots, \alpha_q)'$ is a $q \times 1$ vector of regression coefficients, and $\mathcal{E} = (\boldsymbol{\varepsilon}_1', \cdots, \boldsymbol{\varepsilon}_n')'$ is an $N \times 1$ error vector. In addition, $\mathcal{E}$ is a randomly distributed error vector with mean vector $\mathbf{0} \in \mathbb{R}^N$ and covariance matrix $\sigma^2\boldsymbol{\Sigma} \in \mathbb{R}^{N \times N}$, where $\boldsymbol{\Sigma}$ is an $N \times N$ block diagonal matrix with i-th block $\boldsymbol{\Sigma}^i$. If $p_i = p$ and $\boldsymbol{\Sigma}^i = \boldsymbol{\Sigma}_c$, for $i = 1, \cdots, n$, then $\boldsymbol{\Sigma} = \mathbf{I}_n \otimes \boldsymbol{\Sigma}_c$. In addition, $\mathbf{X}$ is an $N \times q$ matrix with the $(p(i-1) + j)$-th row $(x_{ij1}, \cdots, x_{ijq})$.

In general, there are two possible approaches to estimate unknown parameters. The first approach assumes that the covariance matrix $\boldsymbol{\Sigma}$ is known. Based on (5.7), the generalized least squares estimator of $\boldsymbol{\alpha}$ can be obtained by minimizing the following object function:

$$L(\boldsymbol{\alpha}) = (\mathbb{Y} - \mathbf{X}\boldsymbol{\alpha})'\boldsymbol{\Sigma}^{-1}(\mathbb{Y} - \mathbf{X}\boldsymbol{\alpha}). \tag{5.9}$$

Taking the first derivative of $L(\boldsymbol{\alpha})$ with respect to $\boldsymbol{\alpha}$ and then setting it equal to zero, we obtain that

$$\dot{L}(\boldsymbol{\alpha}) = \mathbf{X}'\boldsymbol{\Sigma}^{-1}(\mathbb{Y} - \mathbf{X}\boldsymbol{\alpha}) = \sum_{i=1}^{n} \mathbf{X}_i'(\boldsymbol{\Sigma}^i)^{-1}(Y_i - \mathbf{X}_i\boldsymbol{\alpha}) = \mathbf{0}, \tag{5.10}$$

where $\mathbf{X}_i = (x_{ijm}) \in \mathbb{R}^{p_i \times q}$. The resulting estimator is

$$\widehat{\boldsymbol{\alpha}} = (\mathbf{X}'\boldsymbol{\Sigma}^{-1}\mathbf{X})^{-1}\mathbf{X}'\boldsymbol{\Sigma}^{-1}\mathbb{Y}. \tag{5.11}$$

The second approach assumes that $\boldsymbol{\Sigma}^i$ is a function of an unknown parameter vector $\boldsymbol{\beta} = (\beta_1, \cdots, \beta_K)'$ and K is relatively small in comparison with p_i for all $i = 1, \cdots, n$. Note that the structured covariance matrices discussed in Sect. 3.1 are in this framework. One popular method to estimate $\boldsymbol{\alpha}$ is the generalized estimating equations (GEE) proposed by Liang and Zeger (1986). Basically, GEE extends the quasi-likelihood method of Wedderburn (1974) and the generalized linear model (McCullagh and Nelder 1989) by allowing repeated measurements and introducing a working correlation matrix. Assume that $E(y_{ij}) = \mu_{ij} = h(x_{ij}'\boldsymbol{\alpha})$, where $h(\cdot)$ is a known inverse link function. To estimate $\boldsymbol{\alpha}$, Liang and Zeger (1986) proposed generalized estimating equations as follows:

$$\sum_{i=1}^{n} (\dot{\boldsymbol{\mu}}_i)'\mathbf{V}_i^{-1}(Y_i - \boldsymbol{\mu}_i) = \mathbf{0}, \tag{5.12}$$

where $\boldsymbol{\mu}_i = (\mu_{i1}, \cdots, \mu_{ip_i})' \in \mathbb{R}^{p_i}$, $\dot{\boldsymbol{\mu}}_i = \partial\boldsymbol{\mu}_i/\partial\boldsymbol{\alpha} \in \mathbb{R}^{p_i \times q}$, $\mathbf{V}_i = \mathbf{G}_i^{1/2}\mathbf{R}(\boldsymbol{\beta})\mathbf{G}_i^{1/2} \in \mathbb{R}^{p_i \times p_i}$, $\mathbf{G}_i \in \mathbb{R}^{p_i \times p_i}$ is the diagonal matrix of the marginal variance of the i-th subject, and $\mathbf{R}(\boldsymbol{\beta}) \in \mathbb{R}^{p_i \times p_i}$ is the working correlation matrix fully characterized by parameter $\boldsymbol{\beta}$.

Based on the above equation, Liang and Zeger (1986) not only presented a detailed iterative procedure to estimate unknown parameters, $\boldsymbol{\alpha}$ and $\boldsymbol{\beta}$, but also demonstrated the asymptotic properties of those estimators. Furthermore, Liang and Zeger (1986) provided several examples to illustrate the estimation of covariance parameter $\boldsymbol{\beta}$. Note that (5.10) is a special case of (5.12). Hence, we can apply GEE to find the parameter estimators of $\boldsymbol{\alpha}$ and $\boldsymbol{\beta}$ if $\boldsymbol{\Sigma}$ in (5.9) is a function of $\boldsymbol{\beta}$. In addition, the generalized estimating equation is applicable for discrete responses such as binary data in logistic regression and count data in Poisson regression. Some useful references of GEE can be found in Liang and Zeger (1986), Zeger et al. (1988), Zeger and Liang (1992), Diggle et al. (2002), and Hardin and Hilbe (2012).

To accommodate the misspecified working correlation matrix, Qu et al. (2000) then linked $\mathbf{R}(\boldsymbol{\beta})$ to a linear combination of basis matrices. That is,

$$\mathbf{R}^{-1}(\boldsymbol{\beta}) = \sum_{k=1}^{K} \beta_k \mathbf{M}_k, \tag{5.13}$$

where $\mathbf{M}_k$ are known matrices for $k = 1, \cdots, K$ and $\boldsymbol{\beta} = (\beta_1 \cdots, \beta_K)'$ is an unknown parameter vector.

They further introduced the following extended score function without estimating the parameter vector $\boldsymbol{\beta}$:

$$g(\boldsymbol{\alpha}) = \frac{1}{n}\sum_{i=1}^{n} g_i(\boldsymbol{\alpha}) = \frac{1}{n}\begin{pmatrix} \sum_{i=1}^{n} (\dot{\boldsymbol{\mu}}_i)'\mathbf{G}_i^{-1/2}\mathbf{M}_1\mathbf{G}_i^{-1/2}(Y_i - \boldsymbol{\mu}_i) \\ \vdots \\ \sum_{i=1}^{n} (\dot{\boldsymbol{\mu}}_i)'\mathbf{G}_i^{-1/2}\mathbf{M}_K\mathbf{G}_i^{-1/2}(Y_i - \boldsymbol{\mu}_i) \end{pmatrix}. \tag{5.14}$$

Afterward, they proposed the quadratic influence function

$$Q_n(\boldsymbol{\alpha}) = g(\boldsymbol{\alpha})'\mathbf{C}_n^{-1} g(\boldsymbol{\alpha}), \tag{5.15}$$

where $\mathbf{C}_n = n^{-2}\sum_{i=1}^{n} g_i(\boldsymbol{\alpha}) g_i(\boldsymbol{\alpha})'$. Thus, we can obtain the estimator $\widehat{\boldsymbol{\alpha}}$ by minimizing $Q_n(\boldsymbol{\alpha})$, and this method is mainly designed for longitudinal data analysis. Some useful references and application can be found in Qu et al. (2000), Qu and Song (2004), and Qu and Li (2006).

In model (5.7), we implicitly assume that $\boldsymbol{\alpha}$, $\boldsymbol{\beta}$, σ^2, and $\boldsymbol{\Sigma}^i$ are fixed and not a function of times t_{ij} or locations s_{ij}. Suppose some of those parameters are a function of t_{ij}, see, e.g., the model given below

$$\mathbb{Y}(t_{ij}) = \mathbf{X}(t_{ij})\boldsymbol{\alpha}(t_{ij}) + \mathcal{E}(t_{ij}), \tag{5.16}$$

where $\mathcal{E}$ is a randomly distributed error vector with mean vector $\mathbf{0} \in \mathbb{R}^N$ and covariance matrix $\sigma^2(t_{ij})\boldsymbol{\Sigma}(t_{ij}) \in \mathbb{R}^{N\times N}$. A similar model is in Fan et al. (2007), and model (5.16) can be viewed as a special application of varying coefficient models (see Hastie and Tibshirani 1993; Fan and Zhang 1999, 2008). Accordingly, model (5.16) can be estimated by nonparametric or semiparametric methods, such as kernel, local polynomial and smoothing spline. Useful references on this subject can be found in Zeger and Diggle (1994), Hoover et al. (1998), Wu et al. (1998), Zhang et al. (1998), Lin et al. (2001), Wu and Zhang (2006), and Wu and Tian (2018).

5.1.4 Multivariate Regression Analysis

In the previous two subsections, we have discussed covariance and correlation matrices, which are associated with weighted regression models and longitudinal data analysis, respectively. Here, we study multivariate regression models, which also intrinsically consist of covariance matrices. As mentioned in Sect. 4.2, the aim of multivariate regression is establishing the relationship between the p components of the mean vector of multivariate responses, $Y = (y_1, \cdots, y_p)' \in \mathbb{R}^p$, and the q components of the explanatory vector, $X = (x_1, \cdots, x_q)' \in \mathbb{R}^q$. The multivariate regression model is

$$Y = \mathbf{A}'X + \boldsymbol{\varepsilon}, \tag{5.17}$$

where $\mathbf{A} = (\boldsymbol{\alpha}_1, \cdots, \boldsymbol{\alpha}_p)$ is a $q \times p$ matrix of regression coefficients, $\boldsymbol{\alpha}_j = (\alpha_{1j}, \cdots, \alpha_{qj})'$, and $\boldsymbol{\varepsilon} = (\varepsilon_1, \cdots, \varepsilon_p)' \in \mathbb{R}^p$ is a random vector with the mean $\mathbf{0} \in \mathbb{R}^p$ and covariance matrix $\boldsymbol{\Sigma} \in \mathbb{R}^{p\times p}$. With n independent samples (or subjects) on Y and the associated explanatory vector X, the multivariate regression model can be expressed as either

$$Y_i = \mathbf{A}'X_i + \boldsymbol{\varepsilon}_i \quad (i = 1, \cdots, n) \quad \text{or} \tag{5.18}$$

$$Y_{(j)} = \mathbf{X}\boldsymbol{\alpha}_j + \boldsymbol{\varepsilon}_{(j)} \quad (j = 1, \cdots, p), \tag{5.19}$$

where $\mathbf{X} = (X_1, \cdots, X_n)' \in \mathbb{R}^{n\times q}$, $\boldsymbol{\varepsilon}_i = (\epsilon_{i1} \cdots, \epsilon_{ip})'$, $Y_{(j)} = (y_{1j}, \cdots, y_{nj})'$ for $j = 1, \cdots, p$, and $\boldsymbol{\varepsilon}_{(j)} = (\epsilon_{1j} \cdots, \epsilon_{nj})'$. The matrix expression of (5.18) becomes

$$\mathbf{Y} = \mathbf{X}\mathbf{A} + \mathbf{E}, \tag{5.20}$$

where $\mathbf{Y} = (Y_1, \cdots, Y_n)' = (Y_{(1)}, \cdots, Y_{(p)}) \in \mathbb{R}^{n\times p}$, $Y_i = (y_{i1}, \cdots, y_{ip})'$ for $i = 1, \cdots, n$, and $\mathbf{E} = (\boldsymbol{\varepsilon}_1, \cdots, \boldsymbol{\varepsilon}_n)' \in \mathbb{R}^{n\times p}$.

Let $\mathbb{Y}^* = \text{vec}(\mathbf{Y}) \in \mathbb{R}^{np}$ and $\mathcal{E}^* = \text{vec}(\mathbf{E}) \in \mathbb{R}^{np}$, where $\mathcal{E}^*$ is a random vector with mean $\mathbf{0} \in \mathbb{R}^{np}$ and covariance $(\boldsymbol{\Sigma} \otimes \mathbf{I}_n) \in \mathbb{R}^{np\times np}$. Using the fact that $\text{vec}(\mathbf{XA}) = (\mathbf{I}_p \otimes \mathbf{X})\text{vec}(\mathbf{A})$, model (5.20) can be re-expressed as

$$\mathbb{Y}^* = (\mathbf{I}_p \otimes \mathbf{X})\text{vec}(\mathbf{A}) + \mathcal{E}^*. \tag{5.21}$$

Then we can obtain the weighted least squares estimator of $\mathbf{A}$ by minimizing the following object function:

$$\{\mathbb{Y}^* - (\mathbf{I}_p \otimes \mathbf{X})\text{vec}(\mathbf{A})\}'(\boldsymbol{\Sigma} \otimes \mathbf{I}_n)^{-1}\{\mathbb{Y}^* - (\mathbf{I}_p \otimes \mathbf{X})\text{vec}(\mathbf{A})\}. \tag{5.22}$$

As a result,

$$\begin{aligned}\text{vec}(\widehat{\mathbf{A}}) &= \{(\mathbf{I}_p \otimes \mathbf{X})'(\boldsymbol{\Sigma} \otimes \mathbf{I}_n)^{-1}(\mathbf{I}_p \otimes \mathbf{X})\}^{-1}(\mathbf{I}_p \otimes \mathbf{X})'(\boldsymbol{\Sigma} \otimes \mathbf{I}_n)^{-1}\mathbb{Y}^* \\ &= (\mathbf{I}_p \otimes (\mathbf{X}'\mathbf{X})^{-1}\mathbf{X}')\mathbb{Y}^*. \end{aligned} \tag{5.23}$$

After algebraic simplification, the weighted least squares estimator of $\mathbf{A}$ is

$$\widehat{\mathbf{A}} = (\mathbf{X}'\mathbf{X})^{-1}\mathbf{X}'\mathbf{Y}, \tag{5.24}$$

which is identical to the least squares estimator. It is worth noting that the weighted regression estimator in (5.3) is different from the OLS estimator, and the longitudinal data estimator in (5.11) is also different from the OLS estimator. Based on (5.23) and (5.24), we have that $\mathrm{vec}(\widehat{\mathbf{A}}') = ((\mathbf{X}'\mathbf{X})^{-1}\mathbf{X}' \otimes \mathbf{I}_p)\mathbb{Y}$ and $\widehat{\mathbf{A}}' = \mathbf{Y}'\mathbf{X}(\mathbf{X}'\mathbf{X})^{-1}$, where $\mathbb{Y} = \mathrm{vec}(\mathbf{Y}')$.

Suppose that the random error vector $\boldsymbol{\varepsilon}$ is independent and identically distributed as $N(0, \boldsymbol{\Sigma})$. Then $\widehat{\mathbf{A}}$ is also the maximum likelihood estimator. In addition, $\mathrm{vec}(\widehat{\mathbf{A}})$ is distributed as $N(\mathrm{vec}(\mathbf{A}), \boldsymbol{\Sigma} \otimes (\mathbf{X}'\mathbf{X})^{-1})$. Subsequently, $\widehat{\mathbf{Y}} = \mathbf{X}\widehat{\mathbf{A}}$, and $\mathrm{vec}(\widehat{\mathbf{Y}})$ is distributed as $N(\mathrm{vec}(\mathbf{XA}), \boldsymbol{\Sigma} \otimes \mathbf{H})$, where $\mathbf{H} = \mathbf{X}(\mathbf{X}'\mathbf{X})^{-1}\mathbf{X}'$. If $\boldsymbol{\Sigma}$ is fully unspecified, then its maximum likelihood estimator and unbiased estimator are $\widehat{\boldsymbol{\Sigma}} = n^{-1}(\mathbf{Y} - \widehat{\mathbf{Y}})'(\mathbf{Y} - \widehat{\mathbf{Y}})$ and $\tilde{\boldsymbol{\Sigma}} = (n - q)^{-1}(\mathbf{Y} - \widehat{\mathbf{Y}})'(\mathbf{Y} - \widehat{\mathbf{Y}})$, respectively. The statistical inference of multivariate regression model can be found in related books; see, e.g., Anderson (2003), Seber (2004), Johnson and Wichern (2007), Izenman (2008), and Cook (2018).

Finally, it is of interest to extend multivariate regression models to seemingly unrelated regressions proposed by Zellner (1962). Specifically, we link each component's response vector $Y_{(j)}$ to $\boldsymbol{\alpha}_j$ in (5.19), with different covariate matrix, X_j, for $j = 1, \cdots, p$. A good reference book on the topic of seemingly unrelated regression model can be found in Srivastava and Giles (1987).

5.2 Covariance-Mean Regression Analysis

5.2.1 Model Structure

The prior section considers four different mean regressions along with their covariance structures. Note that the covariance structure in multivariate linear regression discussed in Sect. 5.1.4 was not specified. Motivated by Anderson (1973), we follow the approach of Zou et al. (2022) and introduce a hybrid model by encompassing multivariate mean regression in Sect. 5.1.4 and covariance regression in Chap. 4. Without loss of generality, we assume that the first covariate in the mean regression model is a $q \times 1$ vector with 1s and the first similarity matrix in the covariance regression model is a $p \times p$ identity matrix.

We first consider the multivariate mean regression model

$$\mathbf{Y} = \mathbf{XA} + \mathbf{E}, \tag{5.25}$$

where $\mathbf{A}$ is defined in (5.17), and $\mathbf{Y}$, $\mathbf{X}$, and $\mathbf{E}$ are defined in (5.20). In addition, we assume that the i-th component of $\mathbf{E}$, $\boldsymbol{\varepsilon}_i$, is a random vector with mean $\mathbf{0} \in \mathbb{R}^p$ and covariance matrix $\boldsymbol{\Sigma}^i \in \mathbb{R}^{p\times p}$ for $i = 1, \cdots, n$. To model covariance structure, we adopt the approach of Zou et al. (2022) and construct the similarity matrix $\mathbf{W}_k^i = \mathbf{W}_k^i(Z_k^i) = \big(w(Z_{kj_1}^i, Z_{kj_2}^i)\big) \in \mathbb{R}^{p\times p}$, where Z_k^i is the k-th $p \times 1$ explanatory vector collected at the i-th observation, $Z_{kj_1}^i$ and $Z_{kj_2}^i$ are the j_1-th and

j_2-th elements of Z_k^i for $j_1, j_2 \in \{1, \cdots, p\}$, respectively. Note that the similarity $w(Z_{kj_1}^i, Z_{kj_2}^i)$ is a decreasing function of the distance $d(Z_{kj_1}^i, Z_{kj_2}^i)$ between $Z_{kj_1}^i$ and $Z_{kj_2}^i$ (e.g., $w(Z_{kj_1}^i, Z_{kj_2}^i) = \exp\{-d^2(Z_{kj_1}^i, Z_{kj_2}^i)\}$). The similarity matrix for a vector of discrete explanatory variables can be defined by referring to Johnson and Wichern (2007). We next adapt the approach of Chap. 4 and consider a linear covariance regression model

$$\mathbf{\Sigma}^i(\boldsymbol{\beta}) = \sum_{k=1}^{K} \beta_k \mathbf{W}_k^i, \tag{5.26}$$

where $\boldsymbol{\beta} = (\beta_1, \cdots, \beta_K)'$ is the vector of covariance regression coefficients, $\mathbf{W}_k^i \in \mathbb{R}^{p\times p}$ is the similarity matrix, $\mathbf{W}_1^i = \mathbf{I}_p$ is the $p \times p$ identity matrix, for $i = 1, \cdots, n$. Accordingly, encompassing (5.25) and (5.26) forms the covariance-mean regression (CMR) model; see Zou et al. (2022).

Let $\mathbb{Y} = \mathrm{vec}(\mathbf{Y}')$ and $\mathcal{E} = \mathrm{vec}(\mathbf{E}')$. Then the CMR model can be expressed as

$$\mathbb{Y} = (\mathbf{X} \otimes \mathbf{I}_p)\mathrm{vec}(\mathbf{A}') + \mathcal{E}, \tag{5.27}$$

where $\mathcal{E}$ is a random vector with the mean $\mathbf{0} \in \mathbb{R}^{np}$ and covariance matrix $\mathbf{\Sigma}(\boldsymbol{\beta}) = \mathrm{diag}(\mathbf{\Sigma}^i(\boldsymbol{\beta})) \in \mathbb{R}^{np\times np}$ for $i = 1, \cdots, n$.

5.2.2 Parameter Estimators and Hypothesis Tests

To estimate unknown parameters of $\boldsymbol{\alpha}$ and $\boldsymbol{\beta}$ in the CMR model, we consider two different approaches proposed by Zou et al. (2022). The first approach is assuming that the vector $\boldsymbol{\varepsilon}_i$ of random errors is independently distributed as $N(\mathbf{0}, \mathbf{\Sigma}^i(\boldsymbol{\beta}))$ for $i = 1, \cdots, n$. Denote the true mean coefficient matrix and the true covariance coefficient vector as $\mathbf{A}_0$ and $\boldsymbol{\beta}_0 = (\beta_{10}, \cdots, \beta_{K0})'$, respectively.

In practice, $\mathbf{A}_0$ is often unknown. Based on the observed $\mathbb{Y} = \mathrm{vec}(\mathbf{Y}')$, we consider the log-likelihood function for both unknown parameters $\mathbf{A}$ and $\boldsymbol{\beta}$, which is

$$\begin{aligned}\ell_{np}(\mathbf{A}, \boldsymbol{\beta}) = &-\frac{np}{2}\log(2\pi) - \frac{1}{2}\log|\mathbf{\Sigma}(\boldsymbol{\beta})| \\ &- \frac{1}{2}\left\{\mathbb{Y} - (\mathbf{X}\otimes\mathbf{I}_p)\mathrm{vec}(\mathbf{A}')\right\}'\mathbf{\Sigma}^{-1}(\boldsymbol{\beta})\left\{\mathbb{Y} - (\mathbf{X}\otimes\mathbf{I}_p)\mathrm{vec}(\mathbf{A}')\right\}.\end{aligned} \tag{5.28}$$

Since $\mathbf{A}$ is a nuisance parameter, we obtain the profile likelihood method by removing parameter $\mathbf{A}$ from (5.28). Specifically, given $\boldsymbol{\beta}$, we maximize $\ell_{np}(\mathbf{A}, \boldsymbol{\beta})$ with respect to $\mathbf{A}$ and have that

$$\text{vec}\left\{\widehat{\mathbf{A}}'(\boldsymbol{\beta})\right\} = \left\{(\mathbf{X}\otimes\mathbf{I}_p)'\boldsymbol{\Sigma}^{-1}(\boldsymbol{\beta})(\mathbf{X}\otimes\mathbf{I}_p)\right\}^{-1}(\mathbf{X}\otimes\mathbf{I}_p)'\boldsymbol{\Sigma}^{-1}(\boldsymbol{\beta})\mathbb{Y}. \tag{5.29}$$

We then substitute $\mathbf{A}'$ in (5.28) with $\widehat{\mathbf{A}}'(\boldsymbol{\beta})$ and obtain the profile log-likelihood function

$$\ell_{np}(\boldsymbol{\beta}) = \ell_{np}(\widehat{\mathbf{A}}(\boldsymbol{\beta}),\boldsymbol{\beta}) = -\frac{np}{2}\log(2\pi) - \frac{1}{2}\log|\boldsymbol{\Sigma}(\boldsymbol{\beta})| - \frac{1}{2}\mathbf{U}(\boldsymbol{\beta})'\mathbf{U}(\boldsymbol{\beta}), \tag{5.30}$$

where $\mathbf{U}(\boldsymbol{\beta}) = \mathbf{M}(\boldsymbol{\beta})\boldsymbol{\Sigma}^{-1/2}(\boldsymbol{\beta})\mathbb{Y}$, $\mathbf{M}(\boldsymbol{\beta}) = \mathbf{I}_{np} - \mathbf{H}(\boldsymbol{\beta})$, and $\mathbf{H}(\boldsymbol{\beta}) = \boldsymbol{\Sigma}^{-1/2}(\boldsymbol{\beta})(\mathbf{X}\otimes\mathbf{I}_p)\{(\mathbf{X}\otimes\mathbf{I}_p)'\boldsymbol{\Sigma}^{-1}(\boldsymbol{\beta})(\mathbf{X}\otimes\mathbf{I}_p)\}^{-1}(\mathbf{X}\otimes\mathbf{I}_p)'\boldsymbol{\Sigma}^{-1/2}(\boldsymbol{\beta})$ is an idempotent matrix.

Based on (5.30), we obtain the following profile score function:

$$\frac{\partial\ell_{np}(\boldsymbol{\beta})}{\partial\beta_k} = -\frac{1}{2}\text{tr}\left\{\boldsymbol{\Sigma}^{-1}(\boldsymbol{\beta})\mathbf{W}_k\right\} + \frac{1}{2}\mathbf{U}(\boldsymbol{\beta})'\boldsymbol{\Sigma}^{-1/2}(\boldsymbol{\beta})\mathbf{W}_k\boldsymbol{\Sigma}^{-1/2}(\boldsymbol{\beta})\mathbf{U}(\boldsymbol{\beta}), \tag{5.31}$$

where $\mathbf{W}_k = \text{diag}\{W_k^{(1)},\cdots,W_k^{(n)}\}$ for $k = 1,\cdots,K$. Define

$$s_{np,k}(\boldsymbol{\beta}) = \frac{\partial\ell_{np}(\boldsymbol{\beta})}{\partial\beta_k} + \frac{1}{2}\text{tr}\left[\boldsymbol{\Sigma}^{-1/2}(\boldsymbol{\beta})\mathbf{W}_k\boldsymbol{\Sigma}^{-1/2}(\boldsymbol{\beta})\mathbf{H}(\boldsymbol{\beta})\right]. \tag{5.32}$$

Then Zou et al. (2022) obtained the Z-estimator $\widehat{\boldsymbol{\beta}}_Z$ (van der Vaart 1998) by solving the estimating equation $s_{np,k}(\boldsymbol{\beta}) = 0$ for all $k = 1,\cdots,K$. Subsequently, we replace $\boldsymbol{\beta}$ in (5.29) with $\widehat{\boldsymbol{\beta}}_Z$, which yields the estimator $\widehat{\mathbf{A}}$.

We next present the theoretical properties of $\widehat{\boldsymbol{\beta}}_Z$ demonstrated by Zou et al. (2022).

Theorem 5.1 *Assume that Conditions* ***(C5.1)–(C5.5)*** *hold in Sect. 5.2.3. Then, as $n\to\infty$ and p is either fixed or goes to infinity*

$$\sqrt{np}\big(\widehat{\boldsymbol{\beta}}_Z - \boldsymbol{\beta}_0\big) \xrightarrow{d} N\left(0, 2\mathcal{M}_{-1/2}^{-1}\right),$$

where $\mathcal{M}_{-1/2}$ is defined in Condition ***(C5.3)****. In addition, under Conditions* ***(C5.1)****,* ***(C5.2)****,* ***(C5.3′)****,* ***(C5.4)****, and* ***(C5.5)*** *in Sect. 5.2.3, we have that, as n is fixed and $p\to\infty$,*

$$\sqrt{np}\big(\widehat{\boldsymbol{\beta}}_Z - \boldsymbol{\beta}_0\big) \xrightarrow{d} N\left(0, 2\left\{\mathcal{M}_{-1/2}(\mathbf{M}_0)\right\}^{-1}\right),$$

where $\mathbf{M}_0 = \mathbf{M}(\boldsymbol{\beta}_0)$ and $\mathcal{M}_{-1/2}(\mathbf{M}_0)$ is defined in Condition (C5.3′).

The above theorem indicates that $\widehat{\boldsymbol{\beta}}_Z$ is a consistent estimator of $\boldsymbol{\beta}_0$ under the assumption that $p\to\infty$. In addition, Zou et al. (2022) showed that $\widehat{\boldsymbol{\beta}}_Z$ is an efficient estimator when $\mathbf{A}_0$ is known.

As mentioned in Chap. 4, the computation of the MLE-type estimator in covariance regression can be challenging when p is large. Hence, we introduce the second approach, i.e., the ordinary least squares (OLS) approach, proposed by Zou et al. (2022). Specifically, we first obtain the mean regression estimator of $\mathbf{A}$ given below.

$$\mathrm{vec}\left\{\tilde{\mathbf{A}}'_{\mathrm{OLS}}\right\} = \left\{(\mathbf{X} \otimes \mathbf{I}_p)'(\mathbf{X} \otimes \mathbf{I}_p)\right\}^{-1} (\mathbf{X} \otimes \mathbf{I}_p)'\mathbb{Y}. \tag{5.33}$$

Accordingly, $\tilde{\mathbf{A}}_{\mathrm{OLS}} = (\mathbf{X}'\mathbf{X})^{-1}\mathbf{X}'\mathbf{Y}$, which is the same as that in (5.24). We then obtain the residual vector below, which can be used to estimate the covariance matrix of the random error vector.

$$\widetilde{\mathcal{E}} = \mathbb{Y} - (\mathbf{X} \otimes \mathbf{I}_p)\mathrm{vec}\left(\tilde{\mathbf{A}}'_{\mathrm{OLS}}\right) = \tilde{\mathbf{Q}}\mathbb{Y} = \tilde{\mathbf{Q}}\mathcal{E}, \tag{5.34}$$

where $\tilde{\mathbf{Q}} = \mathbf{I}_{np} - (\mathbf{X}\otimes\mathbf{I}_p)\left\{(\mathbf{X} \otimes \mathbf{I}_p)'(\mathbf{X} \otimes \mathbf{I}_p)\right\}^{-1} (\mathbf{X}\otimes\mathbf{I}_p)'$ is an idempotent matrix. This, in conjunction with (5.26) and (5.27), leads to

$$\mathrm{E}(\widetilde{\mathcal{E}}\widetilde{\mathcal{E}}') = \tilde{\mathbf{Q}}\boldsymbol{\Sigma}(\boldsymbol{\beta}_0)\tilde{\mathbf{Q}} = \sum_{k=1}^{K} \beta_{k,0}\widetilde{\mathbf{W}}_k,$$

where $\widetilde{\mathbf{W}}_k = \tilde{\mathbf{Q}}\mathbf{W}_k\tilde{\mathbf{Q}}$ and $\mathbf{W}_k$ is defined in (5.31) for $k = 1, \cdots, K$.

Besides, Zou et al. (2022) obtained the unconstrained OLS estimator that minimizes the objective function below:

$$D_{np}(\boldsymbol{\beta}) := \left\|\widetilde{\mathcal{E}}\widetilde{\mathcal{E}}' - \sum_{k=1}^{K} \beta_k\widetilde{\mathbf{W}}_k\right\|_F^2 .$$

After algebraic simplification, the unconstrained OLS estimator of $\boldsymbol{\beta}$ is

$$\tilde{\boldsymbol{\beta}}_{\mathrm{OLS}} = \left(\mathrm{tr}(\widetilde{\mathbf{W}}_{k_1}\widetilde{\mathbf{W}}_{k_2})\right)^{-1}_{K\times K} \left(\widetilde{\mathcal{E}}'\widetilde{\mathbf{W}}_k\widetilde{\mathcal{E}}\right)_{K\times 1}. \tag{5.35}$$

It is the same as that of (4.17) when the mean regression component is not included in the model. The asymptotic property of $\tilde{\boldsymbol{\beta}}_{\mathrm{OLS}}$ demonstrated by Zou et al. (2022) is given below.

Theorem 5.2 *Assume that Conditions* **(C5.1)–(C5.4)** *hold in Sect. 5.2.3. Then*

$$\sqrt{np}\left(\tilde{\boldsymbol{\beta}}_{\mathrm{OLS}} - \boldsymbol{\beta}_0\right) \xrightarrow{d} N\left(0, 2\mathcal{M}_0^{-1}\mathcal{M}_{1/2}\mathcal{M}_0^{-1}\right),$$

as $n \to \infty$ and p is either fixed or goes to infinity, where $\mathcal{M}_l$, for $l = 0$ and $1/2$, is defined in Condition **(C5.3)**. *In addition, under Conditions* **(C5.1)**, **(C5.2)**, **(C5.3′)**, *and* **(C5.4)** *in Sect. 5.2.3, as n is fixed and $p \to \infty$, we obtain that*

$$\sqrt{np}\left(\tilde{\boldsymbol{\beta}}_{\mathrm{OLS}}-\boldsymbol{\beta}_0\right) \xrightarrow{d} N\left(0, 2\left\{\mathcal{M}_0(\tilde{\mathbf{Q}})\right\}^{-1}\mathcal{M}_{1/2}(\tilde{\mathbf{Q}})\left\{\mathcal{M}_0(\tilde{\mathbf{Q}})\right\}^{-1}\right),$$

where $\mathcal{M}_l(\tilde{\mathbf{Q}})$ *is defined in Condition (C5.3′) for* $l = 0$ *and* $1/2$.

The above theorem indicates that $\tilde{\boldsymbol{\beta}}_{\mathrm{OLS}}$ is a consistent estimator. In addition, the asymptotic variance of $\tilde{\boldsymbol{\beta}}_{\mathrm{OLS}}$, $2(np)^{-1}\mathcal{M}_0^{-1}\mathcal{M}_{1/2}\mathcal{M}_0^{-1}$, is greater than that of $\widehat{\boldsymbol{\beta}}_Z$, $2(np)^{-1}\mathcal{M}_{-1/2}^{-1}$; see the proof in Zou et al. (2022). However, the computation of the OLS estimator can be considerably simpler and faster than that of $\widehat{\boldsymbol{\beta}}_Z$, especially when p is large.

Note that the covariance matrix estimator $\boldsymbol{\Sigma}^i(\tilde{\boldsymbol{\beta}}_{\mathrm{OLS}})$ may not be positive definite. To assure its positiveness, we impose the constraint $\boldsymbol{\beta} \in \mathbf{B}^+ := \{\boldsymbol{\beta} : \boldsymbol{\Sigma}^i(\boldsymbol{\beta}) \succ \mathbf{0},\ \text{for } i = 1, \cdots, n\}$, where $\succ$ was defined in Sect. 1.6.1. Under this constraint, we obtain the constrained OLS (OLS+) estimator $\tilde{\boldsymbol{\beta}}_{\mathrm{OLS+}} = \arg\min_{\boldsymbol{\beta}\in\mathbf{B}^+} D_{np}(\boldsymbol{\beta})$. One can show that the constrained OLS estimator is identical to the unconstrained OLS estimator with probability tending to one as $\max\{n, p\} \to \infty$. Consequently, $\tilde{\boldsymbol{\beta}}_{\mathrm{OLS}}$ and $\tilde{\boldsymbol{\beta}}_{\mathrm{OLS+}}$ yield the same asymptotic distribution. The algorithm for obtaining the constrained estimator is in the supplementary material of Zou et al. (2022).

In this section, we have obtained the three consistent parameter estimators, $\widehat{\boldsymbol{\beta}}_Z$, $\tilde{\boldsymbol{\beta}}_{\mathrm{OLS}}$ and $\tilde{\boldsymbol{\beta}}_{\mathrm{OLS+}}$, and denote them by $\tilde{\boldsymbol{\beta}}$ without ambiguity. Then, it can be shown that $\boldsymbol{\Sigma}^i(\tilde{\boldsymbol{\beta}})$ and its inverse, $\{\boldsymbol{\Sigma}^i(\tilde{\boldsymbol{\beta}})\}^{-1}$ are consistent estimators of $\boldsymbol{\Sigma}_0^i$ and $\{\boldsymbol{\Sigma}_0^i\}^{-1}$, respectively, in both spectral norm and Frobenius norm standardized by $p^{-1/2}$ when either p or n goes to infinity, where $\boldsymbol{\Sigma}_0^i = \boldsymbol{\Sigma}_0^i(\boldsymbol{\beta}_0)$. Furthermore, the asymptotic properties in Theorems 5.1 and 5.2 can be extended by allowing the number of similarity matrices (K) to go to infinity. The detailed proofs of the above results can be found in Zou et al. (2022).

It is worth noting that the asymptotic covariance in Theorems 5.1 and 5.2 depends on $\boldsymbol{\Sigma}(\boldsymbol{\beta}_0)$. In practice, one can replace $\boldsymbol{\beta}_0$ by its consistent estimators, $\widehat{\boldsymbol{\beta}}_Z$ and $\tilde{\boldsymbol{\beta}}$, which yields the correspondingly consistent estimators of the asymptotic covariances. Then, we can conduct hypothesis testing and construct confidence intervals for covariance regression coefficients. In hypothesis testing with covariance regression, it is natural to test the significance of auxiliary information. Specifically, one can test the sphericity of $\boldsymbol{\Sigma}_0^i$, namely, $\boldsymbol{\Sigma}_0^i = \beta_{10}\mathbf{I}_p$ for some finite positive constant β_{10}. As a result, the null and alternative hypotheses are

$$H_{0\beta} : \beta_{20} = \cdots = \beta_{K0} = 0 \text{ vs } H_{1\beta} : \beta_{k0} \neq 0 \text{ for some } k = 2, \cdots, K.$$

Using the asymptotic normality property of $\tilde{\boldsymbol{\beta}}$, one can construct a Wald test statistic, which follows a chi-squared distribution asymptotically with $K - 1$ degrees of freedom under the null hypothesis.

In addition to the above hypothesis setting, it is important to assess the adequacy of the covariance regression model in (5.26). To this end, we consider the following hypotheses:

$$H_0 : \mathbf{\Sigma}_0^{(i)} = \sum_{k=1}^{K} \beta_{k0} \mathbf{W}_k^i \text{ for all } i = 1, \cdots, n, \text{ and} \tag{5.36}$$

$$H_1 : \mathbf{\Sigma}_0^i \neq \sum_{k=1}^{K} \beta_{k0} \mathbf{W}_k^i \text{ for some } i = 1, \cdots, n. \tag{5.37}$$

To test the null hypothesis, Zou et al. (2022) proposed two tests, which examine the adequacy of the covariance regression model (5.26) under the heterogeneous covariance setting. They also showed that these two test statistics are asymptotically standard normal under the null hypothesis.

After studying the above models, we can integrate them and propose the generalized covariance-mean regression model. As stated in Sect. 5.1.3, let the mean responses be $E(y_{ij}) = \mu_{ij} = h(x_{ij}'\boldsymbol{\alpha})$, where $h(\cdot)$ is a known inverse link function. In addition, assume that the precision matrix of $\mathbf{R}(\boldsymbol{\beta})$ in the linear combination structure (5.13) is a function of covariates and can change along with the subject i. Denote the precision matrix accordingly by $\mathbf{R}^i(\boldsymbol{\beta})^{-1}$ for $i = 1, \cdots, n$. Following (5.26), one can link it to a set of similarity matrices:

$$\mathbf{R}^i(\boldsymbol{\beta})^{-1} = \sum_{k=1}^{K} \beta_k \mathbf{W}_k^i. \tag{5.38}$$

As a result, (5.38) can broaden the application of Qu et al.'s (2000) quadratic influence function by allowing $\mathbf{R}(\boldsymbol{\beta})$ to be heterogeneous across subjects. It would be valuable to employ iterative procedure to obtain parameter estimators of $\boldsymbol{\alpha}$ and $\boldsymbol{\beta}$ and show their theoretical properties. Furthermore, the extension of the above study to the multivariate version of generalized estimating equations (Brobbey et al. 2022) is worth investigation.

5.2.3 *Technical Conditions*

To study asymptotic properties of parameter estimators, we adopt Zou et al.'s (2022) conditions given below.

(C5.1) For any symmetric matrices in $\{\mathbf{W}_k^i \in \mathbb{R}^{p\times p} : i = 1, \cdots, n, \text{ and } k = 1, \cdots, K\}$, assume that there exists a finite constant $\delta > 0$ such that, for any $n \geq 1$ and $p \geq 1$,

$$\max_{1\leq i\leq n} \left\| \mathbf{W}_k^i \right\|_2 \leq \delta, \; k = 1, \cdots, K. \tag{5.39}$$

(C5.2) Assume that there exists a finite constant $a_{\min} > 0$ such that

$$\min_{1\le i\le n} \lambda_p \left\{\mathbf{\Sigma}_0^i\right\} \ge a_{\min}, \tag{5.40}$$

for any $n \ge 1$ and $p \ge 1$, where $\lambda_p \left\{\mathbf{\Sigma}_0^i\right\}$ is the smallest eigenvalue of $\mathbf{\Sigma}_0^i$.

Note that Conditions (C5.1) and (C5.2) are modified from their corresponding conditions in Sect. 4.2.

(C5.3) Assume that the matrices

$$\frac{1}{np}\left(\mathrm{tr}(\mathbf{\Sigma}_0^{2l}\mathbf{W}_{k_1}\mathbf{\Sigma}_0^{2l}\mathbf{W}_{k_2})\right)_{K\times K} \to \mathcal{M}_l,$$

respectively, for $l = -1, -1/2, 0, 1/2$, as $\max\{n, p\} \to \infty$, where $\mathbf{\Sigma}_0 = \mathbf{\Sigma}(\boldsymbol{\beta}_0) = \mathrm{diag}(\mathbf{\Sigma}_0^i)$, $\mathcal{M}_l \in \mathbb{R}^{K\times K}$ are finite positive definite matrices, $\mathbf{\Sigma}_0^0 := \mathbf{I}_{np}$ and $(\mathbf{\Sigma}_0^i)^0 := \mathbf{I}_p$.

Note that this condition is a generalization of Condition (C4.3) in Sect. 4.2.

(C5.3′) Let $\mathbf{M}_0 = \mathbf{M}(\boldsymbol{\beta})|_{\boldsymbol{\beta}=\boldsymbol{\beta}_0}$, where $\mathbf{M}(\boldsymbol{\beta})$ is defined in (5.30). Assume that, as n is fixed and $p \to \infty$, the matrix

$$\frac{1}{np}\left(\mathrm{tr}(\mathbf{\Sigma}_0^{-1/2}\mathbf{M}_0\mathbf{\Sigma}_0^{-1/2}\mathbf{W}_{k_1}\mathbf{\Sigma}_0^{-1/2}\mathbf{M}_0\mathbf{\Sigma}_0^{-1/2}\mathbf{W}_{k_2})\right)_{K\times K} \to \mathcal{M}_{-1/2}(\mathbf{M}_0),$$

where $\mathcal{M}_{-1/2}(\mathbf{M}_0) \in \mathbb{R}^{K\times K}$ is a finite positive definite matrix. In addition, assume that, as n is fixed and $p \to \infty$, the matrices

$$\frac{1}{np}\left(\mathrm{tr}(\tilde{\mathbf{Q}}\mathbf{\Sigma}_0^{2l}\tilde{\mathbf{Q}}\mathbf{W}_{k_1}\tilde{\mathbf{Q}}\mathbf{\Sigma}_0^{2l}\tilde{\mathbf{Q}}\mathbf{W}_{k_2})\right)_{K\times K} \to \mathcal{M}_l(\tilde{\mathbf{Q}}),$$

respectively, for $l = 0$ and $1/2$, where $\tilde{\mathbf{Q}}$ is defined in (5.34) and $\mathcal{M}_l(\tilde{\mathbf{Q}}) \in \mathbb{R}^{K\times K}$ are finite positive definite matrices.

Note that Condition (C5.3) is insufficient to obtain the asymptotic distribution of estimators when n is fixed. However, Condition (C5.3′) is sufficient in this case.

(C5.4) Assume that $\mathbf{X}'\mathbf{X}$ is a nonsingular matrix when n is fixed. In addition, assume that $n^{-1}\mathbf{X}'\mathbf{X} \to \mathbf{\Sigma}_\mathbf{X}$ as $n \to \infty$, where $\mathbf{\Sigma}_\mathbf{X}$ is a finite positive definite matrix.

This condition is standard in linear regression models.

(C5.5) Let $s_{np}(\boldsymbol{\beta}) = \left(s_{np,1}(\boldsymbol{\beta}), \cdots, s_{np,K}(\boldsymbol{\beta})\right)'$, where $s_{np,k}(\boldsymbol{\beta})$ is defined in (5.32) for $k = 1, \cdots, K$. Assume that $\boldsymbol{\beta}_0$ is an interior point of the compact parameter space $\mathbf{B} \subset \mathbb{R}^K$ and $\boldsymbol{\beta}_0$ is the unique root of the equation $\mathrm{E}\{s_{np}(\boldsymbol{\beta})\} = 0$.

This condition ensures the identification of the parameter $\boldsymbol{\beta}$ when we construct $\widehat{\boldsymbol{\beta}}_Z$ in Sect. 5.2.2.

5.3 Empirical Examples

This section briefly reviews two examples associated with covariance-mean regression models. The first example is from Qu and Song (2004), who studied urinary incontinence based on the data from Preisser and Qaqish (1999). The response variable is 1 if the subject's daily life is bothered by accidental loss of urine, 0 otherwise. In this study, there are 137 patients from 38 practices, and each cluster contains at least 1 patient and at most 8 patients. It is natural to assume that subjects are correlated if they are from the same hospital practice. There are five covariates: gender, age, number of daily leaking accidents, severity of leaking, and number of times using the toilet per day. In the mean model fitting, Qu and Song (2004) considered logistic regression. In addition, they applied the inverse of the equicorrelation matrix to construct the quadratic inference function and generalized estimating equations (see Qu et al. 2000). They found that the quadratic inference function is more robust against outliers than generalized estimating equations. In practice, it would be of interest to fit the data with additional correlation matrices such as AR(1), MA(1), and Toeplitz, and then compare their performance.

The second example is taken from Zou et al. (2022), who studied the relationship between stock return comovement and firms' fundamentals based on the Standard and Poors (S&P) 500 component stocks. The dataset was obtained from the Wharton Research Data Services (WRDS). After eliminating the stocks with missing values in the response or covariates, there are 364 securities in the study. The response vectors contain the monthly excess returns of the $p = 364$ stocks for $n = 24$ months ranging from March 2015 to February 2017. There are $q = 2$ mean regression covariates, intercept, and the monthly excess return on the market. In addition, there are five covariance regression covariates: **industry** (IND); **leverage** computed by liability-to-asset ratio; **cash flow**; **size** measured by the logarithm of market value; and **book-to-market ratio**. Based on these five covariates, the corresponding similarity matrices are constructed. Note that the identity matrix $\mathbf{I}_p$ is included in the covariance regression model.

To study the effect of stock return comovement, Zou et al. (2022) employed the covariance-mean regression model to fit the data. After taking into account the mean regression effect on monthly excess market returns, they considered 31 ($2^5 - 1$) possible candidate models induced by five similarity matrices. The results show that the covariance regression model with $\mathbf{I}_p$ and the similarity matrix induced by IND is adequate. Specifically, IND is positively and significantly related to the covariance of monthly excess returns among 364 securities after including monthly excess market returns in mean responses.

Acknowledgments Section 5.2 and the second example of Sect. 5.3 in this chapter are derived in part from *Inference on Covariance-Mean Regression* published in the Journal of Econometrics, 2022 by Elsevier.

Chapter 6
Fixed and Random Covariance Models

This chapter reviews fixed and random mean regression models and then discusses their differences with respect to their expected test (prediction) errors based on the test data and the parameter estimator obtained from the training data. It is not surprising that the random setting model yields a larger test error than that of the fixed setting model. We then focus on the fixed and random covariance regression models. The theoretical properties of regression parameter estimators are established, which show that the random setting yields larger bias and variance in test errors. In addition, the estimators of the expected test errors under the Fixed-X and Random-X settings are obtained, which lead to their corresponding Mallows-type selection criteria $\widehat{C}_p$ and $\widehat{RC}_p$, respectively. These two criteria can be used to assess the performance of competing covariance regression models. An empirical example is briefly discussed.

6.1 Fixed and Random Mean Regression Models

6.1.1 Model Structure and Explanatory Variables (Covariates)

Regression analysis is one of the most widely and commonly used techniques for studying the relationship between the response and explanatory variables. To begin with the mean regression model, let $y_i \in \mathbb{R}^1$ be the response variable, and $X_i = (x_{i1}, \cdots, x_{iq})' \in \mathbb{R}^q$ be the explanatory vector for $i = 1, \cdots, n$. One usually expresses their relationship in the following form:

$$y_i = g(X_i) + \varepsilon_i, \tag{6.1}$$

where ε_i is i.i.d. from the distribution such as $N(0, \sigma^2)$ with $\sigma > 0$ for $i = 1, \cdots, n$. The explanatory variables can be either fixed or random. The fixed explanatory

W. Lan, C.-L. Tsai, *Covariance Analysis and Beyond*,
https://doi.org/10.1007/978-3-032-08796-6_6

variables are often constructed using dummy variables based on natural categories (e.g., two sexes or four seasons), experimental designs (e.g., the three levels of oil pressures or the four types of vaccines), or counting data with discrete choices or specific intervals (e.g., survey data that uses the Likert scale of ratings 1–5), among others. In contrast, the random explanatory variables are usually continuous, such as prices, scores, densities, and measurements, which have some distributions associated with them.

It is worth noting that fixed and random covariates impact parameter estimations differently. Suppose that $g(X_i)$ is the true model, but the approximately fitted model is $X_i'\boldsymbol{\alpha}$, where $\boldsymbol{\alpha} = (\alpha_1, \cdots, \alpha_q)' \in \mathbb{R}^q$. Under restrictive conditions with random covariates, White (1980b) demonstrated that the ordinary least squares (OLS) estimator of $\boldsymbol{\alpha}$ is consistent. However, this result is not valid with fixed covariates. On the other hand, if the true model is linear, then conditional inferences of $y_i|X_i$ are valid, and the OLS estimator is consistent under either fixed covariates or random covariates, with some conditions.

Note that when the true model is nonlinear and $\{X_i\}$ is random, the OLS estimator can break down; see Buja et al. (2019). This means that X_i cannot be treated as fixed, for the purpose of making conditional inferences, when the underlying process is nonlinear with random covariates. Useful references about conditional inferences and ancillaries can be found in White (1981), Brown (1990), Reid (1995) and Ghosh et al. (2010). A seminal book of Cox and Hinkley (1974) is also a good reference for this subject.

In addition to parameter estimation, Breiman and Spector (1992) studied the predictive capability in the context of model selection. To this end, they introduced prediction error (PE) and mean squared error (ME). Note that PE refers to the expected prediction error. They argued that treating the explanatory vector $X \in \mathbb{R}^q$ as fixed could lead to highly biased estimation of PE. As a result, non-resampling estimates, such as C_p (Mallows 1973) and adjusted R^2, could become highly biased methods of model selection. To this end, they considered the following five methods for random-X covariates: (i) test samples; (ii) cross-validation (Stone 1974); (iii) V-fold cross-validation (Geisser 1975); (iv) bootstrap (Efron 1979); (v) partial cross-validation (Girard 1998). In addition, they conducted extensive simulations to compare those methods. Later, Izenman (2008) employed PE and ME to study prediction accuracy based on similar methods. Recently, some researchers in their high-dimensional and machine learning articles have considered the Random-X setting in predictions; see, e.g., Dobriban and Wager (2018), Hastie et al. (2022), and Bai et al. (2023). Hence, we next study the impact of fixed and random covariates on expected prediction errors. For the sake of clarification, we prefer to use expected test error (see Section 7.2 in Hastie et al. 2009) rather than expected prediction error in the remainder of this chapter.

6.1.2 Expected Test (Prediction) Errors Under Three X-Scenarios

This subsection assumes that the true model is (6.1). As discussed in Sect. 6.1.1, this model can be interpreted under two different scenarios. One scenario is the fixed-X setting, while the other scenario is the Random-X setting. When X_i is fixed, such as a designed vector, the model (6.1) is well defined. Note that we sometimes denote the fixed X_i by x_i to differentiate it from that of the random X_i. As for the scenario with random X_i, we assume that $(X_i', y_i)'$ is i.i.d. and follows a joint distribution P, such as multivariate normal for $i = 1, \cdots, n$. In this case, we also assume that $E(y_i|X_i) = g(X_i)$ and $E(\varepsilon_i|X_i) = 0$ or X_i is independent of ε_i.

The main goal of this chapter is assessing expected test errors. To this end, let $(X_i', y_i)'$ be the training data for $i = 1, \cdots, n$, and let $(X_{0i}', y_{0i})'$ be the test data for $i = 1, \cdots, m$. Here, we assume that $m = n$. We then adopt the approach from a seminal paper of Rosset and Tibshirani (2020) and precisely define the following three settings for evaluating expected test errors, where fixed and random settings have been informally discussed earlier. Note that Rosset and Tibshirani (2020) used the term "prediction error" rather than "expected test error". For the sake of convenience, we sometimes use the term "test error" rather than "expected test error" provided it does not cause confusion.

(I) Fixed-X Setting: X_i is fixed, $X_{0i} = x_i$, and $\tilde{y}_i$ is independently generated from the conditional distribution of $y_i|x_i$;
(II) Same-X Setting: X_i is random, $X_{0i} = X_i$, and y_{0i} is independently generated from the conditional distribution of $y_i|X_i$;
(III) Random-X setting: X_i is random, and $(X_{0i}', y_{0i})'$ is independently generated from the joint distribution P.

Note that we use y_{0i} in the Same-X setting rather than $\tilde{y}$ given in Rosset and Tibshirani (2020). For the sake of simplification, we only consider fitting linear model below.

To assess the expected test (prediction) error, we fit the data with a linear regression model,

$$y_i = X_i'\boldsymbol{\alpha} + \varepsilon_i, \tag{6.2}$$

where $i = 1, \cdots, n$. Let $Y = (y_1, \cdots, y_n)'$, $Y_0 = (y_{01}, \cdots, y_{0n})'$, $\tilde{Y} = (\tilde{y}_1, \cdots, \tilde{y}_n)'$, and $\mathbf{X} = (X_1, \cdots, X_n)' \in \mathbb{R}^{n \times q}$ with full column rank. In addition, denote by $\widehat{\boldsymbol{\alpha}} = (\mathbf{X}'\mathbf{X})^{-1}\mathbf{X}'Y$ the ordinary least squares (OLS) estimator of $\boldsymbol{\alpha}$ (i.e., the maximum likelihood estimator under the normality assumption). We then follow Rosset and Tibshirani's (2020) squared loss function and define the three expected test errors below,

$$
\begin{aligned}
&(1)\ Err_F = E_1\Big\{\frac{1}{n}\sum_{i=1}^{n}(\tilde{y}_i - x_i'\widehat{\boldsymbol{\alpha}})^2\Big\},\\
&(2)\ Err_S = E_2\Big\{\frac{1}{n}\sum_{i=1}^{n}(y_{0i} - X_i'\widehat{\boldsymbol{\alpha}})^2\Big\} = E_2\Big\{(y_{01} - X_1'\widehat{\boldsymbol{\alpha}})^2\Big\}, \text{ and}\\
&(3)\ Err_R = E_3\Big\{(y_{01} - X_{01}'\widehat{\boldsymbol{\alpha}})^2\Big\}.
\end{aligned}
\tag{6.3}
$$

Note that E_1 refers to the expectations with respect to $\tilde{Y}$ and Y, E_2 refers to the expectations with respect to y_{01}, Y, and $\mathbf{X}$, and E_3 refers to the expectations with respect to y_{01}, Y, X_{01}, and $\mathbf{X}$. We will discuss the differences among these three expected errors in the next subsection. For the sake of illustration, we assume that the fitted model is linear. However, the above definitions and their associated results are applicable to the general mean regression function $\widehat{g}$; see Rosset and Tibshirani (2020).

6.1.3 Comparison of Expected Test Errors

After algebraic calculations, Err_F, Err_S, and Err_R can be simplified as follows.

$$
Err_F = \frac{1}{n}\sum_{i=1}^{n}\Big[E(x_i'\widehat{\boldsymbol{\alpha}}) - g(x_i)\Big]^2 + \frac{1}{n}\sum_{i=1}^{n}\text{var}(x_i'\widehat{\boldsymbol{\alpha}}) + \sigma^2; \tag{6.4}
$$

$$
Err_S = B + V + \sigma^2; \quad \text{and}
$$

$$
Err_R = B + V + B^+ + V^+ + \sigma^2,
$$

where $B = E_{\mathbf{X}}[E(X_1'\widehat{\boldsymbol{\alpha}}|\mathbf{X}) - g(X_1)]^2$, $V = E_{\mathbf{X}}[\text{var}(X_1'\widehat{\boldsymbol{\alpha}}|\mathbf{X})]$, $B^+ = E_{\mathbf{X},X_{01}}[E(X_{01}'\widehat{\boldsymbol{\alpha}}|\mathbf{X}, X_{01}) - g(X_{01})]^2 - E_{\mathbf{X}}[E(X_1'\widehat{\boldsymbol{\alpha}}|\mathbf{X}) - g(X_1)]^2$, and $V^+ = E_{\mathbf{X},X_{01}}[\text{var}(X_{01}'\widehat{\boldsymbol{\alpha}})|\mathbf{X}, X_{01})] - E_{\mathbf{X}}[\text{var}(X_1'\widehat{\boldsymbol{\alpha}}|\mathbf{X})]$. Note that the B and B^+ are the squared bias and excess squared bias, respectively, while the V and V^+ correspondingly denote the variance and excess variance. It is known that more complexity of the model usually results in larger variance and smaller bias. Suppose that the true regression model is linear with the OLS estimator, one can show that both $B = 0$ and $B^+ = 0$.

To make comparisons, Rosset and Tibshirani (2020) demonstrated that $B^+ \geq 0$ and $V^+ \geq 0$ and

$$
Err_R - Err_S = B^+ + V^+. \tag{6.5}
$$

Accordingly, $Err_R \geq Err_S$, which implies that the Random-X test error is no smaller than the Same-X test error.

In practice, the three expected test errors are unknown. To estimate them, Rosset and Tibshirani (2020) introduced the concept of "optimism," measured as the difference between the test error and the training error. Based on the three expected test errors, their resulting optimisms are defined below.

$$\begin{aligned}
&(1)\ Opt_F = E_{Y,\tilde{Y}}\left\{\frac{1}{n}\sum_{i=1}^{n}(\tilde{y}_i - x_i'\widehat{\boldsymbol{\alpha}})^2 - \frac{1}{n}\sum_{i=1}^{n}(y_i - x_i'\widehat{\boldsymbol{\alpha}})^2\right\} = \frac{2}{n}\sum_{i=1}^{n}\operatorname{cov}(y_i, x_i'\widehat{\boldsymbol{\alpha}});\\
&(2)\ Opt_S = E_{\mathbf{X},Y,Y_0}\left\{\frac{1}{n}\sum_{i=1}^{n}(y_{0i} - X_i'\widehat{\boldsymbol{\alpha}})^2 - \frac{1}{n}\sum_{i=1}^{n}(y_i - X_i'\widehat{\boldsymbol{\alpha}})^2\right\}\\
&\qquad = \frac{2}{n}\sum_{i=1}^{n}E_{\mathbf{X}}[\operatorname{cov}(y_i, X_i'\widehat{\boldsymbol{\alpha}})|\mathbf{X}];\\
&(3)\ Opt_R = E_{\mathbf{X},Y,X_{01},y_{01}}\left\{(y_{01} - X_{01}'\widehat{\boldsymbol{\alpha}})^2 - (y_1 - X_1'\widehat{\boldsymbol{\alpha}})^2\right\}.
\end{aligned} \tag{6.6}$$

Then, Rosset and Tibshirani (2020) showed that $Opt_R = Opt_S + B^+ + V^+$. Under the linear model with the OLS estimator, $Opt_F = Opt_S = 2\sigma^2 q/n$ and $B^+ = 0$. By the normality assumption of covariates, they further demonstrated that $V^+ = \sigma^2 q(q+1)/[n(n-q-1)]$. Accordingly, $Opt_R = Opt_S + \sigma^2 q(q+1)/[n(n-q-1)] = \frac{\sigma^2 q}{n}[2 + (q+1)/(n-q-1)]$. Using the above results, they obtained unbiased estimators of expected test errors, as given below.

$$\widehat{Err_F} = \widehat{Err_S} = C_P = RSS/n + 2\sigma^2 q/n \quad \text{and}$$

$$\widehat{Err_R} = \widehat{Err_S} + \widehat{V}^+ = RC_p = C_p + \widehat{V}^+ = RSS/n + \frac{\sigma^2 q}{n}[2 + (q+1)/(n-q-1)],$$

where $RSS = \sum_{i=1}^{n}(y_i - x_i'\widehat{\boldsymbol{\alpha}})^2$ and C_p is known as Mallows' C_p, often used in mean regression model selection. Accordingly, RC_p is the random version of C_p that was proposed by Rosset and Tibshirani (2020).

Note that C_p and RC_p can be used for model selections. In this case, one often replaces the unknown σ^2 with $\tilde{\sigma}^2 = \sum_{i=1}^{n}(y_i - x_i^{*\prime}\widehat{\boldsymbol{\alpha}}^*)^2/(n-q^*)$, calculated from the full model ($x_i^* \supseteq x_i$), where the dimension of x_i^* and $\boldsymbol{\alpha}^*$ in the full model is q^* and $q^* \geq q$. In addition, it is of interest to note that C_p is approximately equal to Akaike's (1973) Akaike information criterion. In addition, $\widehat{V}^+$ is one-half of the difference between Hurvich and Tsai's (1989) corrected Akaike information criterion AIC$_c$ and AIC by replacing q with $q+1$ in $\widehat{V}^+$ and then dividing it by a constant n. Moreover, Shen and Huang (2020) suggested employing $\widehat{Err_R}$ to select the tuning parameter in lasso for variable selection. For fitting the general model setting (6.1) or the non-Gaussian distribution of covariates, the reader can refer to Rosset and Tibshirani (2020) for details. In addition to mean regression models, we next analyze fixed and random covariance regression models.

6.2 Fixed and Random Covariance Regression Models

6.2.1 *Model Structure and Parameter Estimators*

After exploring fixed and random mean regression models, we now study covariance regression models. Although these models have been discussed in Chap. 4, we mainly focus on the fixed setting. To extend the application of the covariance regression model, we introduce the random covariance setting. To this end, we follow Zou et al.'s (2025) approach, and assume that Y_i and the $\mathbf{X}_i$ are random variables defined on the probability space $(\Omega, \mathcal{F}, \mathrm{P})$ for $i = 1, \cdots, n$, where Ω is the sample space, $\mathcal{F}$ is the σ-algebra of the subsets of Ω, and P is a probability measure on $(\Omega, \mathcal{F})$. We then introduce the σ-algebra $\mathcal{F}_{\mathcal{X}} = \sigma\langle \mathbf{X}_1, \cdots, \mathbf{X}_n \rangle \subset \mathcal{F}$, which is generated by $\mathbf{X}_1, \cdots, \mathbf{X}_n$. Note that $\mathbf{X}_i$ can be a vector, matrix, or tensor, although we mainly focus on matrices here. For the simplification of notation across both mean regression and covariance regression, we denote the explanatory (or similarity) matrix by $\mathbf{X}$ rather than $\mathbf{W}$ like we did in Chaps. 4 and 5.

To construct the relationship between the covariance of the response vector Y_i and the explanatory variables $\mathbf{X}_i$, Zou et al. (2025) proposed the following random covariance regression model:

$$\mathrm{Cov}[Y_i | \mathcal{F}_{\mathcal{X}}] = \mathbf{G}_{\boldsymbol{\beta}}(\mathbf{X}_i), \quad (i = 1, \cdots, n), \tag{6.7}$$

where $Y_i \in \mathbb{R}^p$ is the response vector, $\mathbf{X}_i = (\mathbf{X}_i^{(1)}, \cdots, \mathbf{X}_i^{(K)})$, $\mathbf{X}_i^{(k)}$s are $p \times p$ matrices, $\mathbf{G}_{\boldsymbol{\beta}}(\cdot)$ is a pre-specified parametric and measurable map characterized by a K-dimensional parameter vector $\boldsymbol{\beta} = (\beta_1, \cdots, \beta_K)' \in \mathbb{R}^K$ such that $\mathbf{G}_{\boldsymbol{\beta}}(\mathbf{X}_i) \in \mathcal{S}_p^+ \subset \mathbb{R}^{p \times p}$ for all $i = 1, \cdots, n$, and $\mathcal{S}_p^+ = \{\mathbf{G} \in \mathbb{R}^{p \times p} : \mathbf{G} \text{ is a positive definite matrix}\}$.

To illustrate the above model, we consider one fixed-**X** example and one random-**X** example below.

Example 6.1 (Fixed-X) Assume that $\mathbf{X}_1 = \mathbf{x}_1, \cdots, \mathbf{X}_n = \mathbf{x}_n$ are all fixed. That is, the $(\mathbf{x}_1, \cdots, \mathbf{x}_n)$ are fixed or designed by users. Under the trivial σ-algebra, $\mathcal{F}_{\mathcal{X}} = \sigma\langle \mathbf{x}_1, \cdots, \mathbf{x}_n \rangle = \{\emptyset, \Omega\}$, the Fixed-$X$ covariance model is $\mathrm{Cov}[Y_i | \mathcal{F}_{\mathcal{X}}] = \mathrm{Cov}(Y_i) = \mathbf{G}_{\boldsymbol{\beta}}(\mathbf{x}_i)$. Accordingly, the compound symmetry covariance matrix stated in Sect. 3.1 has the form $\mathbf{G}_{\boldsymbol{\beta}}(\mathbf{x}_i) = \beta_1 \mathbf{I}_p + \beta_2 \mathbf{x}_i$, where $\boldsymbol{\beta} = (\beta_1, \beta_2)'$, $\mathbf{I}_p$ is the p-dimensional identity matrix, $\mathbf{x}_i = \mathbf{x} = \mathbf{1}\mathbf{1}' - \mathbf{I}_p$, and $\mathbf{1}$ represents a p-dimensional vector of ones. More examples can be found across various fields, such as when $\mathbf{x}_i$

is a set of high-dimensional fixed matrices (see, e.g., Zheng et al. 2019), or when $\mathbf{x}_i$ is a structure of geographic coordinates in spatial statistics (see, e.g., Cressie 1993; Sherman 2011).

Example 6.2 (Random-X) In data analysis, it is common to consider the explanatory variables $\mathbf{X}_i$ as random, along with Y_i. Under this setting, the second example adapts the linear covariance regression model discussed in Chap. 4. Specifically, it is

$$\mathrm{Cov}[Y_i|\mathcal{F}_{\mathcal{X}}] = \mathbf{G}_{\boldsymbol{\beta}}(\mathbf{X}_i) = \sum_{k=1}^{K} \beta_k \mathbf{X}_i^{(k)}, \tag{6.8}$$

where $\mathbf{X}_i = \left(\mathbf{X}_i^{(1)}, \cdots, \mathbf{X}_i^{(K)}\right)$, $\boldsymbol{\beta} = (\beta_1, \cdots, \beta_K)'$, and the $\mathbf{X}_i^{(k)}$s are $p \times p$ similarity matrices constructed by their associated features.

Based on (6.7), we employ the quasi-maximum likelihood estimation (QMLE) approach discussed in Chap. 4 to find the estimator $\widehat{\boldsymbol{\beta}}_{QMLE}$ by maximizing the log-likelihood function

$$\ell(\boldsymbol{\beta}) = -\frac{np}{2}\log(2\pi) - \frac{1}{2}\sum_{i=1}^{n}\log\left|\mathbf{G}_{\boldsymbol{\beta}}(\mathbf{X}_i)\right| - \frac{1}{2}\sum_{i=1}^{n} Y_i'\mathbf{G}_{\boldsymbol{\beta}}^{-1}(\mathbf{X}_i)Y_i. \tag{6.9}$$

In addition to QMLE, Zou et al. (2025) considered the weighted least squares (WLS) approach. Let $\mathcal{Y}_i = \mathrm{vec}(Y_iY_i')$ and $\mathcal{G}_{\boldsymbol{\beta}}(\mathbf{X}_i) = \mathrm{vec}(\mathbf{G}_{\boldsymbol{\beta}}(\mathbf{X}_i))$. For given $\boldsymbol{\beta}$, define a p^2-dimensional error vector $\mathcal{Y}_i - \mathcal{G}_{\boldsymbol{\beta}}(\mathbf{X}_i)$ and a positive semidefinite weight matrix $\mathbf{W}(\mathbf{X}_i) \in \mathbb{R}^{p^2\times p^2}$. Subsequently, introduce the following total error,

$$Tr(\boldsymbol{\beta}) = \frac{1}{n}\sum_{i=1}^{n}\left\{\mathcal{Y}_i - \mathcal{G}_{\boldsymbol{\beta}}(\mathbf{X}_i)\right\}' \mathbf{W}(\mathbf{X}_i)\left\{\mathcal{Y}_i - \mathcal{G}_{\boldsymbol{\beta}}(\mathbf{X}_i)\right\}. \tag{6.10}$$

Accordingly, the WLS estimator $\widehat{\boldsymbol{\beta}}_{WLS}$ can be obtained by minimizing $Tr(\boldsymbol{\beta})$. Suppose that $\mathbf{W}(\mathbf{X}_i) = \mathbf{I}_{p^2}$. Then minimizing (6.10) yields the ordinary least squares (OLS) estimator. To improve the efficiency of the OLS estimator, let $\mathbf{W}(\mathbf{X}_i) = (\mathrm{Cov}[\mathcal{Y}_i|\mathbf{X}_i])^- = \mathbf{V}^-(\mathbf{X}_i)$ be the generalized (Moore–Penrose) inverse of $\mathrm{Cov}[\mathcal{Y}_i|\mathbf{X}_i]$. Then, minimizing (6.10) leads to a generalized least squares (GLS) estimator $\widehat{\boldsymbol{\beta}}_{GLS}$, which is asymptotically the most efficient among all $\widehat{\boldsymbol{\beta}}_{WLS}$; see Zou et al. (2025). Since $\mathrm{Cov}[\mathcal{Y}_i|\mathbf{X}_i]$ is usually unknown in practice, we can replace it by its estimator (see $\widehat{\mathbf{V}}(\mathbf{X}_i)$ in (6.26) of Sect. 6.2.4), and obtain the feasible GLS estimator $\widehat{\boldsymbol{\beta}}_{FGLS}$. To assure that this estimator yields a positive definite matrix $\mathbf{G}_{\boldsymbol{\beta}}(\mathbf{X}_i)$, Zou et al. (2025) proposed a constrained FGLS estimator $\widehat{\boldsymbol{\beta}}^{c}_{FGLS}$. We next study the theoretical properties of $\widehat{\boldsymbol{\beta}}_{QMLE}$ and $\widehat{\boldsymbol{\beta}}_{WLS}$.

6.2.2 Asymptotic Properties of Parameter Estimators

Based on Assumption 1 from Zou et al. (2025) presented in Sect. 6.3.2, we obtain the true covariance regression model,

$$\mathrm{Cov}[Y_i|\mathcal{F}_{\mathcal{X}}] = \mathbf{G}_{\boldsymbol{\beta}_0}(\mathbf{X}_i) \;\; (i = 1, \cdots, n), \tag{6.11}$$

where $\boldsymbol{\beta}_0 = (\beta_{01}, \cdots, \beta_{0K})'$ is the true parameter vector. Before demonstrating asymptotic properties of $\widehat{\boldsymbol{\beta}}_{QMLE}$ and $\widehat{\boldsymbol{\beta}}_{WLS}$, define

$$\mathbf{A}_{i,\mathrm{Q}}^{(k)} = \frac{1}{2}\mathbf{G}_{\boldsymbol{\beta}_0}^{-1/2}(\mathbf{X}_i)\left\{ \left.\frac{\partial \mathbf{G}_{\boldsymbol{\beta}}(\mathbf{X}_i)}{\partial \beta_k}\right|_{\boldsymbol{\beta}=\boldsymbol{\beta}_0} \right\} \mathbf{G}_{\boldsymbol{\beta}_0}^{-1/2}(\mathbf{X}_i) \tag{6.12}$$

and $\Upsilon_i^{(4)} = \mathrm{diag}\{\mu_{i1}^{(4)} - 3, \cdots, \mu_{ip}^{(4)} - 3\} \in \mathbb{R}^{p\times p}$, where $\mu_{ij}^{(4)} = \mathrm{E}[\varepsilon_{ij}^4|\mathcal{F}_{\mathcal{X}}]$. In addition, denote $\mathbf{V}_{np,\mathrm{Q}} = 2(np)^{-1}\sum_{i=1}^n \left(\mathrm{tr}(\mathbf{A}_{i,\mathrm{Q}}^{(k_1)}\mathbf{A}_{i,\mathrm{Q}}^{(k_2)})\right)_{K\times K}$ and $\boldsymbol{\Omega}_{np,\mathrm{Q}} = (np)^{-1}\sum_{i=1}^n \left(\mathrm{tr}(\mathbf{A}_{i,\mathrm{Q}}^{(k_1)} * \Upsilon_i^{(4)} * \mathbf{A}_{i,\mathrm{Q}}^{(k_2)})\right)_{K\times K}$, where $*$ denote the Hadamard product of two matrices defined in Chap. 1. In addition to Assumption 1, we present Assumptions 2, 3, 4 from Zou et al. (2025) in Sect. 6.3.2, which are needed to obtain theoretical results.

Under Assumptions 1, 2, 3, 4, Zou et al. (2025) showed that, as $\max\{n, p\} \to \infty$,

$$\sqrt{np}\left(\widehat{\boldsymbol{\beta}}_{QMLE} - \boldsymbol{\beta}_0\right) \xrightarrow{d} N(\mathbf{0}, \boldsymbol{\Sigma}_{\boldsymbol{\beta}_0}^{Q}), \tag{6.13}$$

where $\boldsymbol{\Sigma}_{\boldsymbol{\beta}_0}^{Q} = \mathbf{V}_{\mathrm{Q}}^{-1} + \mathbf{V}_{\mathrm{Q}}^{-1}\boldsymbol{\Omega}_{\mathrm{Q}}\mathbf{V}_{\mathrm{Q}}^{-1}$, and $\mathbf{V}_{\mathrm{Q}}$ and $\boldsymbol{\Omega}_{\mathrm{Q}}$ are defined in Assumption 6. It can also be shown that, under Assumptions 1, 2, 5 and 6, as $\max\{n, p\} \to \infty$,

$$\sqrt{np}\left(\widehat{\boldsymbol{\beta}}_{WLS} - \boldsymbol{\beta}_0\right) \xrightarrow{d} N(\mathbf{0}, \boldsymbol{\Sigma}_{\boldsymbol{\beta}_0}^{W}), \tag{6.14}$$

where Assumptions 5, 6 are given in Sect. 6.3.2, $\boldsymbol{\Sigma}_{\boldsymbol{\beta}_0}^{W} = \mathbf{M}_{\mathrm{W}}^{-1}\mathbf{V}_{\mathrm{W}}\mathbf{M}_{\mathrm{W}}^{-1} + \mathbf{M}_{\mathrm{W}}^{-1}\boldsymbol{\Omega}_{\mathrm{W}}\mathbf{M}_{\mathrm{W}}^{-1}$, and $\mathbf{M}_{\mathrm{W}}$, $\mathbf{V}_{\mathrm{W}}$ and $\boldsymbol{\Omega}_{\mathrm{W}}$ are defined in Assumption 6. As stated in Sect. 6.2.1, we obtain the estimator $\widehat{\boldsymbol{\beta}}_{GLS}$ when $\mathbf{W}(\mathbf{X}_i) = (\mathrm{Cov}[\mathcal{Y}_i|\mathbf{X}_i])^{-}$. Thus, the above asymptotic result is also valid for the estimator $\widehat{\boldsymbol{\beta}}_{GLS}$. Note that the asymptotic covariance matrices $\boldsymbol{\Sigma}_{\boldsymbol{\beta}_0}^{Q}$ and $\boldsymbol{\Sigma}_{\boldsymbol{\beta}_0}^{W}$ include the unknown parameter vector $\boldsymbol{\beta}_0$, which can be replaced in these covariance estimators with $\widehat{\boldsymbol{\beta}}_{QMLE}$ and $\widehat{\boldsymbol{\beta}}_{WLS}$, respectively.

After obtaining the asymptotic distributions of $\widehat{\boldsymbol{\beta}}_{QMLE}$ and $\widehat{\boldsymbol{\beta}}_{WLS}$, Zou et al. (2025) further demonstrated the asymptotic distributions of $\mathbf{G}_{\widehat{\boldsymbol{\beta}}_{QMLE}}(\mathbf{X}_i)$ and

$\mathbf{G}_{\widehat{\boldsymbol{\beta}}_{WLS}}(\mathbf{X}_i)$ as well as their corresponding precision matrices under the Frobenius norm. Although the above asymptotic properties are developed under the Random-**X** setting, they are automatically valid under the Fixed-**X** setting. This is because the Fixed-**X** setting is a special case of the Random-**X** setting. Based on estimation results, we next study test errors for model assessment.

6.2.3 Expected Test Errors and Model Assessment

It is expected that covariance regression performance relies on the **X** setting. We use the same three settings (Fixed-X setting, Same-X setting and Random-X setting) as defined for the mean regression model in Sect. 6.1.2, except replacing vectors X_i, X_{0i}, and x_i with matrices $\mathbf{X}_i$, $\mathbf{X}_{0i}$, and $\mathbf{x}_i$, respectively, and replacing scalars y_i and y_{0i} with vectors Y_i and Y_{0i}, correspondingly, in those three settings. To assess model fit, Zou et al. (2025) proposed the expected training error and the expected test error, respectively, below.

$$TR = \frac{1}{n}\sum_{i=1}^{n}\left\{\mathcal{Y}_i - \widehat{\mathcal{G}}_{\boldsymbol{\beta}}(\mathbf{X}_i)\right\}' \mathbf{W}(\mathbf{X}_i)\left\{\mathcal{Y}_i - \widehat{\mathcal{G}}_{\boldsymbol{\beta}}(\mathbf{X}_i)\right\} \quad \text{and} \tag{6.15}$$

$$TE = \frac{1}{n}\sum_{i=1}^{n}\left\{\mathcal{Y}_{0i} - \widehat{\mathcal{G}}_{\boldsymbol{\beta}}(\mathbf{X}_{0i})\right\}' \mathbf{W}(\mathbf{X}_{0i})\left\{\mathcal{Y}_{0i} - \widehat{\mathcal{G}}_{\boldsymbol{\beta}}(\mathbf{X}_{0i})\right\}, \tag{6.16}$$

where $\mathcal{Y}_i = \text{vec}(Y_i Y_i')$ as defined in Sect. 6.2.1, $\mathcal{Y}_{0i} = \text{vec}(Y_{0i} Y_{0i}')$, $\widehat{\mathcal{G}}_{\boldsymbol{\beta}}(\mathbf{X}_i) = \text{vec}(\mathbf{G}_{\widehat{\boldsymbol{\beta}}}(\mathbf{X}_i))$, $\widehat{\mathcal{G}}_{\boldsymbol{\beta}}(\mathbf{X}_{0i}) = \text{vec}(\mathbf{G}_{\widehat{\boldsymbol{\beta}}}(\mathbf{X}_{0i}))$, and $\widehat{\boldsymbol{\beta}}$ can be any of the three estimators, $\widehat{\boldsymbol{\beta}}_{QMLE}$, $\widehat{\boldsymbol{\beta}}_{WLS}$, and $\widehat{\boldsymbol{\beta}}_{GLS}$. Since TR does not take into account model complexity, we only focus on TE. Note that the expected test error TE depends on both training and test data. In practice, test observations may not be available, and we can employ the training data.

To evaluate the test error of a covariance regression model, Zou et al. (2025) followed Rosset and Tibshirani's (2020) approach and proposed the following measure,

$$Err = E\{TE\}, \tag{6.17}$$

where the expectation E is taken according to the three settings; see E_1, E_2 and E_3 defined in Sect. 6.1.2. As a result, we have

$$
\begin{aligned}
&(i)\ Err_F = n^{-1}\sum_{i=1}^{n} E\|\mathcal{Y}_{0i}^{*} - \widehat{\mathcal{G}}_{\boldsymbol{\beta}}^{*}(\mathbf{x}_i)\|_2^2;\\
&(ii)\ Err_S = E\|\mathcal{Y}_{01}^{*} - \widehat{\mathcal{G}}_{\boldsymbol{\beta}}^{*}(\mathbf{X}_1)\|_2^2;\ \text{and}\\
&(iii)\ Err_R = E\|\mathcal{Y}_{01}^{*} - \widehat{\mathcal{G}}_{\boldsymbol{\beta}}^{*}(\mathbf{X}_{01})\|_2^2,
\end{aligned}
\tag{6.18}
$$

where $\mathcal{Y}_{0i}^{*} = [\mathbf{W}^{1/2}(\mathbf{X}_{0i})]\mathcal{Y}_{0i}$, $\widehat{\mathcal{G}}_{\boldsymbol{\beta}}^{*}(\mathbf{x}_i) = [\mathbf{W}^{1/2}(\mathbf{x}_i)]\widehat{\mathcal{G}}_{\boldsymbol{\beta}}(\mathbf{x}_i)$, $\widehat{\mathcal{G}}_{\boldsymbol{\beta}}^{*}(\mathbf{X}_1) = [\mathbf{W}^{1/2}(\mathbf{X}_1)]\widehat{\mathcal{G}}_{\boldsymbol{\beta}}(\mathbf{X}_1)$, $\widehat{\mathcal{G}}_{\boldsymbol{\beta}}^{*}(\mathbf{X}_{01}) = [\mathbf{W}^{1/2}(\mathbf{X}_{01})]\widehat{\mathcal{G}}_{\boldsymbol{\beta}}(\mathbf{X}_{01})$, and $\|\cdot\|_2$ is the spectral norm. Note that, for the sake of convenience, we also use $\mathcal{Y}_{0i}$ in the Fixed-**X** setting.

Let $\mathbf{V}(\mathbf{X}_1) = \mathrm{Cov}[\mathbf{Y}_1|\mathbf{X}_1]$, $\mathbf{V}(\mathbf{x}_i) = \mathrm{Cov}[\mathbf{Y}_i|\mathbf{x}_i]$, $\nu^2(\mathbf{X}_1) = tr\{\mathbf{W}(\mathbf{X}_1)\mathbf{V}(\mathbf{X}_1)\}$, $\nu^2(\mathbf{x}_i) = tr\{\mathbf{W}(\mathbf{x}_i)\mathbf{V}(\mathbf{x}_i)\}$, $\mathbb{X} = \{\mathbf{X}_i : i \in \{1, \cdots, n\}\}$ and $(\mathbb{X}, \mathbb{Y}) = \{(\mathbf{X}_i, Y_i) : i \in \{1, \cdots, n\}\}$. Under Assumption 7 in Sect. 6.3.2, Zou et al. (2025) showed the following results:

(i) under the Fixed-**X** setting, $Err_F = n^{-1}\sum_{i=1}^{n}\nu^2(\mathbf{x}_i) + B_{Err_F} + V_{Err_F}$, where

$$
\begin{aligned}
B_{Err_F} &= \frac{1}{n}\sum_{i=1}^{n}\left\|E\left(\widehat{\mathcal{G}}_{\boldsymbol{\beta}}^{*}(\mathbf{x}_i)\right) - \mathcal{G}_{\boldsymbol{\beta}}^{*}(\mathbf{x}_i)\right\|_2^2\\
\text{and } V_{Err_F} &= \frac{1}{n}\sum_{i=1}^{n}\mathrm{tr}\left\{\mathrm{Cov}\left(\widehat{\mathcal{G}}_{\boldsymbol{\beta}}^{*}(\mathbf{x}_i)\right)\right\};
\end{aligned}
$$

(ii) under the Same-**X** setting, $Err_S = E\left\{\nu^2(\mathbf{X}_1)\right\} + B_{Err_S} + V_{Err_S}$, where

$$
\begin{aligned}
B_{Err_S} &= E\left\|E\left[\widehat{\mathcal{G}}_{\boldsymbol{\beta}}^{*}(\mathbf{X}_1)|\mathbb{X}\right] - \widehat{\mathcal{G}}_{\boldsymbol{\beta}}^{*}(\mathbf{X}_1)\right\|_2^2\\
\text{and } V_{Err_S} &= E\,\mathrm{tr}\left\{\mathrm{Cov}\left[\widehat{\mathcal{G}}_{\boldsymbol{\beta}}^{*}(\mathbf{X}_1)|\mathbb{X}\right]\right\};
\end{aligned}
\tag{6.19}
$$

(iii) under the Random-**X** setting, $Err_R = E\left\{\nu^2(\mathbf{X}_{01})\right\} + B_{Err_R} + V_{Err_R}$, where

$$
\begin{aligned}
B_{Err_R} &= E\left\|\mathrm{E}\left[\widehat{\mathcal{G}}_{\boldsymbol{\beta}}^{*}(\mathbf{X}_{01})|\mathbb{X}, \mathbf{X}_{01}\right] - \mathcal{G}_{\boldsymbol{\beta}}^{*}(\mathbf{X}_{01})\right\|_2^2\\
\text{and } V_{Err_R} &= E\,\mathrm{tr}\left\{\mathrm{Cov}\left[\widehat{\mathcal{G}}_{\boldsymbol{\beta}}^{*}(\mathbf{X}_{01})|\mathbb{X}, \mathbf{X}_{01}\right]\right\}.
\end{aligned}
\tag{6.20}
$$

Note that $E\left\{\nu^2(\mathbf{X}_{01})\right\} = E\left\{\nu^2(\mathbf{X}_1)\right\}$. Accordingly,

$$
Err_R - Err_S = B^{(d)} + V^{(d)},
\tag{6.21}
$$

where $B^{(d)} = B_{Err_R} - B_{Err_S}$ and $V^{(d)} = V_{Err_R} - V_{Err_S}$ are the bias and variance differences, respectively. If $B^{(d)} + V^{(d)} \geq 0$, then the Random-**X** error is no smaller than the Same-**X** error.

To estimate those error measures, Zou et al. (2025) adapted the concept of optimism in mean regression models discussed in Sect. 6.1.3 and defined optimism in the covariance regression model context as per below:

$$Opt = E\{TE - TR\} = Err - E\{TR\}. \tag{6.22}$$

Suppose Assumption 7 is satisfied. Then, Zou et al. (2025) demonstrated that the three measures of optimism, under the Fixed-**X**, Same-**X**, and Random-**X** settings, respectively, are given below:

$$\begin{aligned}
&(i)\ Opt_F = 2n^{-1}\sum_{i=1}^{n} \operatorname{tr}\operatorname{Cov}\big(\mathcal{Y}_i^*, \widehat{\mathcal{G}}_{\boldsymbol{\beta}}^*(\mathbf{x}_i)\big);\\
&(ii)\ Opt_S = 2n^{-1}\sum_{i=1}^{n} \operatorname{tr} E\operatorname{Cov}\big[\mathcal{Y}_i^*, \widehat{\mathcal{G}}_{\boldsymbol{\beta}}^*(\mathbf{X}_i)\big|\mathbb{X}\big];\ \ and\\
&(iii)\ Opt_R = Opt_S + B^{(d)} + V^{(d)}.
\end{aligned} \tag{6.23}$$

Suppose that UE is an unbiased estimator of Opt_F. By (6.22), the unbiased estimator of Err_F is $\widehat{Err_F} = TR + UE$. In addition, Zou et al. (2025) showed that $\widehat{Err_S} = \widehat{Err_F}$. Let $\widehat{B}^{(d)}$ and $\widehat{V}^{(d)}$ be the estimators of $B^{(d)}$ and $V^{(d)}$, respectively. Using (6.21), we then have $\widehat{Err_R} = \widehat{Err_F} + \widehat{B}^{(d)} + \widehat{V}^{(d)} = TR + UE + \widehat{B}^{(d)} + \widehat{V}^{(d)}$. This indicates that the difference between the estimator of the Random-**X** error and that of the Fixed-**X** error is due to the sum of the differences between their corresponding biases and variances. Analogously to mean regression models discussed in Sect. 6.1.3, $\widehat{Err_F}$ and $\widehat{Err_R}$ can refer to Mallows C_p and the random version of C_p (i.e., RC_p), respectively, in covariance regression models. We next study C_p and RC_p in linear covariance regression models.

6.2.4 The C_p and RC_p in Linear Covariance Regression Models

Consider the linear regression model, $\mathbf{G}_{\boldsymbol{\beta}}(\mathbf{X}_i) = \sum_{k=1}^{K} \beta_k \mathbf{X}_i^{(k)}$, in (6.8). Substituting it into (6.10), we obtain the weighted least square estimator of $\boldsymbol{\beta}$,

$$\widehat{\beta}_{WLS} = \left\{\sum_{i=1}^{n} \mathcal{X}_i' \mathbf{W}(\mathbf{X}_i)\mathcal{X}_i\right\}^{-1} \sum_{i=1}^{n} \mathcal{X}_i' \mathbf{W}(\mathbf{X}_i)\mathcal{Y}_i, \tag{6.24}$$

where $\mathcal{X}_i = (\mathrm{vec}(\mathbf{X}_i^{(1)}), \cdots, \mathrm{vec}(\mathbf{X}_i^{(K)})) \in \mathbb{R}^{p^2 \times K}$, also defined in Sect. 6.3.1 (III).

Under the linear model structure with $\mathbf{W}(\mathbf{X}_\zeta) = \mathbf{V}^{-1}(\mathbf{X}_\zeta)$, one can employ Proposition 6.2 in Sect. 6.3.3 to show that $Opt_F = Opt_S = 2K/n$, where $\mathbf{V}(\mathbf{X}_\zeta)$ is defined in Proposition 6.1 of Sect. 6.3.3, and $\mathbf{X}_\zeta$ can be either random or fixed. This finding is basically the same as that in mean regression models. Both propositions are in Zou et al. (2025). Suppose that $\mathbf{V}$ is known. We can further apply Proposition 6.2 and obtain the unbiased estimator of Opt_F and Opt_S given below:

$$UE_L = 2n^{-1}\text{tr}\big\{(\mathcal{X}^{*\prime}\mathcal{X}^*)^{-1}\mathcal{X}^{*\prime}\mathbb{V}^*(\mathbb{X})\mathcal{X}^*\big\},$$

where $\mathcal{X}^*$ and $\mathbb{V}^*(\mathbb{X})$ are defined in Sect. 6.3.1 (III). Note that the above expression is also applicable when $\mathbb{X} = \mathbf{x}$ and $\mathcal{X}^* = \chi^*$, where χ^{ast} is defined in Sect. 6.3.1 (III). Accordingly, Zou et al. (2025) defined Mallows' C_p for linear covariance regression models under the Fixed-X and Same-X settings as follows:

$$C_p \triangleq \widehat{Err_F} = \widehat{Err_S} = TR + UE_L. \tag{6.25}$$

In practice, $\mathbf{V}$ is often unknown. In that situation, we can refer to Proposition 6.1 in Sect. 6.3.3 and consider the following estimator proposed by Zou et al. (2025),

$$\begin{aligned}\widehat{\mathbf{V}}(\mathbf{X}_i) = (\widehat{\mu}^{(4)} - 3)\sum_{j=1}^{p}&\left\{\mathbf{G}_{\widehat{\boldsymbol{\beta}}}^{1/2}(\mathbf{X}_i)\mathbf{E}_{jj}\mathbf{G}_{\widehat{\boldsymbol{\beta}}}^{1/2}(\mathbf{X}_i) \otimes \mathbf{G}_{\widehat{\boldsymbol{\beta}}}^{1/2}(\mathbf{X}_i)\mathbf{E}_{jj}\mathbf{G}_{\widehat{\boldsymbol{\beta}}}^{1/2}(\mathbf{X}_i)\right\}\\ &+2\mathbf{N}_p\left\{\mathbf{G}_{\widehat{\boldsymbol{\beta}}}(\mathbf{X}_i) \otimes \mathbf{G}_{\widehat{\boldsymbol{\beta}}}(\mathbf{X}_i)\right\},\end{aligned} \tag{6.26}$$

where $\widehat{\mu}^{(4)} = (np)^{-1}\sum_{i=1}^{n}\sum_{j=1}^{p}\widehat{\varepsilon}_{ij}^4$, $(\widehat{\varepsilon}_{i1},\cdots,\widehat{\varepsilon}_{ip})' \triangleq \widehat{\mathcal{E}}_i = \mathbf{G}_{\widehat{\boldsymbol{\beta}}}^{-1/2}(\mathbf{X}_i)Y_i$, $\mathbf{G}_{\widehat{\boldsymbol{\beta}}}(\mathbf{X}_i) = \sum_{k=1}^{K}\widehat{\beta}_k\mathbf{X}_i^{(k)}$, $\mathbf{E}_{jj}$ and $\mathbf{N}_p$ are defined in Proposition 6.1, and $\widehat{\boldsymbol{\beta}}$ can be any consistent estimator such as $\widehat{\boldsymbol{\beta}}_{WLS}$. Accordingly, we obtain the estimator of UE_L, $\widehat{UE_L} = 2n^{-1}\text{tr}\big\{(\widehat{\mathcal{X}^{*\prime}}\widehat{\mathcal{X}}^*)^{-1}\widehat{\mathcal{X}^{*\prime}}\widehat{\mathbb{V}}^*(\mathbb{X})\widehat{\mathcal{X}}^*\big\}$. This leads to C_p for unknown $\mathbf{V}$, and it is

$$\widehat{C}_p = TR + \widehat{UE_L}. \tag{6.27}$$

To find RC_p under the linear covariance model (6.8) being correctly specified, Zou et al. (2025) showed that $B^{(d)} = 0$. Suppose Assumption 7 holds and $\boldsymbol{\beta} = \widehat{\boldsymbol{\beta}}_{WLS}$, they further demonstrated that $V_{Err_S} = E(UE_L)/2 = Opt_S/2$. Thus, we obtain an unbiased estimator of V_{Err_S},

$$\widetilde{V}_{Err_S} = UE_L/2. \tag{6.28}$$

After algebraic simplification, V_{Err_R} in (6.20) is

$$V_{Err_R} = E\,\mathrm{tr}\left\{\mathrm{Cov}\left[\widehat{\mathcal{G}^*}_{\boldsymbol{\beta}}(\mathbf{X}_{01})\middle|\mathbb{X}, \mathbf{X}_{01}\right]\right\} \tag{6.29}$$
$$= E\,\mathrm{tr}\left\{(\mathcal{X}^{*\prime}\mathcal{X}^*)^{-1}\mathcal{X}^{*\prime}\mathbb{V}^*(\mathbb{X})\mathcal{X}^*\left(\mathcal{X}^{*\prime}\mathcal{X}^*\right)^{-1}\mathcal{X}^{*\prime}{}_{01}\mathcal{X}^*{}_{01}\right\}.$$

Since the test data $\mathbf{X}_{01}$ is not available, Zou et al. (2025) proposed the following approximation:

$$E\,\mathrm{tr}\left\{\left(\mathcal{X}^{*(-i)\prime}\mathcal{X}^{*(-i)}\right)^{-1}\mathcal{X}^{*(-i)\prime}\mathbb{V}^*(\mathbb{X}^{(-i)})\mathcal{X}^{*(-i)}\left(\mathcal{X}^{*(-i)\prime}\mathcal{X}^{*(-i)}\right)^{-1}\mathcal{X}_i^{*\prime}\mathcal{X}_i^*\right\}, \tag{6.30}$$

where $\mathcal{X}^{*(-i)} = (\mathcal{X}_1{}^{*\prime}, \cdots, \mathcal{X}_{i-1}{}^{*\prime}, \mathcal{X}_{i+1}^{*\prime}, \cdots, \mathcal{X}_n{}^{*\prime})'$ and $\mathbb{X}^{(-i)} = (\mathbf{X}_1, \cdots, \mathbf{X}_{i-1}, \mathbf{X}_{i+1}, \cdots, \mathbf{X}_n)$ for $i = 1, \cdots, n$. This leads to a cross-validation estimator of V_{Err_R}, which is

$$\widetilde{V}_{ErrR} = \frac{1}{n}\sum_{i=1}^{n}\mathrm{tr}\left\{\left(\mathcal{X}^{*(-i)\prime}\mathcal{X}^{*(-i)}\right)^{-1}\mathcal{X}^{*(-i)\prime}\mathbb{V}^*(\mathbb{X}^{(-i)})\mathcal{X}^{*(-i)}\right.$$
$$\left.\left(\mathcal{X}^{*(-i)\prime}\mathcal{X}^{*(-i)}\right)^{-1}\mathcal{X}_i^{*\prime}\mathcal{X}_i^*\right\}. \tag{6.31}$$

Accordingly, we obtain the estimator of $V^{(d)}$, $\tilde{V}^{(d)} = \widetilde{V}_{ErrR} - \widetilde{V}_{ErrS}$. This, together with (6.21), (6.25), (6.28), and $B^{(d)} = 0$, yields

$$RC_p \triangleq \widehat{Err_R} = C_p + \tilde{V}^{(d)} = TR + \frac{1}{2}UE_L + \widetilde{V}_{ErrR}. \tag{6.32}$$

Since $\mathbf{V}$ is usually unknown in practice, we can replace it with its estimator $\widehat{\mathbf{V}}$ in UE_L and $\widetilde{V}_{ErrR}$ of the above equation. As a result, we obtain the estimator of RC_p, which is

$$\widehat{RC_p} = TR + \frac{1}{2}\widehat{UE_L} + \widehat{V}_{ErrR}. \tag{6.33}$$

As shown in Zou et al. (2025), $V_{ErrR} \geq V_{ErrS}$. Hence, we expect that $\widehat{RC_p}$ is larger than $\widehat{C_p}$. Analogously to mean regression model selections discussed in Sect. 6.1.3, both $\widehat{RC_p}$ and $\widehat{C_p}$ can be used in linear covariance regression model assessment and selection.

This chapter focuses on fixed and random mean regression and fixed and random covariance regression separately. It would be useful to take into account the setting in Chap. 5 and explore fixed and random covariance-mean regression analysis. To demonstrate theoretical results mentioned in this section, we next present necessary notation and assumptions and provide two useful propositions.

6.3 Notation, Assumptions, and Propositions

6.3.1 *Notation*

(I) Let $\mathbf{Z} = (z_{j_1 j_2})_{l\times m}$ be an $l \times m$ generic matrix $\mathbf{Z}$ with the (j_1, j_2)-th element $z_{j_1 j_2}$, for $j_1 = 1, \cdots, l$ and $j_2 = 1, \cdots, m$. If $\mathbf{Z} = (z_{j_1 j_2})_{l\times m}$ is a random matrix, then, for $1 \le \xi \le \infty$, denote by $\|z_{j_1 j_2}\|_{L^\xi} = (\mathrm{E}|z_{j_1 j_2}|^\xi)^{1/\xi}$ the L^ξ-norm of the random variable $z_{j_1 j_2}$. Then define

$$\|\mathbf{Z}\|_{L^\xi} = \left(\|z_{j_1 j_2}\|_{L^\xi}\right)_{l\times m}, \tag{6.34}$$

which is an $l \times m$ matrix consisting of the L^ξ-norm of $z_{j_1 j_2}$. In addition, define the matrix ξ-norm below:

$$\|\mathbf{Z}\|_\xi = \sup\left\{\frac{\|\mathbf{Z}\boldsymbol{\gamma}\|_\xi}{\|\boldsymbol{\gamma}\|_\xi} : \boldsymbol{\gamma} \in \mathbb{R}^m \text{ and } \boldsymbol{\gamma} \neq \mathbf{0}\right\}. \tag{6.35}$$

Moreover, define

$$|||\mathbf{Z}|||_{L^{\xi_1},\xi_2} = \left\| \|\mathbf{Z}\|_{L^{\xi_1}} \right\|_{\xi_2} \tag{6.36}$$

for $\xi_1, \xi_2 \in [1, \infty]$, where ξ_1 and ξ_2 are associated with the norm power in the L^{ξ_1}-norm (6.34) and the matrix ξ_2-norm (6.35), respectively.

(II) Let $\left(\mathbf{G}_{\boldsymbol{\beta}_0}^{1/2}(\mathbf{X}_i)\right)_j$ be the j-th column of $\mathbf{G}_{\boldsymbol{\beta}_0}^{1/2}(\mathbf{X}_i)$ for $j = 1, \cdots, p$. In addition, define

$$\begin{aligned}
\mathbf{A}_{i,\mathrm{W}}^{(k)} = \frac{1}{2}\Bigg(&\mathrm{vec}'\left\{\left.\frac{\partial \mathbf{G}_{\boldsymbol{\beta}}(\mathbf{X}_i)}{\partial \beta_k}\right|_{\boldsymbol{\beta}=\boldsymbol{\beta}_0}\right\}\mathbf{W}(\mathbf{X}_i)\Big\{\left(\mathbf{G}_{\boldsymbol{\beta}_0}^{1/2}(\mathbf{X}_i)\right)_{j_1} \otimes \left(\mathbf{G}_{\boldsymbol{\beta}_0}^{1/2}(\mathbf{X}_i)\right)_{j_2} \\
&+ \left(\mathbf{G}_{\boldsymbol{\beta}_0}^{1/2}(\mathbf{X}_i)\right)_{j_2} \otimes \left(\mathbf{G}_{\boldsymbol{\beta}_0}^{1/2}(\mathbf{X}_i)\right)_{j_1}\Big\}\Bigg)_{p\times p}, \\
\overline{\mathbf{A}}_{i,\mathrm{W}}^{(k_1,k_2)} = -\frac{1}{2}\Bigg(&\mathrm{vec}'\left\{\left.\frac{\partial^2 \mathbf{G}_{\boldsymbol{\beta}}(\mathbf{X}_i)}{\partial \beta_{k_1}\partial \beta_{k_2}}\right|_{\boldsymbol{\beta}=\boldsymbol{\beta}_0}\right\}\mathbf{W}(\mathbf{X}_i)\Big\{\left(\mathbf{G}_{\boldsymbol{\beta}_0}^{1/2}(\mathbf{X}_i)\right)_{j_1} \\
&\otimes \left(\mathbf{G}_{\boldsymbol{\beta}_0}^{1/2}(\mathbf{X}_i)\right)_{j_2} \\
&+ \left(\mathbf{G}_{\boldsymbol{\beta}_0}^{1/2}(\mathbf{X}_i)\right)_{j_2} \otimes \left(\mathbf{G}_{\boldsymbol{\beta}_0}^{1/2}(\mathbf{X}_i)\right)_{j_1}\Big\}\Bigg)_{p\times p} \quad \text{and}
\end{aligned} \tag{6.37}$$

$$\overline{\overline{\mathbf{A}}}_{i,\mathrm{W}}^{(k_1,k_2,k_3)}(\boldsymbol{\beta}) = \frac{1}{2}\left(\mathrm{vec}'\left\{ \frac{\partial^3 \mathbf{G}_{\boldsymbol{\beta}}(\mathbf{X}_i)}{\partial\beta_{k_1}\partial\beta_{k_2}\partial\beta_{k_3}}\bigg|_{\boldsymbol{\beta}=\boldsymbol{\beta}_0}\right\}\mathbf{W}(\mathbf{X}_i)\left\{\left(\mathbf{G}_{\boldsymbol{\beta}_0}^{1/2}(\mathbf{X}_i)\right)_{j_1} \otimes \left(\mathbf{G}_{\boldsymbol{\beta}_0}^{1/2}(\mathbf{X}_i)\right)_{j_2} + \left(\mathbf{G}_{\boldsymbol{\beta}_0}^{1/2}(\mathbf{X}_i)\right)_{j_2} \otimes \left(\mathbf{G}_{\boldsymbol{\beta}_0}^{1/2}(\mathbf{X}_i)\right)_{j_1}\right\}\right)_{p\times p}$$

where $\otimes$ is the Kronecker product defined in Sect. 1.6, and $i = 1, \cdots, n$, $1 \leq j_1, j_2 \leq p$ and $1 \leq k_1, k_2 \leq K$. Moreover, denote

$$\mathbf{V}_{np,\mathrm{W}} = \frac{2}{np}\sum_{i=1}^{n}\left(\mathrm{tr}(\mathbf{A}_{i,\mathrm{W}}^{(k_1)}\mathbf{A}_{i,\mathrm{W}}^{(k_2)})\right)_{K\times K},$$

$$\boldsymbol{\Omega}_{np,\mathrm{W}} = \frac{1}{np}\sum_{i=1}^{n}\left(\mathrm{tr}(\mathbf{A}_{i,\mathrm{W}}^{(k_1)} * \Upsilon_i^{(4)} * \mathbf{A}_{i,\mathrm{W}}^{(k_2)})\right)_{K\times K} \quad \text{and}$$

$$\mathbf{M}_{np,\mathrm{W}} = \frac{1}{np}\sum_{i=1}^{n}\left(\mathrm{vec}'\left\{\frac{\partial \mathbf{G}_{\boldsymbol{\beta}_0}(\mathbf{X}_i)}{\partial\beta_{k_1}}\right\}\mathbf{W}(\mathbf{X}_i)\mathrm{vec}\left\{\frac{\partial \mathbf{G}_{\boldsymbol{\beta}_0}(\mathbf{X}_i)}{\partial\beta_{k_2}}\right\}\right)_{K\times K}.$$

(III.a) Under the Random-X setting, let $\mathbf{X}_i = (\mathbf{X}_i^{(1)}, \cdots, \mathbf{X}_i^{(K)})$, $\mathcal{X}_i = \left(\mathrm{vec}(\mathbf{X}_i^{(1)}), \cdots, \mathrm{vec}(\mathbf{X}_i^{(K)})\right) \in \mathbb{R}^{p^2\times K}$, $\mathcal{X}_{01} = \mathrm{vec}(\mathbf{X}_{01}) \in \mathbb{R}^{p^2}$, and $\mathbb{X} = (\mathbf{X}_1, \cdots, \mathbf{X}_n)$ defined in Sect. 6.2.3. In addition, let $\mathbb{W}(\mathbb{X}) = \mathrm{diag}\{\mathbf{W}(\mathbf{X}_1), \cdots, \mathbf{W}(\mathbf{X}_n)\} \in \mathbb{R}^{np^2\times np^2}$ be a matrix whose diagonal blocks are $\mathbf{W}(\mathbf{X}_1), \cdots, \mathbf{W}(\mathbf{X}_n)$ and the other elements are zeros.

(III.b) Under the Fixed-X setting, let $\mathbf{X}_i = \mathbf{x}_i = (\mathbf{x}_i^{(1)}, \cdots, \mathbf{x}_i^{(K)})$, $\chi_i = \left(\mathrm{vec}(\mathbf{x}_i^{(1)}), \cdots, \mathrm{vec}(\mathbf{x}_i^{(K)})\right) \in \mathbb{R}^{p^2\times K}$, and $\mathbf{x} = (\mathbf{x}_1, \cdots, \mathbf{x}_n)$. In addition, let $\mathbb{W}(\mathbf{x}) = \mathrm{diag}\{\mathbf{W}(\mathbf{x}_1), \cdots, \mathbf{W}(\mathbf{x}_n)\} \in \mathbb{R}^{np^2\times np^2}$ be a matrix whose diagonal blocks are $\mathbf{W}(\mathbf{x}_1), \cdots, \mathbf{W}(\mathbf{x}_n)$ and the other elements are zeros.

(III.c) Let $\mathcal{X}_i^* = \mathbf{W}^{1/2}(\mathbf{X}_i)\mathcal{X}_i \in \mathbb{R}^{p^2\times K}$, $\mathcal{X}^* = (\mathcal{X}_1^{*\prime}, \cdots, \mathcal{X}_n^{*\prime})' \in \mathbb{R}^{np^2\times K}$, $\mathcal{X}_{01}^* = \mathbf{W}^{1/2}(\mathbf{X}_{01})\mathcal{X}_{01}$, $\chi_i^* = \mathbf{W}^{1/2}(\mathbf{x}_i)\chi_i \in \mathbb{R}^{p^2\times K}$, and $\chi^* = (\chi_1^{*\prime}, \cdots, \chi_n^{*\prime})' \in \mathbb{R}^{np^2\times K}$. In addition, let $\mathbb{V}^*(\mathbb{X}) = \mathbb{W}^{1/2}(\mathbb{X})\mathbb{V}(\mathbb{X})\mathbb{W}^{1/2}(\mathbb{X}) \in \mathbb{R}^{np^2\times np^2}$ and $\mathbb{V}^*(\mathbf{x}) = \mathbb{W}^{1/2}(\mathbf{x})\mathbb{V}(\mathbf{x})\mathbb{W}^{1/2}(\mathbf{x}) \in \mathbb{R}^{np^2\times np^2}$, where $\mathbb{V}(\mathbb{X}) = \mathrm{diag}\{\mathbf{V}(\mathbf{X}_1), \cdots, \mathbf{V}(\mathbf{X}_n)\} \in \mathbb{R}^{np^2\times np^2}$, $\mathbb{V}(\mathbf{x}) = \mathrm{diag}\{\mathbf{V}(\mathbf{x}_1), \cdots, \mathbf{V}(\mathbf{x}_n)\} \in \mathbb{R}^{np^2\times np^2}$, and $\mathbf{V}(\cdot) \in \mathbb{R}^{p^2\times p^2}$ is defined in Proposition 6.1 of Sect. 6.3.3.

6.3.2 Assumptions

The following assumptions correspond to Assumptions 1, 2, 3 and 4, Assumptions S.3'–S.4' and Assumption 6 in Zou et al. (2025).

Assumption 1 Assume that $Y_i = \mathbf{G}_{\boldsymbol{\beta}_0}^{1/2}(\mathbf{X}_i)\mathcal{E}_i$ for $i = 1, \cdots, n$, where $\boldsymbol{\beta}_0 = (\beta_{01}, \cdots, \beta_{0K})'$ is the true parameter vector of $\boldsymbol{\beta}$, $\mathbf{G}_{\beta_0}^{1/2}(\mathbf{X}_i)$ is symmetric and the square root of covariance matrix $\mathbf{G}_{\beta_0}(\mathbf{X}_i)$, $\mathcal{E}_i = (\varepsilon_{i1}, \cdots, \varepsilon_{ip})'$, ε_{ij} is conditionally independent relative to the σ-algebra $\mathcal{F}_{\mathcal{X}}$ in both i and j, $E[\varepsilon_{ij}|\mathcal{F}_{\mathcal{X}}] = 0$, and $E[\varepsilon_{ij}^2|\mathcal{F}_{\mathcal{X}}] = 1$ for all $i = 1, \cdots, n$ and $j = 1, \cdots, p$. In addition, assume that there exists a finite constant $\delta_1 > 0$ such that $\sup_{i,j} E|\varepsilon_{ij}|^{4+\delta_1} < \infty$.

Assumption 2 Assume that $\mathbf{G}_{\boldsymbol{\beta}}(\mathbf{X}_i)$ $(i = 1, \cdots, n)$ denotes an indexed series of nonsingular matrices, uniformly over $\boldsymbol{\beta}$ that lies in a compact parameter space $\mathfrak{B} \subset \mathbb{R}^K$ and $K < \infty$. In addition, assume that $\boldsymbol{\beta}_0$ is an interior point of $\mathfrak{B}$. Moreover, assume that $\mathbf{G}_{\boldsymbol{\beta}}(\mathbf{x}_i)$ is three times differentiable with respect to $\boldsymbol{\beta}$ for $P_{\mathbf{X}_i}$-almost for every $\mathbf{x}_i$ (see van der Vaart 1998) and that $\mathbf{G}_{\boldsymbol{\beta}}(\mathbf{x}_i)$ (and its first three derivatives with respect to $\boldsymbol{\beta}$) are all measurable functions of x_i, where $P_{\mathbf{X}_i} = \mathrm{P} \circ \mathbf{X}_i^{-1}$ (see Chung 2001) is the induced measure from the probability space $(\Omega, \mathcal{F}, \mathrm{P})$.

Assumption 3 Assume that there exists finite $\delta_2 > 8/\delta_1$ such that

$$\max \left\{ \sup_{p\geq 1} \sup_{i} \left\| \left| \mathbf{G}_{\boldsymbol{\beta}_0}^{1/2}(\mathbf{X}_i) \right| \right\|_{L^{\tau_c(2+\delta_2)},1}, \sup_{p\geq 1} \sup_{i} \left\| \left| \sup_{\boldsymbol{\beta}\in\mathfrak{B}} \left| \mathbf{G}_{\boldsymbol{\beta}}^{-1}(\mathbf{X}_i) \right| \right| \right\|_{L^{\tau_0(2+\delta_2)},1} \right. \tag{6.38}$$

$$\sup_{p\geq 1} \sup_{i} \left\| \left| \sup_{\boldsymbol{\beta}\in\mathfrak{B}} \left| \frac{\partial \mathbf{G}_{\boldsymbol{\beta}}(\mathbf{X}_i)}{\partial \beta_k} \right| \right| \right\|_{L^{\tau_1(2+\delta_2)},1}, \sup_{p\geq 1} \sup_{i} \left\| \left| \sup_{\boldsymbol{\beta}\in\mathfrak{B}} \left| \frac{\partial^2 \mathbf{G}_{\boldsymbol{\beta}}(\mathbf{X}_i)}{\partial \beta_{k_1} \partial \beta_{k_2}} \right| \right| \right\|_{L^{\tau_2(2+\delta_2)},1},$$

$$\left. \sup_{p\geq 1} \sup_{i} \left\| \left| \sup_{\boldsymbol{\beta}\in\mathfrak{B}} \left| \frac{\partial^3 \mathbf{G}_{\boldsymbol{\beta}}(\mathbf{X}_i)}{\partial \beta_{k_1} \partial \beta_{k_2} \partial \beta_{k_3}} \right| \right| \right\|_{L^{\tau_3(2+\delta_2)},1} \right\} < \infty,$$

for any $k, k_1, k_2, k_3 \in \{1, \cdots, K\}$, and for some $\tau_c, \tau_0, \tau_1, \tau_2, \tau_3 \in (0, \infty)$ that satisfy

$$\max \left\{ \left(\frac{2}{\tau_c} + \frac{4}{\tau_0} + \frac{3}{\tau_1} \right), \left(\frac{2}{\tau_c} + \frac{3}{\tau_0} + \frac{1}{\tau_1} + \frac{1}{\tau_2} \right), \left(\frac{2}{\tau_c} + \frac{2}{\tau_0} + \frac{1}{\tau_3} \right) \right\} \leq 1,$$

where δ_1 is defined in Assumption 1.

Assumption 4 As $\max\{n, p\} \to \infty$, we have that $\mathbf{V}_{np,\mathrm{Q}} \xrightarrow{p} \mathbf{V}_{\mathrm{Q}}$ and $\boldsymbol{\Omega}_{np,\mathrm{Q}} \xrightarrow{p} \boldsymbol{\Omega}_{\mathrm{Q}}$, where $\mathbf{V}_{\mathrm{Q}}$ and $\boldsymbol{\Omega}_{\mathrm{Q}}$ are finite $K \times K$ matrices. In addition, assume both $\mathbf{V}_{\mathrm{Q}}$ and $\mathbf{V}_{\mathrm{Q}} + \boldsymbol{\Omega}_{\mathrm{Q}}$ are positive definite.

Assumption 5 Assume that there exists finite $\delta_2 > 8/\delta_1$ such that

$$\max\left\{\sup_{p\geq 1}\sup_{i}\left|\left|\left|\mathbf{A}_{i,\mathrm{W}}^{(k)}\right|\right|\right|_{L^{2+\delta_2},1}, \sup_{p\geq 1}\sup_{i}\left|\left|\left|\overline{\mathbf{A}}_{i,\mathrm{W}}^{(k_1,k_2)}\right|\right|\right|_{L^{2+\delta_2},1},\right.$$

$$\left.\sup_{p\geq 1}\sup_{i}\left|\left|\left|\sup_{\boldsymbol{\beta}\in\mathfrak{B}}\left|\overline{\overline{\mathbf{A}}}_{i,\mathrm{W}}^{(k_1,k_2,k_3)}(\boldsymbol{\beta})\right|\right|\right|\right|_{L^{2+\delta_2},1}\right\} < \infty$$

for any $k, k_1, k_2, k_3 \in \{1, \cdots, K\}$, where δ_1 is defined in Assumption 1.

Assumption 6 Assume that $\mathbf{V}_{np,\mathrm{W}} \xrightarrow{p} \mathbf{V}_{\mathrm{W}}$, $\boldsymbol{\Omega}_{np,\mathrm{W}} \xrightarrow{p} \boldsymbol{\Omega}_{\mathrm{W}}$ and $\mathbf{M}_{np,\mathrm{W}} \xrightarrow{p} \mathbf{M}_{\mathrm{W}}$ as $\max\{n, p\} \to \infty$, where $\mathbf{V}_{\mathrm{W}}$, $\boldsymbol{\Omega}_{\mathrm{W}}$, $\mathbf{M}_{\mathrm{W}}$ are finite $K \times K$ matrices. In addition, assume that both $\mathbf{V}_{\mathrm{W}} + \boldsymbol{\Omega}_{\mathrm{W}}$ and $\mathbf{M}_{\mathrm{W}}$ are positive definite.

Assumption 7 Assume that the data are generated by $Y_\zeta = \mathbf{G}_{\boldsymbol{\beta}_0}^{1/2}(X_\zeta)\mathcal{E}_\zeta$ for all $\zeta = 1, \cdots, n$ and $01, \cdots, 0n$, and $\mathcal{E}_\zeta = (\varepsilon_{\zeta 1}, \cdots, \varepsilon_{\zeta p})'$ is independent of $\mathcal{F}_{\mathcal{X}}$, where the $\varepsilon_{\zeta j}$s are i.i.d. in both ζ and j with $E(\varepsilon_{\zeta j}) = 0$, $E(\varepsilon_{\zeta j}^2) = 1$, $E(\varepsilon_{\zeta j}^4) = \mu^{(4)}$, $E|\varepsilon_{\zeta j}|^{4+\delta_1} < \infty$ for $j = 1, \cdots, p$, and $\delta_1 > 0$ is defined in Assumption 1.

6.3.3 Propositions

The following two propositions correspond to Propositions S.1 and 3 in Zou et al. (2025).

Proposition 6.1 *Suppose Assumption 7 is satisfied. Under any one of the Fixed-***X***, Random-***X***, and Same-***X** *settings, we have that, for* $\zeta \in \{1, \cdots, n\} \cup \{01, \cdots, 0n\}$,

$$Cov[\mathcal{Y}_\zeta|\mathbf{X}_\zeta] = Cov[\mathcal{Y}_\zeta|\mathcal{F}_{\mathcal{X}_\zeta}]$$

$$= \{\mu^{(4)} - 3\}\sum_{j=1}^{p}\left\{\mathbf{G}^{1/2}(X_\zeta)\mathbf{E}_{jj}\mathbf{G}^{1/2}(\mathbf{X}_\zeta) \otimes \mathbf{G}^{1/2}(\mathbf{X}_\zeta)\mathbf{E}_{jj}\mathbf{G}^{1/2}(\mathbf{X}_\zeta)\right\}$$

$$+2\mathbf{N}_p\left\{\mathbf{G}(\mathbf{X}_\zeta) \otimes \mathbf{G}(\mathbf{X}_\zeta)\right\} \triangleq \mathbf{V}(X_\zeta),$$

where $\mathcal{F}_{\mathcal{X}_\zeta} = \sigma\langle\mathbf{X}_1, \cdots, \mathbf{X}_n, \mathbf{X}_{01}, \cdots, \mathbf{X}_{0n}\rangle$, $\mu^{(4)}$ *is defined in Assumption 1,* $\mathbf{N}_p = (\mathbf{I}_{p^2} + \mathbf{K}_{pp})/2$, $\mathbf{K}_{pp} = \sum_{j_1,j_2=1}^{p}(\mathbf{E}_{j_1j_2} \otimes \mathbf{E}_{j_2j_1})$, *and* $\mathbf{E}_{j_1j_2}$ *is the* $p \times p$ *matrix with the* (j_1, j_2)*-th element being* 1 *and the other elements being 0 for* $j_1, j_2 \in \{1, \cdots, p\}$.

Proposition 6.2 *Suppose Assumption 7 is satisfied. Consider fitting a linear covariance regression model (6.8) with* $\widehat{\boldsymbol{\beta}} = \widehat{\boldsymbol{\beta}}_{WLS}$ *in (6.24). Under the Fixed-X setting, we have that* $Opt_F = 2n^{-1}\mathrm{tr}\{(\mathcal{X}^{*\prime}\mathcal{X}^{*})^{-1}\mathcal{X}^{*\prime}\mathbb{V}^{*}(\mathbf{x})\mathcal{X}^{*}\}$. *In addition, under the Same-X setting, we have that* $Opt_S = 2n^{-1}\mathrm{E}\left[\mathrm{tr}\{(\mathcal{X}^{*\prime}\mathcal{X}^{*})^{-1}\mathcal{X}^{*\prime}\mathbb{V}^{*}(\mathbf{X})\mathcal{X}^{*}\}\right]$.

The proofs of the above two propositions can be found in the supplementary material of Zou et al. (2025).

6.4 An Empirical Example

We briefly discuss the example presented in Zou et al. (2025). Employing the linear random covariance regression model, Zou et al. (2025) analyzed the monthly returns of $p = 5$ stocks collected from the US stock market. In this study, there are five stocks, and they are UnitedHealth Group Incorporated, Goldman Sachs Group Inc., Home Depot Inc., Microsoft Corporation, and Salesforce.com Inc., which were the five largest index component weights of stocks in the Dow Jones Industrial Average (DJIA). For each month i, the response vector Y_i consists of the monthly returns of these five stocks, where $i = 1, \cdots, n$, and $n = 24$ months of training data were collected from June 2015 to May 2017. In addition, $i = 01, \cdots, 0n$, and $n = 24$ months of testing data were collected from June 2017 to May 2019. Moreover, there are five covariance regression covariates: (i) IND: industry or sector classified by the Global Industry Classification Standard, GICS; (ii) LEV: total liabilities divided by total tangible assets in the quarter before month i; (iii) CF: cash flow in the quarter before month i; (iv) SIZE: measured by the logarithm of the market value in the quarter before month i; and (v) BM: book-to-market ratio in the month before month i. Based on these five covariates, the corresponding similarity matrices are constructed. In addition, the identity matrix $\mathbf{I}_p$ is included in the covariance regression model. Note that the data are obtained from Wharton Research Data Services (WRDS).

To compare the covariance regression models in terms of their expected test error estimates $\widehat{C_p}$ and $\widehat{RC_p}$, we consider $2^5 - 1 = 31$ possible candidate models, where each model contains $\mathbf{I}_p$ and a subset of five covariates. In addition, $\widehat{\mathbf{V}}$ is obtained by using the full model with all covariates $\mathbf{I}_p$, IND, LEV, CF, SIZE and BM. The selection results show that $\widehat{C_p}$ chooses the model with variables $\mathbf{I}_p$ and LEV (call this Model I), while $\widehat{RC_p}$ selects the model with variables $\mathbf{I}_p$, LEV, SIZE and BM (call this Model II). Note that the constrained GLS estimates are used in this example.

To assess the performance of Models I and II, Zou et al. (2025) computed their portfolios, and named them Portfolio I and Portfolio II, respectively. The results show that Portfolio II not only has a lower risk level, but also earns a higher average return in comparison with Portfolio I (overall). Hence, it is not surprising that the Sharpe ratio of Portfolio II is much higher than that of Portfolio I. We conclude that $\widehat{RC_p}$ is a better choice than $\widehat{C_p}$ for portfolio management in this empirical example.

Acknowledgments Sections 6.2–6.4 in this chapter are derived in part from *Fixed and Random Covariance Regression Analysis* and its supplementary materials, published in The Annals of Statistics, 2025, Institute of Mathematical Statistics.

Chapter 7
Spatial Autoregressive and Network Autocorrelation Models

This chapter introduces the spatial autoregressive model, which has a specific covariance structure so that it is able to take into account the spatial dependence and heterogeneity of responses by leveraging spatial information. The quasi-maximum likelihood estimation method is employed to estimate unknown parameters. We then allow the adjacency matrix to exhibit network interactions and present two adaptive network autocorrelation models for identifying influential nodes. Furthermore, screening large-scale networks in a network autocorrelation model for selecting relevant nodes is discussed. In addition to models with univariate structure, multivariate network autocorrelation models are analyzed. Moreover, four additional types of network autocorrelation models are provided to discuss the mutual influence, functional varying coefficient, quantile, and community effects, respectively. Finally, three examples are given to briefly illustrate empirical applications.

7.1 Spatial Autoregressive Models

7.1.1 Model Structure

In the last half-century, spatial autoregressive (SAR) models have attracted considerable attention across many fields such as behavior and social science, economics, ecology, medicine, psychology, and finance. Useful references for the SAR model are Whittle (1954), Ord (1975), Cliff and Ord (1981), Anselin (1988), Cressie (1993), Lee (2004), LeSage and Pace (2009), Banerjee et al. (2015), and Lee (2023). The purpose of the SAR model is to describe spatial dependence and heterogeneity in data collected from regions (or locations) in space. In particular, spatial dependence is associated with the neighboring observations at nearby locations. Note that the term "regions in space" can be analogously applied to different subjects such as "actors in a social network."

W. Lan, C.-L. Tsai, *Covariance Analysis and Beyond*,
https://doi.org/10.1007/978-3-032-08796-6_7

To construct the SAR model, let y_i be the response collected at the i-th region (or location) for $i = 1, \cdots, n$. In addition, denote its associated explanatory vector by $X_i = (x_{i1}, \cdots, x_{iq})' \in \mathbb{R}^q$. To measure the degree of interaction of location j on location i, we adopt the approach of Ord (1975) and LeSage and Pace (2009) to construct a non-negative weight w_{ij} for $i, j = 1, \cdots, n$. We assume that $\sum_{j=1}^{n} w_{ij} = 1$ and $w_{ii} = 0$ for $i = 1, \cdots, n$, to ensure that the weights are row-stochastic (i.e., row-standardized). To incorporate spatial dependence and heterogeneity among the response variables, we consider the following SAR model; see, e.g., Ord (1975) and LeSage and Pace (2009),

$$y_i = \rho \sum_{j=1}^{n} w_{ij} y_j + X_i' \boldsymbol{\alpha} + \varepsilon_i, \tag{7.1}$$

where $|\rho| < 1$, $\boldsymbol{\alpha} = (\alpha_1, \cdots, \alpha_q)'$, and the errors ε_i are i.i.d. random variables with mean 0 and variance σ^2 for $i = 1, \cdots, n$. The term $\sum_{j=1}^{n} w_{ij} y_j$ refers to a "spatial lag," and it represents a linear combination of neighboring responses around the i-th location, and ρ is the spatial autoregression parameter (or social interaction coefficient).

The above model can be expressed in matrix form,

$$Y = \rho \mathbf{W} Y + \mathbf{X} \boldsymbol{\alpha} + \boldsymbol{\varepsilon}, \tag{7.2}$$

where $Y = (y_1, \cdots, y_n)'$, $\mathbf{W} = (w_{ij})$ is an $n \times n$ spatial weight matrix, $\mathbf{X} = (X_1, \cdots, X_n)' \in \mathbb{R}^{n \times q}$, and $\boldsymbol{\varepsilon} = (\varepsilon_1, \cdots, \varepsilon_n)'$ is an $n \times 1$ random vector with mean $\mathbf{0}$ and variance $\sigma^2 \mathbf{I}$, and $\mathbf{I}$ is an $n \times n$ identity matrix. The spatial autoregression parameter ρ can be viewed as the partial correlation between the response Y and the spatial lag vector $\mathbf{W}Y$ conditional on covariates X. Note that $\mathbf{W}$ need not be a symmetric matrix.

Using the fact that $|\rho| < 1$, $(\mathbf{I} - \rho \mathbf{W})$ is invertible (see Banerjee et al. 2015). Then model (7.2) can be rewritten as follows:

$$Y = (\mathbf{I} - \rho \mathbf{W})^{-1} \mathbf{X} \boldsymbol{\alpha} + (\mathbf{I} - \rho \mathbf{W})^{-1} \boldsymbol{\varepsilon}. \tag{7.3}$$

Note that this is a dependent and heterogeneous model, and its covariance matrix is $\text{Cov}(Y) = \boldsymbol{\Sigma} = \sigma^2 (\mathbf{I} - \rho \mathbf{W})^{-1} (\mathbf{I} - \rho \mathbf{W}')^{-1}$, which is more complicated than the weighted regression model discussed in Sect. 5.1.1.

7.1.2 Parameter Estimation

A natural approach to estimating unknown parameters in model (7.2) is ordinary least squares (OLS). However, the resulting estimators are not consistent due to the endogeneity of the spatial lag; see Ord (1975), Anselin (1988), and Lee (2002). Hence, we next study the quasi-maximum likelihood estimator (QMLE) proposed by Lee (2004). Specifically, this estimator is derived from a normal likelihood but the random errors in model (7.2) need not be normally distributed.

Define $\mathbf{S}(\rho) = \mathbf{I} - \rho\mathbf{W}$. Then $\boldsymbol{\varepsilon} = \boldsymbol{\varepsilon}(\boldsymbol{\alpha}, \rho) = \mathbf{S}(\rho)Y - \mathbf{X}\boldsymbol{\alpha}$, and the quasi-log-likelihood function of (7.2) is

$$\ell(\boldsymbol{\theta}) = -\frac{n}{2}\log 2\pi - \frac{n}{2}\log\sigma^2 - \frac{1}{2\sigma^2}(\mathbf{S}(\rho)Y - \mathbf{X}\boldsymbol{\alpha})'(\mathbf{S}(\rho)Y - \mathbf{X}\boldsymbol{\alpha}) + \log|\mathbf{S}(\rho)|, \tag{7.4}$$

where $\boldsymbol{\theta} = (\boldsymbol{\alpha}', \rho, \sigma^2)'$. We subsequently employ the concentrated quasi-maximum likelihood method to estimate the parameters. For the given ρ, we maximize $\ell(\boldsymbol{\theta})$ with respect to $\boldsymbol{\alpha}$ and σ^2, which leads to

$$\widehat{\boldsymbol{\alpha}}(\rho) = (\mathbf{X}'\mathbf{X})^{-1}\mathbf{X}'\mathbf{S}(\rho)Y \quad \text{and} \tag{7.5}$$

$$\widehat{\sigma}^2(\widehat{\boldsymbol{\alpha}}(\rho), \rho) = n^{-1}\boldsymbol{\varepsilon}'(\widehat{\boldsymbol{\alpha}}(\rho), \rho)\boldsymbol{\varepsilon}(\widehat{\boldsymbol{\alpha}}(\rho), \rho), \tag{7.6}$$

where $\boldsymbol{\varepsilon}(\widehat{\boldsymbol{\alpha}}(\rho), \rho) = \mathbf{MS}(\rho)Y$ and $\mathbf{M} = \mathbf{I} - \mathbf{X}(\mathbf{X}'\mathbf{X})^{-1}\mathbf{X}'$. The resulting concentrated quasi-log-likelihood function (omitting constants) is

$$\ell_c(\rho) = -\frac{n}{2}\log\widehat{\sigma}^2(\widehat{\boldsymbol{\alpha}}(\rho), \rho) + \log|\mathbf{S}(\rho)|. \tag{7.7}$$

Employing the quasi-Newton method to maximize the above equation with respect to ρ, we get the QMLE, $\widehat{\rho} = \text{argmax}\,\ell_c(\rho)$. Accordingly, we obtain the QMLEs of $\boldsymbol{\alpha}$ and σ^2, which are $\widehat{\boldsymbol{\alpha}} = \widehat{\boldsymbol{\alpha}}(\widehat{\rho})$ and $\widehat{\sigma}^2 = \widehat{\sigma}^2(\widehat{\boldsymbol{\alpha}}, \widehat{\rho})$. Under some regularity assumptions, Lee (2004) demonstrated that $\widehat{\boldsymbol{\theta}} = (\widehat{\boldsymbol{\alpha}}', \widehat{\rho}, \widehat{\sigma}^2)'$ is a consistent estimator and also converges to a normal distribution.

It is worth noting that the computational complexity of QMLE is $O(n^3)$ when computing $|\mathbf{S}(\rho)|$, which can become a bottleneck when n is large. Hence, one may employ the fast approach proposed by Smirnov and Anselin (2001) to reduce the computational cost. In addition, $\widehat{\boldsymbol{\alpha}}(\rho)$ is the same as the generalized least squares estimator (see, e.g., Seber and Lee 2003). It is worth noting that QMLE is MLE if the random errors are normally distributed. Since QMLE and MLE may be sensitive to outliers, Tho et al. (2023) proposed the robust estimation of SAR models, which can be resistant to outliers.

7.2 Adaptive Network Autocorrelation Models

In the last three decades, studying network relationships has become a popular and important area for both practical applications and theoretical developments. To this end, we adapt the SAR setting by changing the term “region (or location)” to “nodes (or actors)." Then, we construct an adjacency matrix $\mathbf{A}^* = (a_{ij}^*) \in \mathbb{R}^{n\times n}$ of the network with n nodes, and denote $a_{ij}^* = 1$ if a direct connection leads from node i to node j and $a_{ij}^* = 0$ otherwise. To exclude self-connections, we set $a_{jj}^* = 0$ for any $1 \le j \le n$. Subsequently, we define $w_{ij} = a_{ij}^*/\sum_{j=1}^n a_{ij}^*$ to be the row-stochastic version of a_{ij}^*, which leads to $\sum_{j=1}^n w_{ij} = 1$ for $i = 1, \cdots, n$ as discussed in Sect. 7.1. Accordingly, the matrix $\mathbf{W} = (w_{ij})_{n\times n} \in \mathbb{R}^{n\times n}$ with $i, j = 1, \cdots, n$, describes the network relationships among the n nodes. In network analysis, $\mathbf{W}$ is called the normalized adjacency matrix that provides an information map describing the relationship between any two adjacent nodes. By replacing the spatial weight matrix $\mathbf{W}$ with the normalized adjacency matrix, model (7.2) is also referred to as the network autocorrelation model (NAM) in social network analysis; see, for example, Neuman and Mizruchi (2010), Wang et al. (2014), and Li and Sewell (2021). Later, we will study various types of network autocorrelation models and their extensions.

Although NAM focuses on networks, it basically has the same model structure as that of SAR. Hence, we exchange NAM and SAR thereafter. In addition, we can analogously obtain the QMLE and its theoretical properties as discussed in Sect. 7.1. Note that the QMLE requires us to compute the determinant of $\mathbf{S}(\rho)$ in (7.4), which results in the computational complexity being $O(n^3)$ as mentioned in Sect. 7.1. To reduce computational complexity, Ma et al. (2020) introduced a naive least squares method and the resulting estimator is consistent and asymptotically normal under some conditions. In a pure NAM model with the sampled network data selected from a large-scale network, Zhou et al. (2017) proposed the approximated and paired maximum likelihood estimators. They also demonstrated that those two estimators are consistent and asymptotically normal under certain conditions. In addition, Huang et al. (2019) proposed a least squares type estimation algorithm for a large-scale network. This is a fast algorithm, and its computational complexity is $O(n)$ for extremely sparse networks. The above estimation methods can play an important role in assessing actors' properties in a big network. In the remainder of this section, we will introduce two adaptive network autocorrelation models for studying actors' network influence.

7.2.1 Network Influence Model

The SAR model (7.2) has been widely used in the literature to characterize the relationship among the Y_is. However, it is not able to identify influential nodes since model (7.2) assumes that all of the nodes have the same influential power measured by the parameter ρ. In practice, node i can have more (or less) influence than node j for any two connected nodes i and j. Accordingly, the influence parameter can be different across nodes. To this end, we consider the model given below.

7.2.1.1 Model Structure

Let ρ_j $(j = 1, \cdots, n)$ be the influence measure of node j in the network. Then the information of node i received from node j is $y_j \rho_j w_{ij}$. This, together with node i's own attributes X_i, yields the following adaptive SAR model proposed by Zou et al. (2021),

$$y_i = \sum_{j=1}^{n} (\rho_j w_{ij}) y_j + X_i' \boldsymbol{\alpha} + \varepsilon_i \ (i = 1, \cdots, n), \tag{7.8}$$

where X_i, $\boldsymbol{\alpha}$, and ε_i are defined in (7.1) and $|\rho_j| < 1$. The matrix expression of (7.8) is given below:

$$Y = \mathbf{WD}Y + \mathbf{X}\boldsymbol{\alpha} + \boldsymbol{\varepsilon}, \tag{7.9}$$

where $\mathbf{D} = \text{diag}(\rho_i)$ is an $n \times n$ diagonal matrix, $\mathbf{X}$ and $\boldsymbol{\varepsilon}$ are defined in (7.2). Based on this model, one can identify influential nodes via their associated influence measures, the ρ_j's. It is obvious that model (7.9) reduces to model (7.2) when the ρ_j's are equal.

Since model (7.9) characterizes the influential power of each node, Zou et al. (2021) named the vector $\mathbf{D1}_n = (\rho_1, \cdots, \rho_n)' \in \mathbb{R}^n$ network influence index, where $\mathbf{1}_n = (1, \cdots, 1)' \in \mathbb{R}^n$. Accordingly, we also refer to the associated adaptive SAR model as the network influence (NIF) model. We next study the parameter estimation of this model.

7.2.1.2 Parameter Estimation

In the NIF model (7.9), it is infeasible to estimate the network influence index when n is sufficiently large. Hence, Zou et al. (2021) parameterized the network influence measure ρ_i by $\rho_i(\boldsymbol{\beta}) = g(Z_i'\boldsymbol{\beta})$, where $g(\cdot)$ is a strictly monotone and known function, $Z_i = (z_{i1}, \cdots, z_{id})' = (z_{i1}, Z_{-1,i}')' \in \mathbb{R}^d$, $Z_{-1,i} = (z_{i2}, \cdots, z_{id})'$ refers to the $d - 1$ possible attributes that may affect the influential power of node i, and $\boldsymbol{\beta} = (\beta_1, \cdots, \beta_d)' \in \mathbb{R}^d$ is an unknown regression coefficient vector. Accordingly, $z_{i1} \equiv 1$ is associated with the intercept β_1 for $i = 1, \cdots, n$. If $\beta_2 = \cdots = \beta_d = 0$, then $\rho_i = g(\beta_1)$, that is, the ρ_i's are equal. Since $\mathbf{D}$ is a function of $\boldsymbol{\beta}$, we further express (7.9) as

$$Y = \mathbf{WD}(\boldsymbol{\beta})Y + \mathbf{X}\boldsymbol{\alpha} + \boldsymbol{\varepsilon}. \tag{7.10}$$

Accordingly, the covariance matrix of Y is $\boldsymbol{\Sigma} = \sigma^2(\mathbf{I} - \mathbf{WD}(\boldsymbol{\beta}))^{-1}(\mathbf{I} - \mathbf{D}'(\boldsymbol{\beta})\mathbf{W}')^{-1}$.

It is of interest to note that there are four known functions (described below) are considered in Zou et al. (2021). They are: (i) $g(Z_i'\boldsymbol{\beta}) = e^{Z_i'\boldsymbol{\beta}}/(1 + e^{Z_i'\boldsymbol{\beta}})$;

(ii) $g(Z_i'\boldsymbol{\beta}) = \Phi(Z_i'\boldsymbol{\beta})$, where $\Phi(\cdot)$ is the cumulative distribution function of the standard normal distribution; (iii) $g(Z_i'\boldsymbol{\beta}) = 1 - e^{-e^{Z_i'\boldsymbol{\beta}}}$; and (iv) $g(Z_i'\boldsymbol{\beta}) = e^{Z_i'\boldsymbol{\beta}}$.

As discussed in Sect. 7.1.2, Zou et al. (2021) also followed Lee's (2004) quasi-maximum likelihood estimation (QMLE) approach to estimate the unknown parameters in model (7.10). To this end, define $\mathbf{S}(\boldsymbol{\beta}) = \mathbf{I} - \mathbf{WD}(\boldsymbol{\beta})$ and assume that it is invertible. Then replacing $\mathbf{S}(\rho)$ in (7.4) with $\mathbf{S}(\boldsymbol{\beta})$, we obtain the quasi-log-likelihood function of (7.10),

$$\ell(\boldsymbol{\theta}) = -\frac{n}{2}\log(2\pi) - \frac{n}{2}\log\sigma^2 - \frac{1}{2\sigma^2}\{\mathbf{S}(\boldsymbol{\beta})Y - \mathbf{X}\boldsymbol{\alpha}\}'\{\mathbf{S}(\boldsymbol{\beta})Y - \mathbf{X}\boldsymbol{\alpha}\} + \log|\mathbf{S}(\boldsymbol{\beta})|,$$

where $\boldsymbol{\theta} = (\boldsymbol{\alpha}', \boldsymbol{\beta}', \sigma^2)'$. For given $\boldsymbol{\beta}$, we maximize $\ell(\boldsymbol{\theta})$ with respect to $\boldsymbol{\alpha}$ and σ^2, which leads to

$$\widehat{\boldsymbol{\alpha}}(\boldsymbol{\beta}) = (\mathbf{X}'\mathbf{X})^{-1}\mathbf{X}'\mathbf{S}(\boldsymbol{\beta})Y \quad \text{and}$$

$$\widehat{\sigma}^2(\widehat{\boldsymbol{\alpha}}(\boldsymbol{\beta}), \boldsymbol{\beta}) = n^{-1}\boldsymbol{\varepsilon}'(\widehat{\boldsymbol{\alpha}}(\boldsymbol{\beta}), \boldsymbol{\beta}))\boldsymbol{\varepsilon}(\widehat{\boldsymbol{\alpha}}(\boldsymbol{\beta}), \boldsymbol{\beta}),$$

where $\boldsymbol{\varepsilon}(\widehat{\boldsymbol{\alpha}}(\boldsymbol{\beta}), \boldsymbol{\beta}) = \mathbf{MS}(\boldsymbol{\beta})Y$ and $\mathbf{M}$ is defined below (7.6). Then the concentrated quasi-log-likelihood function (omitting constants) is

$$\ell_c(\boldsymbol{\beta}) = -\frac{n}{2}\log\widehat{\sigma}^2(\widehat{\boldsymbol{\alpha}}(\boldsymbol{\beta}), \boldsymbol{\beta}) + \log|\mathbf{S}(\boldsymbol{\beta})|. \tag{7.11}$$

Employing the quasi-Newton method to maximize the above equation with respect to $\boldsymbol{\beta}$, we get QMLE, $\widehat{\boldsymbol{\beta}} = \text{argmax}\ell_c(\boldsymbol{\beta})$. Accordingly, we obtain the QMLEs of $\boldsymbol{\alpha}$ and σ^2, which are $\widehat{\boldsymbol{\alpha}} = \widehat{\boldsymbol{\alpha}}(\widehat{\boldsymbol{\beta}})$ and $\widehat{\sigma}^2 = \widehat{\sigma}^2(\widehat{\boldsymbol{\alpha}}, \widehat{\boldsymbol{\beta}})$. Under some conditions, Zou et al. (2021) demonstrated that $\widehat{\boldsymbol{\theta}} = (\widehat{\boldsymbol{\alpha}}', \widehat{\boldsymbol{\beta}}', \widehat{\sigma}^2)'$ converges to a normal distribution.

It is worth noting that the estimator of $\boldsymbol{\beta}$ constitutes the estimated network influence index $\widehat{\mathbf{D}}\mathbf{1}_n = (\widehat{\rho}_1, \cdots, \widehat{\rho}_n)'$. This allows us to identify the most influential nodes. In addition, Zou et al. (2021) proposed three test statistics, i.e., the quasi-likelihood ratio test, the quasi-score test, and the quasi-Wald test, to assess the significance of the NIF model versus the classical NAM model with a constant network correlation. Accordingly, one can determine the contribution of the influence index. To illustrate the NIF model, Zou et al. (2021) presented an empirical example to study the spillover effects in Chinese mutual fund cash flows and effectively identify influential funds with insightful findings.

7.2.2 *Inward-and-Outward Network Influence Model*

In the previous section, we employed the NIF model to examine heterogeneous influence in networks. However, we only discussed assessing the influence of a node on other nodes. Recently, Wu et al. (2022) proposed an Inward-and-Outward Network Influence (IONI) model to assess nodal heterogeneity in both directions. Specifically, this model allows for two types of influence; the first type measures the magnitude of influence that each node exerts on other (outward influence), the second type measures the receptivity of each node to being influenced by other (inward influence). Accordingly, these two types of influence measures classify all nodes into four quadrants (high inward and high outward, low inward and high outward, low inward and low outward, high inward and low outward). We next introduce the IONI model.

7.2.2.1 Model Structure

Let y_i, w_{ij}, X_i, $\boldsymbol{\alpha}$, and ε_i be defined as in model (7.1). Then Wu et al. (2022) introduced the Inward-and-Outward Network Influence (IONI) model as follows:

$$y_i = \sum_{j=1}^{n} (\delta_i w_{ij} \rho_j) y_j + X_i' \boldsymbol{\alpha} + \varepsilon_i, \tag{7.12}$$

where $0 \leq \delta_i < 1$ and $0 \leq \rho_j < 1$. Note that δ_i measures the degree of node i's receptivity to being influenced by others and ρ_j determines the influence magnitude of node j on others. Thus, Wu et al. (2022) named δ_i the inward influence index and ρ_j the outward influence index.

As discussed in Sect. 7.2.1, we assume that the influence indices δ_i and ρ_j are correspondingly related to the nodes' attributes Z_i and V_j. In addition, let $Z_i = (z_{i1}, \cdots, z_{id_1})' \in \mathbb{R}^{d_1}$ with $z_{i1} \equiv 1$ for $i = 1, \cdots, n$ and $V_j = (v_{j1}, \cdots, v_{jd_2})' \in \mathbb{R}^{d_2}$ with $v_{j1} \equiv 1$ for $j = 1, \cdots, n$, and let their associated unknown parameter vectors be $B_1 = (b_{11}, \cdots, b_{1d_1})' \in \mathbb{R}^{d_1}$ and $B_2 = (b_{21}, \cdots, b_{2d_2})' \in \mathbb{R}^{d_2}$, respectively. The above information, together with (7.12), leads to the IONI model,

$$Y = \mathbf{D}_1(B_1)\mathbf{W}\mathbf{D}_2(B_2)Y + \mathbf{X}\boldsymbol{\alpha} + \boldsymbol{\varepsilon}, \tag{7.13}$$

where $\mathbf{D}_1(B_1) = \text{diag}(g_1(Z_i' B_1))$ and $\mathbf{D}_2(B_2) = \text{diag}(g_2(V_j' B_2))$ are $n \times n$ matrices for $i, j = 1, \cdots, n$, g_1, and g_2 are two strictly monotone and known functions, and $\mathbf{X}$, $\boldsymbol{\alpha}$, and $\boldsymbol{\varepsilon}$ are defined in (7.9). Accordingly, the covariance matrix of Y is $\boldsymbol{\Sigma} = \sigma^2(\mathbf{I} - \mathbf{D}_1(B_1)\mathbf{W}\mathbf{D}_2(B_2))^{-1}(\mathbf{I} - \mathbf{D}_2'(B_2)\mathbf{W}'\mathbf{D}_1'(B_1))^{-1}$.

There are three known functions of g_1 recommended by Wu et al. (2022), and they are: (i) $g_1(Z_i' B_1) = e^{Z_i' B_1}/(1 + e^{Z_i' B_1})$; (ii) $g_1(Z_i' B_1) = 1 - e^{-e^{Z_i' B_1}}$; and (iii) $g_1(Z_i' B_1) = \Phi(Z_i' B_1)$, where $\Phi(\cdot)$ is the standard normal cumulative distribution function. These three functions can also be used for g_2. We next estimate unknown parameters.

7.2.2.2 Parameter Estimation

In order to estimate the unknown parameters in model (7.13), we assume that $b_{11} = 0$ in order to ensure identifiability. As discussed in Sects. 7.1.2 and 7.2.1, Wu et al. (2022) adopted the quasi-maximum likelihood estimation (QMLE) method to estimate unknown parameters. Define $\boldsymbol{\theta} = (\boldsymbol{\alpha}', B_1', B_2', \sigma^2)'$, $\mathbf{S}(B_1, B_2) = \mathbf{I} - \mathbf{D}_1(B_1)\mathbf{W}\mathbf{D}_2(B_2)$, and $\boldsymbol{\varepsilon} = \boldsymbol{\varepsilon}(\boldsymbol{\alpha}, B_1, B_2) = \mathbf{S}(B_1, B_2)Y - \mathbf{X}\boldsymbol{\alpha}$. We obtain the quasi-log-likelihood function of (7.13), which is

$$\ell(\boldsymbol{\theta}) = -\frac{n}{2}\log 2\pi - \frac{n}{2}\log\sigma^2 - \frac{1}{2\sigma^2}(\mathbf{S}(B_1, B_2)Y - \mathbf{X}\boldsymbol{\alpha})'(\mathbf{S}(B_1, B_2)Y - \mathbf{X}\boldsymbol{\alpha}) + \log|\mathbf{S}(B_1, B_2)|. \tag{7.14}$$

For given B_1 and B_2, we maximize $\ell(\boldsymbol{\theta})$ with respect to $\boldsymbol{\alpha}$ and σ^2, which leads to $\widehat{\boldsymbol{\alpha}}(B_1, B_2) = (\mathbf{X}'\mathbf{X})^{-1}\mathbf{X}'\mathbf{S}(B_1, B_2)Y$, $\widehat{\sigma}^2(\widehat{\boldsymbol{\alpha}}(B_1, B_2), B_1, B_2) = n^{-1}\boldsymbol{\varepsilon}'(\widehat{\boldsymbol{\alpha}}(B_1, B_2), B_1, B_2)\boldsymbol{\varepsilon}(\widehat{\boldsymbol{\alpha}}(B_1, B_2), B_1, B_2)$, where $\boldsymbol{\varepsilon}(\widehat{\boldsymbol{\alpha}}(B_1, B_2), B_1, B_2) = \mathbf{M}\mathbf{S}(B_1, B_2)Y$, and $\mathbf{M}$ is defined below (7.6). The resulting concentrated quasi-log-likelihood (omitting constants) is

$$\ell_c(B_1, B_2) = -\frac{n}{2}\log\widehat{\sigma}^2(\widehat{\boldsymbol{\alpha}}(B_1, B_2), B_1, B_2) + \log|\mathbf{S}(B_1, B_2)|. \tag{7.15}$$

Accordingly, we obtain the QMLEs of B_1 and B_2, i.e., $(\widehat{B}_1', \widehat{B}_2')' = \text{argmax}\,\ell_c(B_1, B_2)$. Subsequently, we have that $\widehat{\boldsymbol{\alpha}} = \widehat{\boldsymbol{\alpha}}(\widehat{B}_1, \widehat{B}_2)$ and $\widehat{\sigma}^2 = \widehat{\sigma}^2(\widehat{\boldsymbol{\alpha}}, \widehat{B}_1, \widehat{B}_2)$.

Under some conditions, Wu et al. (2022) showed the asymptotic normality of QMLEs. To assess the significance of indices, they presented two test statistics to examine the inward and outward influences. These tests allow us to classify all nodes into the four-quadrant categories via inward and outward influence indices. In addition, they proposed three more tests to check the homogeneity of levels of influences. To illustrate the usefulness of the Inward-and-Outward Network Influence model, they presented an empirical example (see the first example in Sect. 7.6) of classifying customers into four segments so that practitioners can design marketing strategies to improve the efficacy of a target product promotion or opinion dissemination.

7.3 Network Autocorrelation Model for Large-Scale Network Data

For estimating model parameters discussed in Sects. 7.2.1 and 7.2.2, the computational complexity is $O(n^3)$, which is not likely applicable for large-scale network data. Due to research and empirical demands, however, social media sites, such as Facebook, YouTube, Twitter, and Instagram, have become important resources for collecting large-scale social network data that can be studied using big data techniques. Hence, this section mainly focuses on screening large-scale social networks to identify relevant nodes based on network autocorrelation models, when repeated observations are available.

7.3.1 Model Structure

Consider a network that has a large number of nodes, say n. To construct a network model, let $Y_t = (y_{1t}, \cdots, y_{nt})'$ be the response vector collected at time t across n nodes, where $t = 1, \cdots, T$. In addition, let $X_{it} = (x_{i1t}, \cdots, x_{iqt})'$ be a $q \times 1$ explanatory vector collected from node i at time t for $i = 1, \cdots, n$ and $t = 1, \cdots, T$. Then Zhu et al. (2019) proposed the following network autoregression (i.e., autocorrelation) model,

$$y_{it} = \sum_{j=1}^{n} \rho_j w_{ij} y_{jt} + X_{it}'\boldsymbol{\alpha} + \varepsilon_{it} \ (i = 1, \cdots, n \text{ and } t = 1, \cdots, T), \tag{7.16}$$

where ρ_j, w_{ij}, and $\boldsymbol{\alpha}$ are defined in (7.8), and the ε_{it} are i.i.d. random errors with mean 0 and variance σ^2 for $i = 1, \cdots, n$ and $t = 1, \cdots, T$. The matrix expression of (7.16) is given below:

$$Y_t = \mathbf{WD}Y_t + \mathbf{X}_t\boldsymbol{\alpha} + \boldsymbol{\varepsilon}_t, \tag{7.17}$$

where $\mathbf{W}$ and $\mathbf{D} = \text{diag}(\rho_i)$ are defined in Sect. 7.2.1.1, $\mathbf{X}_t = (X_{1t}, \cdots, X_{nt})'$ is the $n \times q$ matrix, and $\boldsymbol{\varepsilon}_t = (\varepsilon_{1t}, \cdots, \varepsilon_{nt})'$. For $t = 1, \cdots, T$, let $\mathbf{Z}_t = \mathbf{W}\text{diag}(Y_t) = (W_{.1}Y_{1t}, \cdots, W_{.n}Y_{nt}) \in \mathbb{R}^{n \times n}$, where $W_{.i}$ is the i-th column of $\mathbf{W}$ for $i = 1, \cdots, n$. In addition, let $\mathbf{Z} = (\mathbf{Z}_1', \cdots, \mathbf{Z}_T')' \equiv (\tilde{Z}_1, \cdots, \tilde{Z}_n) \in \mathbb{R}^{nT \times n}$, $\mathbb{Y}^* = (Y_1', \cdots, Y_T')' \in \mathbb{R}^{nT}$, $\mathbf{X} = (\mathbf{X}_1', \cdots, \mathbf{X}_T')' \equiv (\tilde{X}_1, \cdots, \tilde{X}_q) \in \mathbb{R}^{nT \times q}$, and $\mathcal{E}^* = (\varepsilon_1', \cdots, \varepsilon_T')' \in \mathbb{R}^{nT}$, where $\tilde{Z}_i$ is an $nT \times 1$ vector for $i = 1, \cdots, n$ and $\tilde{X}_j$ is an $nT \times 1$ vector for $j = 1, \cdots, q$. Accordingly, we can re-express (7.17) as follows:

$$\mathbb{Y}^* = \mathbf{ZD1}_n + \mathbf{X}\boldsymbol{\alpha} + \mathcal{E}^*, \tag{7.18}$$

where $\mathbf{D1}_n = (\rho_1, \cdots, \rho_n)'$ is the network influence index as defined in Sect. 7.2.1.1.

7.3.2 Network Influence Screening

In a large-scale network, it is critical to screen the network influence index. To this end, Zhu et al. (2019) proposed the following R^2-type measure for index screening,

$$R_i^2 = \frac{\tilde{Z}_i'\tilde{\mathbf{Y}}(\tilde{\mathbf{Y}}'\tilde{\mathbf{Y}})^{-1}\tilde{\mathbf{Y}}'\tilde{Z}_i}{\tilde{Z}_i'\tilde{Z}_i}, \tag{7.19}$$

where $\tilde{\mathbf{Y}} = (\mathbb{Y}^*, \mathbf{X}) \in \mathbb{R}^{nT\times(q+1)}$. Without covariate information $\mathbf{X}$, R_i^2 measures the correlation between the response Y_i and its connected responses $\mathbf{Y}'W_{.i}$ for $i = 1, \cdots, n$, where $\mathbf{Y} = (Y_1, \cdots, Y_T) \in \mathbb{R}^{n\times T}$. Based on the measure of R_i^2, for given thresholding value c, one can obtain a family of relevant nodes, $\widehat{A}$, which is

$$\widehat{A} = \{1 \le i \le n : R_i^2 > c\}.$$

Under some conditions, Zhu et al. (2019) demonstrated the screening consistency of $\widehat{A}$. After algebraic simplification, we have that $R_i^2 = R_{iX}^2 + \tilde{R}_{iY}^2$, where $R_{iX}^2 = \tilde{Z}_i'\tilde{\mathbf{H}}\tilde{Z}_i/(\tilde{Z}_i'\tilde{Z}_i)$, $\tilde{\mathbf{H}} = \mathbf{X}(\mathbf{X}'\mathbf{X})^{-1}\mathbf{X}'$, $\tilde{R}_{iY}^2 = \tilde{Z}_i'\tilde{\mathbb{Y}}^*(\tilde{\mathbb{Y}}^{*'}\tilde{\mathbb{Y}}^*)^{-1}\tilde{\mathbb{Y}}^{*'}\tilde{Z}_i/(\tilde{Z}_i'\tilde{Z}_i)$, $\tilde{\mathbb{Y}}^* = \tilde{\mathbf{M}}\mathbb{Y}^*$, and $\tilde{\mathbf{M}} = \mathbf{I} - \tilde{\mathbf{H}}$. Note that R_{iX}^2 is the coefficient of determination between $\tilde{Z}_i$ and $\mathbf{X}$, while $\tilde{R}_{iY}^2$ is similar to the coefficient of partial determination, $\tilde{R}_{iY}^{*2}$, between $\tilde{Z}_i$ and $\tilde{\mathbf{Y}}$ by controlling the contribution of covariates $\mathbf{X}$, where $\tilde{R}_{iY}^{*2} = \tilde{R}_{iY}^2(\tilde{Z}_i'\tilde{Z}_i)/(\tilde{Z}_i^{*'}\tilde{Z}_i^*)$ and $\tilde{Z}_i^* = \tilde{\mathbf{M}}\tilde{Z}_i$. Hence, it is of interest to employ $\tilde{R}_{iY}^2$ or $\tilde{R}_{iY}^{*2}$ to screen network nodes.

7.3.3 Parameter Estimation

Without loss of generality, we assume that the first a_1 nodes are selected into $\widehat{A}$ during the screening process. Then we assign 0 as the level of influence for the other a_2 nodes, where $a_2 = n - a_1$. Thus, the resulting influence index vector is $\boldsymbol{\gamma} = (\rho_1, \cdots, \rho_{a_1}, 0, \cdots, 0)' \in \mathbb{R}^n$, where the number of zeros in $\boldsymbol{\gamma}$ is $n - a_1$. Let $\boldsymbol{\theta}_1 = (\rho_1, \cdots, \rho_{a_1})'$ and $\boldsymbol{\theta} = (\boldsymbol{\alpha}', \boldsymbol{\theta}_1', \sigma^2)'$. The quasi-log-likelihood function (omitting a constant) is

$$\ell(\boldsymbol{\theta}) = -\frac{nT}{2}\log\sigma^2 - \frac{1}{2\sigma^2}\sum_{t=1}^{T}(\mathbf{S}(\boldsymbol{\gamma})Y_t - \mathbf{X}_t\boldsymbol{\alpha})'(\mathbf{S}(\boldsymbol{\gamma})Y_t - \mathbf{X}_t\boldsymbol{\alpha}) + T\log|\mathbf{S}(\boldsymbol{\gamma})|, \tag{7.20}$$

where $\mathbf{S}(\boldsymbol{\gamma}) = \mathbf{I} - \mathbf{W}\{\text{diag}(\boldsymbol{\gamma})\}$, and it is assumed to be invertible. Applying the same approach discussed in Sect. 7.2, we can obtain the QMLE of $\boldsymbol{\theta} = (\boldsymbol{\alpha}', \boldsymbol{\theta}_1', \sigma^2)'$ and denote it $\widehat{\boldsymbol{\theta}} = (\widehat{\boldsymbol{\alpha}}', \widehat{\boldsymbol{\theta}}_1', \widehat{\sigma}^2)'$. Under some conditions, Zhu et al. (2019) showed the asymptotic normality of $\widehat{\boldsymbol{\alpha}}$ and $\widehat{\boldsymbol{\theta}}_1$.

In addition to QMLE, Zhu et al. (2019) proposed the penalized quasi-likelihood estimator of $\boldsymbol{\theta}_M = (\boldsymbol{\alpha}', \boldsymbol{\theta}_1')'$. Specifically, for given penalized parameter λ, one can obtain the penalized estimator of $(\boldsymbol{\alpha}', \boldsymbol{\theta}_1')'$ by minimizing the following function,

$$-\ell(\boldsymbol{\theta}_M) + \sum_{j=1}^{a_1} p_\lambda(|\rho_j|), \tag{7.21}$$

where $p_\lambda(|\cdot|)$ is the penalty function such as lasso (Tibshirani 1996), SCAD (Fan and Li 2001), and MCP (Zhang 2010); see also the penalty function defined in (2.12). The estimation algorithm and the oracle property of estimators can be found in Zhu et al. (2019). They also used a Sina Weibo dataset to illustrate the application of their proposed method and identify the most influential users; see the second example in Sect. 7.6.

It is worth noting that one can extend the screening method of Zhu et al. (2019) by allowing the temporal dependence of responses. Specifically, one can employ the GEE method mentioned in Sect. 5.1.3 to estimate unknown parameters and then explore the theoretical properties of resulting estimators.

7.4 Multivariate Models

7.4.1 *Multivariate Network Autocorrelation Model*

7.4.1.1 Model Structure

After discussing the network autocorrelation model for large-scale network data, we next follow the estimation approach in Sect. 5.1.4 and study the multivariate network autocorrelation model. To this end, let $Y_i = (y_{i1}, \cdots, y_{ip})' \in \mathbb{R}^p$ be the multivariate responses collected at node i for $i = 1, \cdots, n$. Accordingly, $\mathbf{Y} = (Y_1, \cdots, Y_n)' \in \mathbb{R}^{n\times p}$ is the response matrix. Its associated explanatory matrix is $\mathbf{X} = (X_1, \cdots, X_n)' \in \mathbb{R}^{n\times q}$, where $X_i = (x_{i1}, \cdots, x_{iq})' \in \mathbb{R}^q$ for $i = 1, \cdots, n$. In addition, let the normalized network adjacency matrix $\mathbf{W} \in \mathbb{R}^{n\times n}$ be defined in Sect. 7.2, and let $\boldsymbol{\rho} = (\rho_{jk}) \in \mathbb{R}^{p\times p}$ be an activity matrix, which measures the activity effects (i.e., social interactions) among the p components of multi-responses. Then the multivariate network autocorrelation (MNAM) model proposed by Zhu et al. (2020) is below:

$$\mathbf{Y} = \mathbf{W}\mathbf{Y}\boldsymbol{\rho} + \mathbf{X}\mathbf{A} + \mathbf{E}, \tag{7.22}$$

where $\mathbf{A} = (\alpha_{lm}) \in \mathbb{R}^{q\times p}$ is an unknown coefficient matrix, $\mathbf{E} = (\boldsymbol{\varepsilon}_1, \cdots, \boldsymbol{\varepsilon}_p) \in \mathbb{R}^{n\times p}$ is a random error matrix, and $\boldsymbol{\varepsilon}_j = (\varepsilon_{1j}, \cdots, \varepsilon_{nj})' \in \mathbb{R}^n$ for $j = 1, \cdots, p$. In addition, we assume that the $\boldsymbol{\varepsilon}_i^* = (\varepsilon_{i1}, \cdots, \varepsilon_{ip})' \in \mathbb{R}^p$ are i.i.d. random vectors with mean $\mathbf{0}$ and covariance $\boldsymbol{\Sigma} \in \mathbb{R}^{p\times p}$ for $i = 1, \cdots, n$. Moreover, let $\mathbb{Y}^* = \text{vec}(\mathbf{Y}) \in \mathbb{R}^{np}$ and $\mathcal{E}^* = \text{vec}(\mathbf{E}) \in \mathbb{R}^{np}$, where $\mathcal{E}^*$ is a random vector with mean $\mathbf{0} \in \mathbb{R}^{np}$ and covariance $(\boldsymbol{\Sigma} \otimes \mathbf{I}_n) \in \mathbb{R}^{np\times np}$. We then adapt the approach of Sect. 5.1.4 and re-express (7.22) as follows:

$$\mathbb{Y}^* = (\boldsymbol{\rho}' \otimes \mathbf{W})\mathbb{Y}^* + (\mathbf{I}_p \otimes \mathbf{X})\text{vec}(\mathbf{A}) + \mathcal{E}^*. \tag{7.23}$$

7.4.1.2 Parameter Estimation

To estimate unknown parameters in (7.23), Zhu et al. (2020) developed an approach for obtaining the QMLE. Let $\boldsymbol{\theta} = (\text{vec}'(\mathbf{A}), \text{vec}'(\boldsymbol{\rho}), \text{vec}^{*'}(\boldsymbol{\Sigma}))'$, where $\text{vec}^*(\boldsymbol{\Sigma})$ only transforms the upper triangular parameters of $\boldsymbol{\Sigma}$ into a column vector. In addition, define $\mathbf{X}^* = \mathbf{I}_p \otimes \mathbf{X} \in \mathbb{R}^{np \times qp}$, and $\mathbf{S}(\boldsymbol{\rho}) = \mathbf{I}_{np} - (\boldsymbol{\rho}' \otimes \mathbf{W}) \in \mathbb{R}^{np \times np}$ and assume that it is invertible. Based on (7.23), the quasi-log-likelihood function (omitting a constant) is

$$\ell(\boldsymbol{\theta}) = -\frac{n}{2}\log|\boldsymbol{\Sigma}| - \frac{1}{2}\left\{\mathbf{S}(\boldsymbol{\rho})\mathbb{Y}^* - \mathbf{X}^*\text{vec}(\mathbf{A})\right\}'(\boldsymbol{\Sigma} \otimes \mathbf{I}_n)^{-1}\left\{\mathbf{S}(\boldsymbol{\rho})\mathbb{Y}^* - \mathbf{X}^*\text{vec}(\mathbf{A})\right\}$$
$$+ \log|\mathbf{S}(\boldsymbol{\rho})|.$$

For given $\boldsymbol{\rho}$, we maximize $\ell(\boldsymbol{\theta})$ with respect to $\mathbf{A}$ and $\boldsymbol{\Sigma}$, which leads to

$$\begin{aligned} \text{vec}(\widehat{\mathbf{A}}(\boldsymbol{\rho})) &= \{\mathbf{X}^{*'}(\boldsymbol{\Sigma} \otimes \mathbf{I}_n)^{-1}\mathbf{X}^*\}^{-1}\mathbf{X}^{*'}(\boldsymbol{\Sigma} \otimes \mathbf{I}_n)^{-1}\mathbf{S}(\boldsymbol{\rho})\mathbb{Y}^* \\ &= (\mathbf{I}_p \otimes (\mathbf{X}'\mathbf{X})^{-1}\mathbf{X}')\mathbf{S}(\boldsymbol{\rho})\mathbb{Y}^*. \end{aligned} \tag{7.24}$$

After algebraic simplification, we have that $\widehat{\mathbf{A}}(\boldsymbol{\rho}) = (\mathbf{X}'\mathbf{X})^{-1}\mathbf{X}'(\mathbf{Y} - \mathbf{W}\mathbf{Y}\boldsymbol{\rho})$. In addition, we obtain that $\widehat{\boldsymbol{\Sigma}}(\boldsymbol{\rho}) = n^{-1}(\mathbf{Y} - \mathbf{W}\mathbf{Y}\boldsymbol{\rho})\mathbf{M}(\mathbf{Y} - \mathbf{W}\mathbf{Y}\boldsymbol{\rho})'$, where $\mathbf{M}$ is defined below (7.6). Accordingly, the concentrated quasi-log-likelihood function is

$$\ell_c(\boldsymbol{\rho}) = -\frac{n}{2}\log|\widehat{\boldsymbol{\Sigma}}(\boldsymbol{\rho})| + \log|\mathbf{S}(\boldsymbol{\rho})|. \tag{7.25}$$

We then obtain the QMLE of $\boldsymbol{\rho}$; i.e., $\widehat{\boldsymbol{\rho}} = \text{argmax}\ell_c(\boldsymbol{\rho})$. Thus, the resulting QMLEs of $\mathbf{A}$ and $\boldsymbol{\Sigma}$ are $\widehat{\mathbf{A}} = \widehat{\mathbf{A}}(\widehat{\boldsymbol{\rho}})$ and $\widehat{\boldsymbol{\Sigma}} = \widehat{\boldsymbol{\Sigma}}(\widehat{\boldsymbol{\rho}})$, respectively.

Under some conditions, Zhu et al. (2020) showed the asymptotic properties of $\widehat{\boldsymbol{\theta}} = (\text{vec}'(\widehat{\mathbf{A}}), \text{vec}'(\widehat{\boldsymbol{\rho}}), \text{vec}^{*'}(\widehat{\boldsymbol{\Sigma}}))'$. Since the computation of QMLE can be complex, Zhu et al. (2020) introduced a least squares type estimator, LSE, which is similar to that of Huang et al. (2019). They also demonstrated the asymptotic properties of LSE under some conditions. Furthermore, they illustrated the usefulness of their model with a Sina Weibo dataset.

7.4.2 *Network Vector Autoregressive Model*

In multivariate time series analysis, the vector autoregressive (VAR) model is one of the most widely used types of models; see, e.g, Lütkepohl (2007), Tsay (2014), and Kilian and Lütkepohl (2017). This section mainly focuses on network vector autoregressive models of order 1, NAR(1), proposed by Zhu et al. (2017).

7.4.2.1 Model Structure

In networks, the multivariate responses can depend on those occurred in the previous period. Hence, it is useful to take into account vector autoregressive (VAR) models for studying social interactions. As discussed in Sect. 7.3.1, let $Y_t = (y_{1t}, \cdots, y_{nt})' \in \mathbb{R}^n$ be the response vector collected across n nodes for time t, where $t = 0, \cdots, T$. In addition, let $X_i = (x_{i1}, \cdots, x_{iq})'$ be a $q \times 1$ explanatory vector collected from node i, and assume that X_i is an i.i.d. random vector with mean $\mathbf{0}$ and covariance matrix $\mathbf{\Sigma}_x$ for $i = 1, \cdots, n$. Moreover, let the normalized adjacency matrix $\mathbf{W} = (w_{ij}) \in \mathbb{R}^{n\times n}$ be defined in Sect. 7.2. Then Zhu et al. (2017) proposed a network vector autoregressive (NAR) model with order 1 given below.

$$y_{it} = \alpha_0 + X_i'\boldsymbol{\alpha} + \rho_1 \sum_{j=1}^{n} w_{ij} y_{j(t-1)} + \rho_2 y_{i(t-1)} + \varepsilon_{it} \ (i = 1, \cdots, n \text{ and } t = 1, \cdots, T), \tag{7.26}$$

where α_0, ρ_1 and ρ_2 are unknown coefficients, $\boldsymbol{\alpha} \in \mathbb{R}^q$ is defined in (7.8), and ε_{it} is an i.i.d. random error with mean 0 and variance σ^2 for $i = 1, \cdots, n$ and $t = 1, \cdots, T$. To express the above model in matrix form, define $Y_t = (y_{1t}, \cdots, y_{nt})' \in \mathbb{R}^n$, $\mathbf{X} = (X_1, \cdots, X_n)' \in \mathbb{R}^{n\times q}$, $\boldsymbol{\alpha}^* = (\alpha_0 \mathbf{1}_n + \mathbf{X}\boldsymbol{\alpha}) \in \mathbb{R}^n$, $\mathbf{W}^* = (\rho_1 \mathbf{W} + \rho_2 \mathbf{I}_n) \in \mathbb{R}^{n\times n}$, and $\boldsymbol{\varepsilon}_t = (\varepsilon_{1t}, \cdots, \varepsilon_{nt})' \in \mathbb{R}^n$ for $t = 1, \cdots, T$. Then, (7.26) can be re-expressed as follows:

$$Y_t = \boldsymbol{\alpha}^* + \mathbf{W}^* Y_{t-1} + \boldsymbol{\varepsilon}_t \ (t = 1, \cdots, T). \tag{7.27}$$

As a result, (7.27) has the VAR(1) model structure with a network coefficient matrix, which was named NAR(1) model by Zhu et al. (2017).

7.4.2.2 Parameter Estimation

Denote the unknown parameters in (7.27) as $\boldsymbol{\theta} = (\alpha_0, \rho_1, \rho_2, \boldsymbol{\alpha}')' \in \mathbb{R}^{q+3}$. To estimate unknown parameters, define $\tilde{X}_{i(t-1)} = (1, \mathbf{w}_{i.}' Y_{t-1}, y_{i(t-1)}, X_i')'$ and $\tilde{\mathbf{X}}_t = (\tilde{X}_{1t}, \cdots, \tilde{X}_{nt})' \in \mathbb{R}^{n\times(q+3)}$, where $\mathbf{w}_{i.} = (w_{i1}, \cdots, w_{in})' \in \mathbb{R}^n$. Then (7.27) can be re-expressed as follows

$$Y_t = \tilde{\mathbf{X}}_{t-1}\boldsymbol{\theta} + \boldsymbol{\varepsilon}_t \ (t = 1, \cdots, T). \tag{7.28}$$

Accordingly, Zhu et al. (2017) obtained the OLS type estimator of $\boldsymbol{\theta}$, which is $\widehat{\boldsymbol{\theta}} = (\sum_{t=1}^T \tilde{\mathbf{X}}_{t-1}' \tilde{\mathbf{X}}_{t-1})^{-1} \sum_{t=1}^T \tilde{\mathbf{X}}_{t-1}' Y_t$. Under some conditions, they further showed the asymptotic normality of $\widehat{\boldsymbol{\theta}}$. In addition, they presented practical applications of the NAR model based on a Sina Weibo dataset. Although we only consider NAR with one lag, the NAR model with multiple lags can be found in Zhu et al. (2017). It is of interest to note that the influence index in NAR(1) is an average measure. Hence, one can incorporate the NIF and IONI indices discussed in Sect. 7.2 into model (7.26), which can enhance and broaden the applications of NAR.

7.5 Additional Network Autocorrelation Models

This section introduces the four additional types of network autocorrelation models that discuss the mutual influence, functional varying coefficient, quantile, and community effects, respectively. To save space, we only present their model structures. Parameter estimations and theoretical properties can be found in their corresponding references.

7.5.1 Mutual Influence Regression Model

When the observed responses in a network are changing with time, it is critical to construct dynamic influence indices. To this end, Fan et al. (2024b) introduced the mutual influence regression (MIR) model to link the mutual influence matrix of actors to a set of similarity matrices induced by their associated attributes. As discussed in Sect. 7.3.1, let $Y_t = (y_{1t}, \cdots, y_{nt})' \in \mathbb{R}^n$ be the response vector collected across n nodes for time t, where $t = 1, \cdots, T$. In addition, let $\boldsymbol{\varepsilon}_t = (\varepsilon_{1t}, \cdots, \varepsilon_{nt})' \in \mathbb{R}^n$, where ε_{it} is an i.i.d. random error with mean 0 and variance σ^2 for $i = 1, \cdots, n$ and $t = 1, \cdots, T$.

To describe the mutual influence among the n actors, Fan et al. (2024b) introduced the following mutual influence model,

$$Y_{it} = b_{i1t}Y_{1t} + \cdots + b_{i(i-1)t}Y_{(i-1)t} + b_{i(i+1)}Y_{(i+1)t} + \cdots + b_{int}Y_{nt} + \varepsilon_{it}, \quad (7.29)$$

where b_{ijt} is the influence effect of Y_{jt} on Y_{it} for $i, j = 1, \cdots, n$ and $t = 1, \cdots, T$. Define $\mathbf{B}_t = (b_{ijt}) \in \mathbb{R}^{n\times n}$ with $b_{iit} = 0$. Then the matrix expression of (7.29) is

$$Y_t = \mathbf{B}_t Y_t + \boldsymbol{\varepsilon}_t, \quad (7.30)$$

where $\mathbf{B}_t$ is the mutual influence matrix, and it characterizes the degree of mutual influence among the n actors at time t.

To build up the relationship between the mutual influence matrix $\mathbf{B}_t$ and a set of similarity matrices, we can adapt the approach of Chap. 4 to construct similarity matrices $\mathbf{W}_k^{(t)}$. To this end, let $Z_k^{(t)}$ be the k-th $n \times 1$ continuous attribute vector for $k = 1, \cdots, K$ and $t = 1, \cdots, T$. In addition, let $Z_{kj_1}^{(t)}$ and $Z_{kj_2}^{(t)}$ be the j_1-th and j_2-th elements of $Z_k^{(t)}$, respectively. Then, define $\mathbf{W}_k^{(t)} = (w(Z_{kj_1}^{(t)}, Z_{kj_2}^{(t)}))_{n\times n}$, where

$w(Z_{kj_1}^{(t)}, Z_{kj_2}^{(t)}) = a(Z_{kj_1}^{(t)} - Z_{kj_2}^{(t)}) / \sum_{j_2=1}^{n} a(Z_{kj_1}^{(t)} - Z_{kj_2}^{(t)})$ and $a(\cdot)$ is the exponential function. Accordingly, the mutual influence regression (MIR) model is

$$Y_t = \mathbf{B}_t(\boldsymbol{\beta})Y_t + \boldsymbol{\varepsilon}_t = \left(\beta_1 \mathbf{W}_1^{(t)} + \cdots + \beta_K \mathbf{W}_K^{(t)}\right)Y_t + \boldsymbol{\varepsilon}_t, \tag{7.31}$$

where $\boldsymbol{\beta} = (\beta_1, \cdots, \beta_K)'$ is an unknown vector of regression coefficients. This model is able to analyze the effect of the mutual influence matrix $\mathbf{B}_t$ at each time t via a set of similarity matrices $\mathbf{W}_k^{(t)}$ and their associated influence parameters β_k. The covariance matrix of Y_t is $\sigma^2(\mathbf{I}_n - \mathbf{B}_t)^{-1}(\mathbf{I}_n - \mathbf{B}_t)^{-1}$, which is used in parameter estimation. The detailed structure of the MIR model and parameter estimators can be found in Fan et al. (2024b).

7.5.2 Network Functional Varying Coefficient Model

In the last three decades, functional data analysis (FDA) has been widely used to study functional data and examine the relationship between functional random variables. Useful references can be found in Ramsay and Dalzell (1991), Ramsay and Silverman (2005), Yao et al. (2005), and Wang et al. (2016), among others. As mentioned in Chap. 5, varying coefficient models also attract considerable attention. Hence, integrating both methods together to form a functional varying coefficient model can broaden practical applications; see, e.g., Sentürk and Müller (2010), Wu et al. (2010), Chiou et al. (2012), and Zhu et al. (2012). Recently, Zhu et al. (2022) incorporated the network structure into this hybrid model and named it network functional varying coefficient (NFVC) model.

Before introducing NFVC, let $y_i(t)$, $\eta_i(t)$, and $\varepsilon_i(t)$, respectively, be the response, the time-dependent effect, and the independent and identically distributed Gaussian random error with mean zero and constant variance σ^2 for node i ($i = 1, \cdots, n$) at time t, where $0 \le t \le 1$. In addition, let $X_i = (x_{i1}, \cdots, x_{iq})'$, as defined in Sect. 7.1.1, and let its associated time-varying coefficient vector be $\boldsymbol{\alpha}(t) \in \mathbb{R}^q$. Moreover, let the normalized adjacency matrix $\mathbf{W} = (w_{ij})$ be defined as in Sect. 7.2, and let $\rho(t)$ be the dynamic influence parameter. Then, Zhu et al. (2022) proposed the following NFVC model,

$$y_i(t) = \rho(t)\sum_{j=1}^{n} w_{ij} y_j(t) + X_i'\boldsymbol{\alpha}(t) + \eta_i(t) + \varepsilon_i(t). \tag{7.32}$$

They also presented parameter estimators and demonstrated their asymptotic properties. Moreover, an empirical example (see also the third example in Sect. 7.6) was presented to illustrate the application of the NFVC model. Finally, it would be of interest to extend the dynamic approach to the NIF and IONI models.

7.5.3 Network Quantile Autoregressive Model

In data analysis, quantile regression has been widely used since it is robust to non-Gaussian or heavy-tailed data. In addition, it allows practitioners to provide more insightful interpretation of parameter estimates obtained via quantiles $\tau \in (0, 1)$. Useful references are Koenker and Bassett (1978), Koenker (2005), Hao and Naiman (2007), Davino et al. (2013), Koenker (2017), and He et al. (2023).

To broaden the application of NAR model discussed in Sect. 7.4.2, Zhu et al. (2019) built on the approach of Koenker and Xiao (2006) and proposed the network quantile autoregressive (NQAR) model given below.

$$y_{it} = \alpha_0(u_{it}) + X_i'\boldsymbol{\alpha}(u_{it}) + \rho_1(u_{it}) \sum_{j=1}^{n} w_{ij} y_{j(t-1)} + \rho_2(u_{it}) y_{i(t-1)} + \varepsilon_{it}, \tag{7.33}$$

for $i = 1, \cdots, n$ and $t = 1, \cdots, T$, where y_{it}, X_i, and w_{ij} are defined in (7.26) and u_{it} is an i.i.d. random variable with the standard uniform distribution on the interval [0,1], for $i = 1, \cdots, n$ and $t = 1, \cdots, T$. In addition, the coefficients α_0, $\boldsymbol{\alpha}(u_{it})$, $\rho_1(u_{it})$ and $\rho_2(u_{it})$ are unknown functions mapping from (0,1) to $\mathbb{R}^1$ and their dimensions are the same as those defined in (7.26). Accordingly, the τ-th conditional quantile function of y_{it} for the given X_i and $Y_{t-1} = (y_{1,t-1}, \cdots, y_{n,t-1})'$, for $i = 1, \cdots, n$ and $t = 1, \cdots, T$, is

$$Q_{y_{it}}(\tau | X_i, Y_{t-1}) = \alpha_0(\tau) + X_i'\boldsymbol{\alpha}(\tau) + \rho_1(\tau) \sum_{j=1}^{n} w_{ij} y_{j(t-1)} + \rho_2(\tau) y_{i(t-1)}. \tag{7.34}$$

The quantile parameter estimators of the above model and their asymptotic properties are in Zhu et al. (2019). When the dimension of covariates X_i is large, one can consider employing Ma et al.'s (2017) quantile partial correlation screening method to identify relevant covariates.

7.5.4 Community Network Autoregressive Model

Due to artificial intelligence, automation, machine learning, and other advancements in information technology as well as a huge number of social media users, a large-scale network dataset can be collected or generated effectively across various resources or fields. To retrieve important and useful information from a big network data set, it is natural to classify such types data into different communities (or clusters or groups). Although finding communities in large networks is a challenging task, the stochastic block model (SBM) proposed by Holland et al. (1983) can be a promising approach. In the area of SBM, there are many good references such as Nowicki and Snijders (2001), Newman and Girvan (2004), Bickel and Chen (2009), Rohe et al. (2011), Choi et al. (2012), Jin (2012), Lei and Rinaldo (2015), Bickel and Sarkar (2016), Zhang and Zhou (2016), Lee and Wilkinson (2019), Hu et al. (2021), and Wu et al. (2025).

Recently, Chen et al. (2023) incorporated the community structure into the network vector autoregressive model (NAR) discussed in Sect. 7.4.2. Before introducing their model, consider an n-node network with K communities. For each node $i \in \{1, \cdots, n\}$, $k(i) \in \{1, \cdots, K\}$ represents the community to which node i belongs. Then define a membership matrix $\boldsymbol{\Theta} = (\theta_{ij}) \in [0, 1]^{n\times K}$. Specifically, we have $\theta_{ij} = 1$ if $j = k(i)$ and 0 otherwise. In addition, define a symmetric connective matrix $\mathbf{P} = (p_{k(i)k(j)}) \in \mathbb{R}^{K\times K}$. For given $\boldsymbol{\Theta}$ and $\mathbf{P}$, let $\mathbf{A}^* = (a^*_{ij}) \in \mathbb{R}^{n\times n}$ be a symmetric adjacency matrix generated by SBM, where $a^*_{ij} = 1$ if two nodes i and j are connected and $a^*_{ij} = 0$ otherwise. For the sake of completeness, the diagonal elements of $\mathbf{A}^*$ are set to zero. We then assume that each a^*_{ij}, for $i > j$, is independently generated from the Bernoulli distribution with the probability $p_{k(i)k(j)}$, where $0 \le p_{k(i)k(j)} \le 1$ is the connection probability of nodes i and j. Accordingly, $E(\mathbf{A}^*) = \boldsymbol{\Theta}\mathbf{P}\boldsymbol{\Theta}' - \text{diag}(\boldsymbol{\Theta}\mathbf{P}\boldsymbol{\Theta}')$ (see Lei and Rinaldo 2015).

As defined in Sect. 7.4.2, let $Y_t = (y_{1t}, \cdots, y_{nt})'$ be the response vector collected at time t across n nodes for $t = 0, \cdots, T$. In addition, let $\mathbf{X}_t = (X_{1t}, \cdots, X_{nt})' \in \mathbb{R}^{n\times q}$, where $X_{it} = (x_{i1t}, \cdots, x_{iqt})'$ be a $q \times 1$ explanatory vector collected from node i at time t for $i = 1, \cdots, n$ and $t = 0, \cdots, T$. Then Chen et al. (2023) proposed the following community-augmented network autoregression (CNAR) model,

$$Y_t = (\boldsymbol{\Theta}\mathbf{P}\boldsymbol{\Theta}' + \rho\mathbf{I}_n)Y_{t-1} + \mathbf{X}_{t-1}\boldsymbol{\alpha} + \boldsymbol{\varepsilon}_t, \tag{7.35}$$

where $\mathbf{P}$, $\rho \in \mathbb{R}$, and $\boldsymbol{\alpha} \in \mathbb{R}^{q\times 1}$ are unknown parameters. In addition, $\boldsymbol{\varepsilon}_t = \boldsymbol{\Gamma}\mathbf{f}_t + \mathbf{e}_t$, where $\boldsymbol{\Gamma} \in \mathbb{R}^{n\times M}$ is a loading matrix, $\mathbf{f}_t \in \mathbb{R}^{M\times 1}$ is the unknown latent factor, and $\mathbf{e}_t \in \mathbb{R}^{n\times 1}$ is the normally distributed random vector with mean $\mathbf{0}$ and variance $\sigma^2\mathbf{I}_n$, and σ^2 is an unknown scaled parameter. The detailed parameter estimators of the CNAR model and their theoretical properties can be found in Chen et al. (2023).

In addition to the above four models, there are various models that can be applied to networks, such as spatial autoregressive model with spatial autoregressive disturbances (Lee and Liu 2010), banded spatiotemporal autoregressive model (Gao et al. 2019), network GARCH model (Zhou et al. 2020), subnetwork estimation for spatial autoregressive models in large-scale networks (Li et al. 2023), transfer learning for SAR models (Li and Song 2024; Zeng et al. 2024), spatial autoregressive model with measurement error in covariates (Paul and Nath 2024), and distributed estimation and inference for spatial autoregression model with large-scale networks (Ren et al. 2024).

7.6 Empirical Examples

This section briefly reviews three examples to discuss the applications of network autocorrelation and spatial autoregressive models. The first example is taken from

Wu et al. (2022) who studied inward-and-outward network influence based on social network data from Sina Weibo, which is the biggest social media platform in China. This dataset is a snapshot taken in November 2013 and contains 2,580 online users' activity and their corresponding attribute variables between January 2013 and November 2013. The adjacency matrix $\mathbf{A}^* = (a_{ij}^*) \in \mathbb{R}^{2,580\times 2,580}$ is defined as follows: $a_{ij}^* = 1$ if user i is following user j and $a_{ij}^* = 0$ otherwise.

The response variable Y_i is the i-th user's activity level, which is proxied by the number of Sina Weibo posts made by user i over the entire study period. In addition, the five attributes are: In-degree (the number of each user's followers in the network); Out-degree (the number of users that i is following in the network); Duration (the time elapsed since registering on the Sina Weibo platform); Self-Introduction (the length of a user's description about himself/herself); Gender: (male = 1 and female = 0). The all five attributes are assigned to each of the three covariates Z, V, and X; see Sect. 7.2.2.

After fitting the IONI model in (7.12), Wu et al. (2022) obtained the four-quadrant classification. Accordingly, practitioners are able to segment Sina Weibo's users into the four groups to improve product promotion or opinion dissemination. For example, if practitioners need to promote a new product on Weibo, they could target the 4.0% who are "early adopters." In contrast, if practitioners want to conduct a promotion for a mature product, they should target the 7.7% who are the "early majority." To effectively leverage the above two types of users, decision-makers can adopt the "fan economy" strategy to raise their conversion rate. As a result, they should target those 15.2% who are "opinion leaders" to make such promotions. Lastly, removing or ignoring the 73.1% who are prolonged "inactive users" can cut promotion costs and improve marketing efficiency. Consequently, this example demonstrates that obtaining the inward and outward influence indices can play an important role in social network applications.

The second example is from Zhu et al. (2019), who employed their portal nodes screening and estimation method to identify the influential users in their Sina Weibo dataset. This finding could assist marketing professionals to attract active users, launch marketing campaigns, and increase advertising profits. The dataset consists of a total number of $N = 842$ active followers of an MBA program account for consecutive $T = 30$ days.

Let the response Y_{it} be defined as the $\log(1 + \text{length})$-transformed Weibo post length for the i-th user on the t-th day. To take into account users' posting behaviors, five covariates X_{it} are considered. The first one is the lag-1 term $Y_{i(t-1)}$, which takes into account the autoregressive effect of Y_{it}. The remainders are the nodal covariates, and they are: gender (male = 1, female = 0); tenure (i.e., the time length since registration with Sina Weibo); the number of personal labels; and the user's description length. Note that the labels and descriptions are created by the users themselves to describe their lifestyles and characteristics, among other things. In addition, the adjacency matrix $\mathbf{A}^* = (a_{ij}^*) \in \mathbb{R}^{842\times 842}$ is defined as follows: $a_{ij}^* = 1$ if user i is following user j and $a_{ij}^* = 0$ otherwise.

After applying the screening procedure via R_j^2 in (7.19) and then subsequently employing the four penalized estimation methods (lasso, adaptive lasso, MCP, and SCAD), Zhu et al. (2019) found that there are fifteen users being identified as portal nodes by at least one of these four methods. Among those users, nine of them are famous online social media accounts, who release latest news about business, economics, and finance. The other six users are famous celebrities, who are also associated with these related fields. The estimated lag-1 autoregressive coefficient is around 0.46, which indicates a positive momentum effect for user behavior. For the nodal covariates, the only significant variable is the length of the user's description, which is positively related to the user's activity level. In sum, identifying those 15 influential users is critical since they can lead to making more profits.

To further study the behavior of users on the Weibo social media platform, we consider the third example from Zhu et al. (2022). The dataset consists of the posting behavior of $N = 850$ active users from January 2nd, 2014 to January 9th, 2014. The response $Y_i(t_{ik})$ is the number of posts of the i-th user at the k-th hour, where $Y_i(t)$ denotes the activeness of the i-th social network user, which is assumed to have a smooth trajectory at time t. This study contains two nodal covariates, and they are the gender (male = 1, female = 0) and the user's number of personal labels. In addition, the adjacency matrix $\mathbf{A}^* = (a_{ij}^*) \in \mathbb{R}^{850\times 850}$ is defined as follows: $a_{ij}^* = 1$ if user i is following user j and $a_{ij}^* = 0$ otherwise.

After fitting the data with the NFVC model (7.32), Zhu et al. (2022) had the following findings. First, the estimated functions depict peaks during the day and troughs at night, which is consistent with human nature. Second, the estimated network effect $\rho(t)$ is positive, which implies that users' posting behaviors are positively related to the posting behavior of their following friends. In addition, the strength of network activity seems to be much lower on Saturdays, possibly because many people spend Saturdays on other aspects of their life rather on than social networking. The estimated $\widehat{\alpha}_1(t)$ indicates that the male users tend to be less active than the female users on weekends, while they show more activity on Thursdays and Fridays. Moreover, the estimate $\widehat{\alpha}_2(t)$ reveals that the users with more self-created labels lean toward being more active on weekdays and less active on weekends. A possible explanation is that the users with more active personal characters tend to have more highly diversified personal lives on weekends.

The above three examples use three different datasets from Weibo to illustrate the inward-and-outward network influence indices, identify the most influential users, and characterize the dynamic influence, respectively. These empirical analyses can be analogously applied to other social media platforms for understanding their users' behavior and increasing revenues.

Chapter 8
Factor Models and Covariance Matrices

Factor models involve a reduction of parameters in the covariance matrix, and they are highly related to principal components. Since these models have played a very important role in data analysis, we not only introduce classical factor models but also include four extended factor models, namely approximate factor models, factor-augmented regression models, dynamic factor models, and matrix factor models. In addition, three examples are presented to briefly illustrate empirical applications.

8.1 Factor Models

Factor analysis has a long history in exploring the underlying structure of observed variables. In 1904, Charles Spearman published a celebrated article, "General Intelligence, Objectively Determined and Measured," which opens an important and promising avenue, factor analysis, for studying latent variables. Specifically, factor analysis is a technique for modeling observed and correlated variables along with their covariance structure via unobserved variables (i.e., factors). The resulting factors with rotations are not only easier to interpret but also uncorrelated with each other. In principal, factor analysis is a dimension-reduction technique that can also be classified as an unsupervised learning method (see Sect. 9.3). In addition, this chapter only focuses on exploratory, rather than confirmatory, factor analysis. Useful references for reviewing factor models can be found in Bartholomew (2005), Finch (2020), and Fan et al. (2021). Moreover, we consider different factor models from those used in studying market efficiency in finance, such as the Sharpe's (1964) single-factor model (i.e., the capital asset pricing model, CAPM) and Fama and French's (1993, 2015) three-factor and five-factor models. This is because the factors in those models are observable.

W. Lan, C.-L. Tsai, *Covariance Analysis and Beyond*,
https://doi.org/10.1007/978-3-032-08796-6_8

8.1.1 Model Structure

In multivariate analysis, factor models are designed to build a linear relationship between observable and latent variables for the purpose of dimension reduction. Let x_{it} be the observed variables for $i = 1, \cdots, n$ and $t = 1, \cdots, T$. This section assumes that n is fixed and T can go to infinity. To reduce the dimension from n to a lower dimension m, we consider the following factor model,

$$x_{it} = \alpha_i + \boldsymbol{\beta}_i' \mathbf{f}_t + \epsilon_{it} \; (i = 1, \cdots, n \text{ and } t = 1, \cdots, T), \tag{8.1}$$

where α_i is an unknown coefficient, $\boldsymbol{\beta}_i = (\beta_{i1}, \cdots, \beta_{im})' \in \mathbb{R}^m$ is an unknown factor loading, $\mathbf{f}_t = (f_{1t}, \cdots, f_{mt})' \in \mathbb{R}^m$ is an unknown common factor, and the ϵ_{it} are independent random errors with mean 0 and variance $d_i > 0$ for $i = 1, \cdots, n$ and $t = 1, \cdots, T$. For the sake of parameter identifiability, we assume that $E(\mathbf{f}_t) = 0$ and $\text{Cov}(\mathbf{f}_t) = \mathbf{I}_m$ for $t = 1, \cdots, T$; see, e.g., Anderson (2003). Let $\boldsymbol{\varepsilon}_t = (\epsilon_{1t}, \cdots, \epsilon_{nt})' \in \mathbb{R}^n$, and we also assume that $\text{Cov}(\mathbf{f}_t, \boldsymbol{\varepsilon}_t) = \mathbf{0}_{m \times n}$ for $t = 1, \cdots, T$. Note that some articles or books use the notation p (or d) and n (or N) in (8.1) rather than n and T; see, e.g., Anderson and Amemiya (1988) and Johnson and Wichern (2007).

Let $X_t = (x_{1t}, \cdots, x_{nt})' \in \mathbb{R}^n$, $\boldsymbol{\alpha} = (\alpha_1, \cdots, \alpha_n)' \in \mathbb{R}^n$ and $\mathbf{B} = (\boldsymbol{\beta}_1, \cdots, \boldsymbol{\beta}_n)' \in \mathbb{R}^{n \times m}$. Then (8.1) can be re-expressed as follows:

$$X_t = \boldsymbol{\alpha} + \mathbf{B}\mathbf{f}_t + \boldsymbol{\varepsilon}_t. \tag{8.2}$$

Note that $\text{Cov}(\boldsymbol{\varepsilon}_t) = \mathbf{D} = \text{diag}(d_i) \in \mathbb{R}^{n \times n}$ and $\text{Cov}(X_t) = \boldsymbol{\Sigma} = \mathbf{B}\mathbf{B}' + \mathbf{D} \in \mathbb{R}^{n \times n}$ for $t = 1, \cdots, T$. In addition, let $\mathbf{X} = (X_1, \cdots, X_T)' \in \mathbb{R}^{T \times n}$, $\mathbf{F} = (\mathbf{f}_1, \cdots, \mathbf{f}_T)' \in \mathbb{R}^{T \times m}$, and $\boldsymbol{\varepsilon} = (\boldsymbol{\varepsilon}_1, \cdots, \boldsymbol{\varepsilon}_T)' \in \mathbb{R}^{T \times n}$. Then the matrix form of (8.1) is

$$\mathbf{X} = \mathbf{A} + \mathbf{F}\mathbf{B}' + \boldsymbol{\varepsilon}, \tag{8.3}$$

where $\mathbf{A} = \mathbf{1}_T \otimes \boldsymbol{\alpha}'$ and $\mathbf{1}_T = (1, \cdots, 1)' \in \mathbb{R}^T$. To ensure that the matrix of factor loadings $\mathbf{B}$ is uniquely determined, we assume that $\mathbf{B}'\mathbf{D}^{-1}\mathbf{B}$ is diagonal with distinct diagonal elements; see, e.g., Anderson (2003) and Bai and Li (2012). It is worth noting that there are alternative approaches to impose identifying restrictions; see Bai and Li (2012) for detailed discussion.

8.1.2 Parameter Estimation

This section considers three parameter estimations. The first is the least squares estimation, the second is the maximum likelihood estimation, and the third is the penalized estimation. Before introducing least squares estimation, we briefly review principal components.

8.1.2.1 Principal Components and Least Squares Estimation

Principal component analysis (PCA) has been widely used across many different fields such as psychology, education, sociology, statistics, biological science, engineering, and economics. Basically, principal component analysis is a dimension-reduction technique that transforms the original variables into a set of linearly uncorrelated variables. Accordingly, it identifies a smaller number of uncorrelated variables (namely principal components) from a larger set of variables. The empirical applications and theoretical foundations of PCA can be found in Jolliffe (2002) and Vidal et al. (2016).

Suppose the random vector $X \in \mathbb{R}^n$ has mean vector $\boldsymbol{\alpha}$ and covariance matrix $\boldsymbol{\Sigma} \in \mathbb{R}^{n\times n}$. To find the least squares estimator of the factor via principal components, we consider the following three steps.

(i) Take the spectral decomposition of $\boldsymbol{\Sigma}$, i.e., $\boldsymbol{\Sigma} = \boldsymbol{\Gamma}\boldsymbol{\Lambda}\boldsymbol{\Gamma}'$, where $\boldsymbol{\Gamma} = (\boldsymbol{\gamma}_1, \cdots, \boldsymbol{\gamma}_n) \in \mathbb{R}^{n\times n}$, $\boldsymbol{\gamma}_i \in \mathbb{R}^n$, $\boldsymbol{\Gamma}\boldsymbol{\Gamma}' = \boldsymbol{\Gamma}'\boldsymbol{\Gamma} = \sum_{i=1}^n \boldsymbol{\gamma}_i\boldsymbol{\gamma}_i' = \mathbf{I}_n$, $\boldsymbol{\Lambda} = \text{diag}(\lambda_1, \cdots, \lambda_n) \in \mathbb{R}^{n\times n}$. Accordingly, λ_i is the i-th largest eigenvalue of $\boldsymbol{\Sigma}$ and $\boldsymbol{\gamma}_i$ is its corresponding eigenvector. Note that the transformation $Z = \boldsymbol{\Gamma}'X$ results in principal components, and $\text{Cov}(Z) = \boldsymbol{\Lambda}$.

(ii) For given m, let $\mathbf{B} = (b_{ij}) = \left(\lambda_1^{1/2}\boldsymbol{\gamma}_1, \cdots, \lambda_m^{1/2}\boldsymbol{\gamma}_m\right) \in \mathbb{R}^{n\times m}$. Accordingly, $\mathbf{B}\mathbf{B}' = \sum_{i=1}^m \lambda_i\boldsymbol{\gamma}_i\boldsymbol{\gamma}_i'$ and $\mathbf{B}'\mathbf{B} = \text{diag}\left(\lambda_1, \cdots, \lambda_m\right) \in \mathbb{R}^{m\times m}$. In addition, let $\mathbf{D} = \text{diag}(d_i)$, where $d_i = \sigma_{ii} - \sum_{j=1}^m b_{ij}^2$, and σ_{ii} is the i-th diagonal element of $\boldsymbol{\Sigma}$ for $i = 1, \cdots, n$. Based on (8.2), we obtain the population least squares estimator $\mathbf{f}_t^* = (\mathbf{B}'\mathbf{B})^{-1}\mathbf{B}'\left(X_t - \boldsymbol{\alpha}\right) = \text{diag}(\lambda_1^{-1}, \cdots, \lambda_m^{-1})\mathbf{B}'(X_t - \boldsymbol{\alpha})$, where $t = 1, \cdots, T$.

(iii) In practice, $\boldsymbol{\alpha}$ and $\boldsymbol{\Sigma}$ are unknown. One can obtain their corresponding estimators by the sample mean and the sample covariance of the data. As a result, $\widehat{\boldsymbol{\alpha}} = \bar{X} = T^{-1}\sum_{t=1}^T X_t$ and $\widehat{\boldsymbol{\Sigma}}_s = T^{-1}\sum_{t=1}^T (X_t - \bar{X})(X_t - \bar{X})'$. Subsequently, the estimators of $\lambda_i, \boldsymbol{\gamma}_i, \sigma_{ii}, b_{ij}, \boldsymbol{\Gamma}$, and $\mathbf{B}$ are followed, and we denote them $\widehat{\lambda}_i, \widehat{\boldsymbol{\gamma}}_i, \widehat{\sigma}_{ii}, \widehat{b}_{ij}, \widehat{\boldsymbol{\Gamma}}$, and $\widehat{\mathbf{B}}_p$, for $i, j = 1, \cdots, n$. Then, the least squares estimator of $\mathbf{f}_t$ is

$$\widehat{\mathbf{f}}_{pt} = \left(\widehat{\mathbf{B}}_p'\widehat{\mathbf{B}}_p\right)^{-1}\widehat{\mathbf{B}}_p'(X_t - \widehat{\boldsymbol{\alpha}}) = \text{diag}\left(\widehat{\lambda}_1^{-1}, \cdots, \widehat{\lambda}_m^{-1}\right)\widehat{\mathbf{B}}_p'(X_t - \widehat{\boldsymbol{\alpha}}),$$
$$t = 1, \cdots, T, \tag{8.4}$$

where $\widehat{\mathbf{B}}_p = \left(\widehat{\lambda}_1^{1/2}\widehat{\boldsymbol{\gamma}}_1, \cdots, \widehat{\lambda}_m^{1/2}\widehat{\boldsymbol{\gamma}}_m\right)$ is the estimator of the factor loadings. As a result, the estimators of $\mathbf{F}'$ and $\mathbf{D}$ are $\widehat{\mathbf{F}}_p' = \left(\widehat{\mathbf{f}}_{p1}, \cdots, \widehat{\mathbf{f}}_{pT}\right) = \left(\widehat{\mathbf{B}}_p'\widehat{\mathbf{B}}_p\right)^{-1}\widehat{\mathbf{B}}_p'\left(\mathbf{X}' - \mathbf{1}_T' \otimes \widehat{\boldsymbol{\alpha}}\right)$ and $\widehat{\mathbf{D}}_p = \text{diag}\left(\widehat{\boldsymbol{\Sigma}}_s - \widehat{\mathbf{B}}_p\widehat{\mathbf{B}}_p'\right)$, respectively. For the sake of simplicity, we denote $\widehat{\mathbf{F}}_p'$ the PC estimator, and define $\widehat{\mathbf{B}}_p$ and $\widehat{\mathbf{D}}_p$ as the PC-associated estimators.

It is worth noting that factor analysis focuses on covariance structure rather than variance, which is the focus of PCA. In addition, factor analysis transforms observed variables into a linear combination of hidden factors, while PCA transforms observed variables into the principal components. We next study the maximum likelihood estimation method.

8.1.2.2 Maximum Likelihood Estimation

In this section, we assume that X_t is identical and independently generated from a multivariate normal distribution with mean vector $\boldsymbol{\alpha}$ and covariance matrix $\boldsymbol{\Sigma} = \mathbf{BB}' + \mathbf{D}$ for $t = 1, \cdots, T$. The resulting log-likelihood function (omitting a constant) is

$$\ell(\boldsymbol{\alpha}, \boldsymbol{\Sigma}) = -\frac{T}{2}\log|\boldsymbol{\Sigma}| - \frac{1}{2}\sum_{t=1}^{T}(X_t - \boldsymbol{\alpha})'\boldsymbol{\Sigma}^{-1}(X_t - \boldsymbol{\alpha}). \tag{8.5}$$

It can be shown that the maximum likelihood estimator (MLE) of $\boldsymbol{\alpha}$ is $\bar{X}$. Note that $\boldsymbol{\Sigma}$ is a function of $\mathbf{B}$ and $\mathbf{D}$. After replacing $\boldsymbol{\alpha}$ in (8.5) with $\bar{X}$, we obtain that

$$\begin{aligned}\ell(\widehat{\boldsymbol{\alpha}}, \boldsymbol{\Sigma}) &= -\frac{T}{2}\log|\boldsymbol{\Sigma}| - \frac{T}{2}tr(\boldsymbol{\Sigma}^{-1}\widehat{\boldsymbol{\Sigma}}_s) \\ &= -\frac{T}{2}\log|\mathbf{BB}' + \mathbf{D}| - \frac{T}{2}tr\big((\mathbf{BB}' + \mathbf{D})^{-1}\widehat{\boldsymbol{\Sigma}}_s\big). \end{aligned} \tag{8.6}$$

Then, one can maximize (8.6) with respect to $\mathbf{B}$ and $\mathbf{D}$ and obtain the corresponding maximum likelihood estimators $\tilde{\mathbf{B}}$ and $\tilde{\mathbf{D}}$. The detailed procedure can be found in Lawley and Maxwell (1962), Jöreskog (1967), and Khamaru and Mazumder (2019). It is worth noting that the estimated loadings may not lead to reasonable interpretations. In this case, factor rotations can be used to obtain sensible and insightful findings; see, e.g., Jolliffe (2002) and Johnson and Wichern (2007).

Under some conditions, Anderson (2003) demonstrated the asymptotic properties of $\tilde{\mathbf{B}}$ and $\tilde{\mathbf{D}}$. In addition, he presented a likelihood ratio test with Bartlett's (1950) correction to test the adequacy of the m common factor model versus the zero-factor model. Specifically, if $[T - 1 - (2n + 4m + 5)/6]\log\big[|\tilde{\mathbf{B}}\tilde{\mathbf{B}}' + \tilde{\mathbf{D}}|/|\widehat{\boldsymbol{\Sigma}}_s|\big] > \chi^2_{[(n-m)^2/2-(n+m)/2],\alpha}$, then the m factor model is adequate at the α level of significance.

After obtaining the estimators of $\boldsymbol{\alpha}$, $\mathbf{B}$ and $\mathbf{D}$, we can adopt the weighted regression model (5.1) and obtain the weighted least squares estimator of $\mathbf{f}_t$ when $\mathbf{f}_t$ is non-stochastic. That is,

$$\widehat{\mathbf{f}}_t^* = (\tilde{\mathbf{B}}'\tilde{\mathbf{D}}^{-1}\tilde{\mathbf{B}})^{-1}\tilde{\mathbf{B}}'\tilde{\mathbf{D}}^{-1}(X_t - \bar{X}). \tag{8.7}$$

Suppose that $\mathbf{f}_t$ is random, and the joint distribution of $(X_t', \mathbf{f}_t')'$ is normally distributed for $t = 1, \cdots, T$, where $E(X_t) = \boldsymbol{\alpha}$, $E(\mathbf{f}_t) = \mathbf{0}$, $\text{Var}(X_t) = \mathbf{BB}' + \mathbf{D}$,

$\text{Cov}(X_t, \mathbf{f}_t) = \mathbf{B}$, $\text{Cov}(\mathbf{f}_t, X_t) = \mathbf{B}'$, and $\text{Var}(\mathbf{f}_t) = \mathbf{I}_m$. Then the conditional mean, $E(\mathbf{f}_t|X_t)$, yields the following estimator of $\mathbf{f}_t$,

$$\tilde{\mathbf{f}}_t^* = \tilde{\mathbf{B}}'(\tilde{\mathbf{B}}\tilde{\mathbf{B}}' + \tilde{\mathbf{D}})^{-1}(X_t - \bar{X}). \tag{8.8}$$

Apparently, the above two estimators are different from the PC estimator in (8.4).

8.1.2.3 Penalized Estimation

Without loss of generality, we assume that $\boldsymbol{\alpha} = \mathbf{0}$ in the rest of this chapter. Suppose there are a large number of factors in exploratory analysis, but only a few factors are relevant. Hence, it is natural to conduct sparse factor analysis. To this end, define $p_\tau(\cdot)$ a non-negative penalty function with a regularization (penalized) parameter τ. Then Choi et al. (2010) obtained lasso (Tibshirani 1996) estimators of $\mathbf{B}$ and $\mathbf{D}$ by minimizing the following penalized negative log-likelihood function with the L_1 penalty,

$$\ell(\mathbf{B}, \mathbf{D}) = \log|\mathbf{B}\mathbf{B}' + \mathbf{D}| + tr\left\{(\mathbf{B}\mathbf{B}' + \mathbf{D})^{-1}\widehat{\boldsymbol{\Sigma}}_s\right\} + \frac{1}{2T}\tau\sum_{i=1}^{n}\sum_{j=1}^{m}(|b_{ij}|), \tag{8.9}$$

where $\tau > 0$ is a penalized parameter. Denote the resulting lasso estimators by $\widehat{\mathbf{B}}_L$ and $\widehat{\mathbf{D}}_L$. To improve the performance of lasso, Choi et al. (2010) applied the approach of Zou (2006) and obtained the adaptive lasso (Alasso) estimators of $\mathbf{B}$ and $\mathbf{D}$, which can enjoy the oracle properties. Specifically, one first obtains the lasso estimators $\widehat{\mathbf{B}}_L$ and $\widehat{\mathbf{D}}_L$ based on (8.9) and then minimizes the following equation to find the Alasso estimators,

$$\ell(\mathbf{B}, \mathbf{D}) = \log|\mathbf{B}\mathbf{B}' + \mathbf{D}| + tr\left\{(\mathbf{B}\mathbf{B}' + \mathbf{D})^{-1}\widehat{\boldsymbol{\Sigma}}_s\right\} + \frac{1}{2T}\tau\sum_{i=1}^{n}\sum_{j=1}^{m}(|b_{ij}|/|\widehat{b}_{Lij}|), \tag{8.10}$$

where $\widehat{b}_{Lij}$ is the ij-th element of $\widehat{\mathbf{B}}_L$. The algorithm for computing the lasso estimator can be found in Choi et al. (2010). Although they focused on the L_1 penalty, the other penalty functions such as SCAD (Fan and Li 2001) and MCP (Zhang 2010) can be considered; see Sect. 2.2.3.

8.1.3 Factor Selection

In factor analysis, the determination of the number of factors is a critical issue. In addition to the penalized approach, Akaike (1987) and Sclove (1987) introduced the Akaike information criterion (AIC) to select the number of factors. That is,

$$\text{AIC} = T \log|\tilde{\boldsymbol{\Sigma}}| + Ttr(\tilde{\boldsymbol{\Sigma}}^{-1}\widehat{\boldsymbol{\Sigma}}_s) + 2C, \tag{8.11}$$

where $C = n(m + 1) - m(m - 1)/2$ is the number of free parameters, the number of factors m ranges from 1 to M and $M < n(n + 1)/2$, and $\tilde{\boldsymbol{\Sigma}}$ is the maximum likelihood covariance estimator calculated from a candidate model with m factors.

To avoid overfitting, one can extend Hurvich and Tsai's (1989) corrected Akaike information criterion (AICc) from regression models to factor models. It is

$$\text{AICc} = T \log|\tilde{\boldsymbol{\Sigma}}| + Ttr(\tilde{\boldsymbol{\Sigma}}^{-1}\widehat{\boldsymbol{\Sigma}}_s) + 2C + 2C(C + 1)/(T - C - 2). \tag{8.12}$$

In contrast to AIC, one can employ Schwarz's (1978) Bayesian information criterion (BIC) to select the number of factors; see Preacher et al. (2013). It is

$$\text{BIC} = T \log|\tilde{\boldsymbol{\Sigma}}| + Ttr(\tilde{\boldsymbol{\Sigma}}^{-1}\widehat{\boldsymbol{\Sigma}}_s) + \log(T)C. \tag{8.13}$$

More discussion on factor selection can be found in Preacher et al. (2013).

The above three selection criteria are based on the standard information-based criteria, $-2\log f + rC$, where $\log f$ is the log-likelihood function of the candidate model, r is a tuning parameter, and C is the number of free parameters in the candidate model. It is worth noting that Choi and Jeong (2019) employed the results in Burnham and Anderson (2004) to obtain the three corresponding criteria given below:

$$\begin{aligned}
(i)\ \text{AIC}_{CJ} &= T\sum_{i=1}^{n}\log(\widehat{\sigma}_i^2) + 2[m(n+T)+n], \\
(ii)\ \text{AICc}_{CJ} &= T\sum_{i=1}^{n}\log(\widehat{\sigma}_i^2) + nT(m+T)/(T-m-2), \quad \text{and} \\
(iii)\ \text{BIC}_{CJ} &= T\sum_{i=1}^{n}\log(\widehat{\sigma}_i^2) + \log(nT)[m(n+T)+n],
\end{aligned} \tag{8.14}$$

where $\widehat{\sigma}_i^2$ is the i-th diagonal element of $T^{-1}\sum_{t=1}^{T}\left(X_t - \widehat{\mathbf{B}\mathbf{f}_t}\right)\left(X_t - \widehat{\mathbf{B}\mathbf{f}_t}\right)'$, and $\widehat{\mathbf{B}}$ and $\widehat{\mathbf{f}}_t$ are the conditional maximum likelihood estimators obtained by maximizing the conditional log-likelihood function of $\boldsymbol{\varepsilon}_t|\mathbf{F} \sim N(\mathbf{0}, \boldsymbol{\Omega})$ for $t = 1, \cdots, T$ (see Choi 2012). Note that $\boldsymbol{\Omega}$ is a positive definite matrix, and the number of factors in the candidate model is m.

Although the above criteria are developed for standard factor models, Choi and Jeong (2019) made intensive comparisons between their criteria and those designed for approximate factor models discussed in the next section.

8.2 Approximate Factor Models

8.2.1 Model Structure

In the previous section, we assumed that the covariance matrix of random errors is diagonal. To relax this assumption, Chamberlain and Rothschild (1983) proposed an approximate factor model, which has the same model setting as that of (8.2) with weak assumptions on random (idiosyncratic) errors. Specifically, the approximate factor (AF) model is

$$X_t = \boldsymbol{\alpha} + \mathbf{B}\mathbf{f}_t + \boldsymbol{\varepsilon}_t, \tag{8.15}$$

where $\mathrm{Cov}(\boldsymbol{\varepsilon}_t) = \mathbf{D}^* = (d_{ij}^*) \in \mathbb{R}^{n\times n}$, the eigenvalues of $\mathbf{D}^*$ are uniformly bounded and $\mathrm{Cov}(X_t) = \boldsymbol{\Sigma} = \mathbf{B}\mathbf{B}' + \mathbf{D}^* \in \mathbb{R}^{n\times n}$ for $t = 1, \cdots, T$. Accordingly, $\mathbf{D}^*$ is not necessarily a diagonal matrix; see Poignard and Terada (2020). In addition, Bai and Ng (2002) allowed the random errors ϵ_{it} to be heteroscedastic and dependent across all i and t for $i = 1, \cdots, n$ and $t = 1, \cdots, T$. Thus, the resulting factor model is more general than the AF model. With a slight abuse of terminology, we still call it the AF model. We next study parameter estimators of factors and factor loadings under this model.

8.2.2 Parameter Estimation

In factor model settings, we assume that the factors, $\mathbf{f}_t$, are fixed rather than random and the intercept term is not included in model (8.1). In addition, we allow both n and $T \to \infty$. Suppose that the number of factors $m < \min\{n, T\}$. Then Bai and Ng (2002) and Bai (2003) proposed a method to estimate $\mathbf{F}$ and $\mathbf{B}$ by solving the following optimization problem:

$$Q(m) = \min_{\mathbf{F},\mathbf{B}} \frac{1}{nT} \sum_{i=1}^{n} \sum_{t=1}^{T} \left[(x_{it} - \boldsymbol{\beta}_i'\mathbf{f}_t)^2\right], \tag{8.16}$$

subject to the constraint $\mathbf{F}'\mathbf{F}/T = \mathbf{I}_m$. The above optimization problem is the same as minimizing $tr(\mathbf{F}'(\mathbf{X}\mathbf{X}')\mathbf{F})$. Hence, the resulting estimator of $\mathbf{F}$, $\widehat{\mathbf{F}}$, is $\sqrt{T}$ times the eigenvectors associated with the m largest eigenvalues of $\mathbf{X}\mathbf{X}'$, which is also a PC estimator. In addition, the estimator of the factor loading matrix is $\widehat{\mathbf{B}}' = \left(\widehat{\mathbf{F}}'\widehat{\mathbf{F}}\right)^{-1}\widehat{\mathbf{F}}'\mathbf{X} = \widehat{\mathbf{F}}'\mathbf{X}/T$. As a result, the estimator of the common component matrix is $\widehat{\mathbf{F}\mathbf{B}}'$. Moreover, the covariance matrix estimator of $\mathbf{D}^*$ is $\widehat{\mathbf{D}}^* = \left(\mathbf{X} - \widehat{\mathbf{F}\mathbf{B}}'\right)\left(\mathbf{X} - \widehat{\mathbf{F}\mathbf{B}}'\right)'/T$. Under some conditions, Bai (2003) demonstrated the consistency and asymptotic normality of the estimators $\widehat{\mathbf{f}}_t$ and $\widehat{\boldsymbol{\beta}}_i$ (up to a rotation matrix) as well as $\widehat{\boldsymbol{\beta}}_i'\widehat{\mathbf{f}}_t$, as n and $T \to \infty$, for $i = 1, \cdots, n$ and $t = 1, \cdots, T$.

It is worth noting that there is an alternative solution of (8.16) by assuming that $\mathbf{B}'\mathbf{B}/n = \mathbf{I}_m$. Then the resulting estimator of $\mathbf{B}$, $\check{\mathbf{B}}$, is $\sqrt{n}$ times the eigenvectors associated with the m largest eigenvalues of $\mathbf{X}'\mathbf{X}$. In addition, the estimator of $\mathbf{F}$ is $\check{\mathbf{F}} = \mathbf{X}\check{\mathbf{B}}/n$. Define the scaled estimator $\widehat{\mathbf{F}} = \check{\mathbf{F}}(\check{\mathbf{F}}'\check{\mathbf{F}}/T)^{1/2} = (\widehat{\mathbf{f}}_1, \cdots, \widehat{\mathbf{f}}_T)'$. Under some conditions, Bai and Ng (2002) showed the asymptotic property of $\widehat{\mathbf{F}}$. Note that parameter estimators $\check{\mathbf{F}}$ and $\check{\mathbf{B}}$ are the same as those obtained by maximizing the conditional log-likelihood function of $\boldsymbol{\varepsilon}|\mathbf{F}$ in Choi (2012), where $\boldsymbol{\Omega}$ defined in the text below (8.14) is $\mathbf{I}_n$. In addition to PC estimation approach, Daniele et al. (2025) employed Bai and Liao's (2016) quasi log-likelihood function to obtain the estimator of the AF model. To allow both n and $T \to \infty$, we next impose penalties to some parameters and obtain estimators.

8.2.3 *Penalized Estimation*

In high-dimensional approximate factor models, it is critical to consider penalized estimators. To this end, Bai and Liao (2016) proposed the following twice of the negative penalized quasi-log-likelihood function,

$$\ell(\mathbf{B}, \mathbf{D}^*) = \frac{1}{n} \log |\mathbf{B}\mathbf{B}' + \mathbf{D}^*| + \frac{1}{n} tr\big((\mathbf{B}\mathbf{B}' + \mathbf{D}^*)^{-1}\widehat{\boldsymbol{\Sigma}}_s\big) + p_\tau(\mathbf{D}^*), \tag{8.17}$$

where $p_\tau(\mathbf{D}^*)$ is a non-negative penalty function of $\mathbf{D}^*$ with a regularization (penalized) parameter τ. It is worth noting that the above penalty function focuses on $\mathbf{D}^*$ rather than $\mathbf{B}$ in (8.9). This is because $\mathbf{D}^*$ in the AF model consists of a large number of parameters. To obtain the three penalized estimators (lasso, adaptive lasso, and SCAD), Bai and Liao (2016) presented the penalty functions for the first two estimators and provided the first derivative of the penalty function of the third estimator given below:

$$\begin{aligned}
&(i)\ p_\tau(\mathbf{D}^*) = \frac{\tau}{n}\sum_{i=1}^{n}\sum_{j=1, j\neq i}^{n} \big(|d_{ij}^*|\big);\\
&(ii)\ p_\tau(\mathbf{D}^*) = \frac{\tau}{n}\sum_{i=1}^{n}\sum_{j=1, j\neq i}^{n} \big(|d_{ij}^*|/|\widehat{d}_{ij}^*|\big);\\
&(iii)\ p_\tau'(\mathbf{D}^*) = \frac{\tau}{n}\sum_{i=1}^{n}\sum_{j=1, j\neq i}^{n} \big(w_{ij}^*|d_{ij}^*|\big),
\end{aligned} \tag{8.18}$$

where $\widehat{d}^*_{ij}$ is a consistent estimator of d^*_{ij}, $w^*_{ij} = \Big[I_{\{|\widehat{\rho}^*_{ij}|\le\tau\}} + \{(a - \widehat{\rho}^*_{ij}/\tau)_+/(a - 1)\}I_{\{|\widehat{\rho}^*_{ij}|>\tau\}}\Big](\widehat{d}^*_{ii}\widehat{d}^*_{jj})^{-1/2}$, $a > 2$ (e.g., $a = 3.7$), and $b_- = b$ if $b > 0$, 0 otherwise, and $\widehat{\rho}^*_{ij} = \widehat{d}^*_{ij}/(\widehat{d}^*_{ii}\widehat{d}^*_{jj})^{1/2}$ for $i, j = 1, \cdots, n$.

A possible consistent estimator of d^*_{ij} is $T^{-1}\sum_{t=1}^{T} e_{it}e_{jt}$, where $e_{it} = y_{it} - \tilde{\boldsymbol{\beta}}_i'\tilde{\mathbf{f}}_t$, and $\tilde{\boldsymbol{\beta}}_i$ and $\tilde{\mathbf{f}}_t$ are the consistent estimators of $\boldsymbol{\beta}_i$ and $\mathbf{f}_t$ given in Sect. 8.1.2.1. Denote the resulting penalized estimator of $\mathbf{D}^*$ by $\widehat{\mathbf{D}}^*_p$. Under some conditions, Bai and Liao (2016) demonstrated the consistency of $\widehat{\mathbf{D}}^*_p$ in terms of the weighted Frobenius norm. In addition, they employed the majorize-minimize EM algorithm to implement the penalized estimators. Moreover, they applied K-fold cross-validations to choose the tuning parameter.

In approximate factor models, the covariance of random errors $\mathbf{D}^*$ is usually not diagonal. Then we can employ the covariance regression method discussed in Chap. 4 to link $\mathbf{D}^*$ to a set of similarity matrices, which not only yields more insightful findings but also broadens the usefulness of AF models. It is also worth noting that approximate factor models have attracted increasing attention across various subjects in factor analysis. In addition to the above topics discussed in this section, the readers may explore the following subjects that are related to AF models: high-dimensional covariance matrix estimation (Fan et al. 2011); maximum likelihood estimation and inference (Bai and Liao 2016); robust covariance estimation (Fan et al. 2018); rank-regularized estimation (Bai and Ng 2019); factor-adjusted regularized model selection (Fan et al. 2020); weaker loadings (Bai and Ng 2023); and high-dimensional portfolio selection (Daniele et al. 2025).

8.2.4 Factor Selection

In approximate factor models, Bai and Ng (2002) introduced the following three different factor selection criteria:

$$
\begin{aligned}
&(i)\ IC_1 = \log(Q(m)) + m\left(\frac{n+T}{nT}\right)\log\left(\frac{nT}{n+T}\right);\\
&(ii)\ IC_2 = \log(Q(m)) + m\left(\frac{n+T}{nT}\right)\log\left(C_{nT}^2\right);\\
&(ii)\ IC_3 = \log(Q(m)) + m\left(\frac{\log(C_{nT}^2)}{C_{nT}^2}\right),
\end{aligned}
\tag{8.19}
$$

where $Q(m)$ is defined in (8.16) for the given order m, and $C_{nT}^2 = \min\{n, T\}$. By minimizing any of these three criteria, one can obtain the estimator of the factor order, and denote it $\widehat{m}$. Bai and Ng (2002) showed that $\widehat{m}$ is a consistent estimator of the true number of factors, m_0, i.e., $P(\widehat{m} = m_0) \to 1$ as n and $T \to \infty$.

8.3 Factor-Augmented Regression Models

In high-dimensional regression analysis, there are many explanatory variables (including lag variables) available for exploring relationships with response variables. Hence, it is natural to employ the factor model approach to reduce the explanatory variables to the low-dimension factors and then fit the responses vector versus the resulting factors in combination with the remaining explanatory variables. Accordingly, the factor-augmented regression (FAR) model has been introduced and studied; see, e.g., Stock and Watson (2002a), Bai and Ng (2006), Pesaran (2006), Bai (2009), Greenaway-McGrevy et al. (2012), Wang (2012, 2022), Goncalves and Perron (2014), Cheng and Hansen (2015), Juodis and Sarafidis (2022), and Fan et al. (2024).

One important goal of the FAR model is forecasting. To this end, we follow Stock and Watson's (2002a) model setting to introduce the model given below. For $t = 1, \cdots, T$, consider that

$$X_t = \mathbf{B}\mathbf{f}_t + \boldsymbol{\varepsilon}_t \quad \text{and} \tag{8.20}$$

$$y_{t+h} = \boldsymbol{\alpha}'_f \mathbf{f}_t + \boldsymbol{\alpha}'_v V_t + u_{t+h}, \tag{8.21}$$

where (8.20) is the same as (8.15) when the intercept term is omitted, $y_t \in \mathbb{R}$ is the response variable, y_{t+h} is the forecast with horizon h, $\boldsymbol{\alpha}_f \in \mathbb{R}^m$ and $\boldsymbol{\alpha}_v \in \mathbb{R}^\nu$ are unknown parameter vectors, $V_t \in \mathbb{R}^\nu$ is the vector of observed explanatory variables, $\boldsymbol{\varepsilon}_t$ is a random error in (8.15) and u_{t+h} is the forecast error. The goal of this model is to forecast y_{T+h}.

Before making forecasts, we need to study the properties of parameter estimators. Under appropriate assumptions, Bai (2003) showed that the estimator $\widehat{\mathbf{f}}_t$ obtained from (8.16) is consistent for $t = 1, \cdots, T$. Subsequently, we can fit the linear regression of y_{t+h} on $\widehat{\mathbf{f}}_t$ and V_t with $t = 1, \cdots, T-h$. Denote the resulting least squares estimators of $\boldsymbol{\alpha}_f$ and $\boldsymbol{\alpha}_v$ by $\widehat{\boldsymbol{\alpha}}_f$ and $\widehat{\boldsymbol{\alpha}}_v$, respectively. Let $\widehat{\boldsymbol{\alpha}}_{fv} = (\widehat{\boldsymbol{\alpha}}'_f, \widehat{\boldsymbol{\alpha}}'_v)'$. Under some conditions, Stock and Watson (2002a) demonstrated the consistency of $\widehat{\boldsymbol{\alpha}}_{fv}$, as n and $T \to \infty$. Later, Bai and Ng (2006) showed that if $\sqrt{T}/n \to 0$, both n and $T \to \infty$, and some conditions hold, then $\sqrt{T}(\widehat{\boldsymbol{\alpha}}_{fv} - \boldsymbol{\alpha}_{fv0})$ is asymptotically normal, where $\boldsymbol{\alpha}_{fv0}$ is the true parameter.

It is worth noting that the above parameter estimators and their associate theoretical properties rely on the number of factors m. Although m is unknown in practice, we can employ the selection methods in Sect. 8.2.4 to determine the number of factors. After obtaining regression parameter estimators, we next study predictions. Let $Z_t = (\mathbf{f}'_t, V'_t)' \in \mathbb{R}^{(m+\nu)}$, and assume that the mean of u_{t+h} conditional on past observations is zero. Based on the mean-squared criterion, the optimal prediction at period $T + h$ is

$$y_{T+h|T} = E(y_{T+h}|Z_T, Z_{T-1}, \cdots) = \boldsymbol{\alpha}'_f \mathbf{f}_T + \boldsymbol{\alpha}'_v V_T = \boldsymbol{\alpha}'_{fv} Z_T. \tag{8.22}$$

Accordingly, the estimator of the optimal prediction at period $T + h$ is

$$\widehat{y}_{T+h|T} = \widehat{\boldsymbol{\alpha}}'_f \tilde{\mathbf{f}}_T + \widehat{\boldsymbol{\alpha}}'_v V_T = \widehat{\boldsymbol{\alpha}}'_{f\iota} \tilde{Z}_T, \tag{8.23}$$

where $\tilde{Z}_T = (\tilde{\mathbf{f}}'_T, V'_T)'$. In order to make inferences regarding predictions, Bai and Ng (2006) demonstrated that if $\sqrt{T}/n \to 0$, n and $T \to \infty$, and some conditions hold, then $(\widehat{y}_{T+h|T} - y_{T+h|T})/\sqrt{\text{var}(\widehat{y}_{T+h|T})}$ is asymptotically standard normal. The detailed expression of $\text{var}(\widehat{y}_{T+h|T})$ can be found in Bai and Ng (2006), and the overall convergence rate of $\widehat{y}_{T+h|T}$ is $\min\{\sqrt{n}, \sqrt{T}\}$. Based on their findings, we can construct prediction intervals that can help practitioners to make better decisions.

Instead of using FAR model to make forecasts, Fan et al. (2024) introduced the following factor augmented sparse linear regression (FASR) model to incorporate sparse explanatory variables into the FAR model.

$$X_t = \mathbf{B}\mathbf{f}_t + \boldsymbol{\varepsilon}_t \quad \text{and} \tag{8.24}$$

$$y_t = \boldsymbol{\alpha}'_f \mathbf{f}_t + \boldsymbol{\alpha}'_\varepsilon \boldsymbol{\varepsilon}_t + u_t \tag{8.25}$$

for $t = 1, \cdots, T$, where the definitions of y_t, X_t, $\mathbf{B}$, f_t, $\boldsymbol{\alpha}_f$, and $\boldsymbol{\varepsilon}_t$ are the same as those defined in (8.20) and (8.21), $\boldsymbol{\alpha}_\varepsilon \in \mathbb{R}^n$ is an unknown parameter vector, and u_t is the random noise. In addition, the number of variables n can be much larger than the sample size T. We can re-express the FASR model in the vector and matrix form below:

$$\mathbf{X} = \mathbf{F}\mathbf{B}' + \boldsymbol{\varepsilon} \quad \text{and} \tag{8.26}$$

$$Y = \mathbf{F}\boldsymbol{\alpha}_f + \boldsymbol{\varepsilon}\boldsymbol{\alpha}_\varepsilon + U, \tag{8.27}$$

where $\mathbf{X}$, $\mathbf{F}$ and $\boldsymbol{\varepsilon} = (\boldsymbol{\varepsilon}_1, \cdots, \boldsymbol{\varepsilon}_T)'$ are defined before (8.3), $Y = (y_1, \cdots, y_T)'$, and $U = (u_1, \cdots, u_T)' \in \mathbb{R}^T$. In fact, the above model can be re-expressed as follows:

$$Y = \mathbf{F}\boldsymbol{\alpha}_d + \mathbf{X}\boldsymbol{\alpha}_\varepsilon + U, \tag{8.28}$$

where $\boldsymbol{\alpha}_d = \boldsymbol{\alpha}_f - \mathbf{B}'\boldsymbol{\alpha}_\varepsilon$. Note that if $\boldsymbol{\alpha}_d = \mathbf{0}$ or $\boldsymbol{\alpha}_\varepsilon = \mathbf{0}$, then we have the sparse linear model or the factor model, respectively. Accordingly, the FASR model encompasses both the sparse regression and dimension reduction.

To estimate unknown parameters $\boldsymbol{\alpha}_f$ and $\boldsymbol{\alpha}_\varepsilon$, Fan et al. (2024) proposed to first estimate $\mathbf{F}$, $\mathbf{B}$, and $\boldsymbol{\varepsilon}$ by minimizing the following objective function under two constraints, $\mathbf{F}'\mathbf{F}/T = \mathbf{I}_m$ and $\mathbf{B}'\mathbf{B}$ is diagonal. Accordingly,

$$\left(\widehat{\mathbf{F}}, \widehat{\mathbf{B}}\right) = \arg\min_{\mathbf{F},\mathbf{B}} ||\mathbf{X} - \mathbf{F}\mathbf{B}'||_F^2 \quad \text{and} \tag{8.29}$$

$$\widehat{\boldsymbol{\varepsilon}} = \mathbf{X} - \widehat{\mathbf{F}}\widehat{\mathbf{B}}'. \tag{8.30}$$

Specifically, the columns of $\widehat{\mathbf{F}}$ are $\sqrt{T}$ times the eigenvectors associated with the largest m eigenvalues of $\mathbf{X}\mathbf{X}'$. In addition, $\widehat{\mathbf{B}}' = \left(\widehat{\mathbf{F}}'\widehat{\mathbf{F}}\right)^{-1}\widehat{\mathbf{F}}'\mathbf{X} = \widehat{\mathbf{F}}'\mathbf{X}/T$ and $\widehat{\boldsymbol{\varepsilon}} =$

$\left(\mathbf{I}_T - \widehat{\mathbf{F}}\widehat{\mathbf{F}}'/T\right)\mathbf{X}$. To extend this model for high-dimensional data, one can adopt the penalized approach as discussed in Sects. 8.1.2.3 and 8.2.3 and obtain the penalized (regularized) estimator of $\boldsymbol{\alpha}_f$ and $\boldsymbol{\alpha}_\varepsilon$ via the penalized objective function below.

$$\left(\widehat{\boldsymbol{\alpha}}_f, \widehat{\boldsymbol{\alpha}}_{\varepsilon,\tau}\right) = \arg\min_{\boldsymbol{\alpha}_f, \boldsymbol{\alpha}_\varepsilon} \frac{1}{2T} \|Y - \widehat{\mathbf{F}}\boldsymbol{\alpha}_f - \widehat{\boldsymbol{\varepsilon}}\boldsymbol{\alpha}_\varepsilon\|_2^2 + p_\tau\left(\boldsymbol{\alpha}_\varepsilon\right), \tag{8.31}$$

where $p_\tau\left(\boldsymbol{\alpha}_\varepsilon\right)$ is a non-negative penalty function of $\boldsymbol{\alpha}_\varepsilon$ with a penalized parameter $\tau > 0$. As mentioned in Sect. 8.2.3, we can consider the three known penalized estimators (lasso, adaptive lasso, and SCAD) although Fan et al. (2024) focused on lasso. After algebraic simplification, the solution of (8.31) under the L_1 penalty is equivalent to

$$\widehat{\boldsymbol{\alpha}}_{\varepsilon,\tau} = \arg\min_{\boldsymbol{\alpha}_\varepsilon} \frac{1}{2T} ||\tilde{Y} - \widehat{\boldsymbol{\varepsilon}}\boldsymbol{\alpha}_\varepsilon||_2^2 + \tau||\boldsymbol{\alpha}_\varepsilon||_1 \text{ and} \tag{8.32}$$

$$\widehat{\boldsymbol{\alpha}}_f = (\widehat{\mathbf{F}}'\widehat{\mathbf{F}})^{-1}\widehat{\mathbf{F}}'Y = \widehat{\mathbf{F}}'Y/T, \tag{8.33}$$

where $\tilde{Y} = (\mathbf{I}_T - \widehat{\mathbf{F}}\widehat{\mathbf{F}}'/T)Y$.

Under appropriate conditions, Fan et al. (2024) demonstrated the asymptotic properties of estimators $\widehat{\mathbf{F}}$, $\widehat{\mathbf{B}}$, $\widehat{\boldsymbol{\varepsilon}}$, $\widehat{\boldsymbol{\alpha}}_f$, and $\widehat{\boldsymbol{\alpha}}_{\varepsilon,\tau}$. In addition, they proposed a test to assess whether $\boldsymbol{\alpha}_\varepsilon = 0$ and developed a test to examine the adequacy of the FASR model (8.28) by testing whether $\boldsymbol{\alpha}_d = 0$. In practice, the true number of factors, m, is unknown. We can employ either the Bai and Ng's (2002) approach discussed in Sect. 8.2.4 or the eigenvalue ratio method (see, e.g., Lam and Yao 2012; Wang 2012; Ahn and Horenstein 2013; Fan et al. 2024) to select the number of factors.

In addition to the above extension of the FAR model, we can integrate the network structure in Sect. 7.3 to factor model settings discussed before. We can also employ the covariance regression method discussed in Chap. 4 to link the covariance matrix $\mathbf{D}^*$ to a set of similarity matrices. Furthermore, there are some refinements and extensions of the factor-augmented regression models being proposed, such as quadratic principal components (Bai and Ng 2008a), bootstrap prediction intervals (Goncalves and Perron 2014), vector autoregressions (Bai et al. 2016), sufficient forecasts (Fan et al. 2017), functional coefficients (Li et al. 2020), threshold effects (Yan and Cheng 2022), structural changes (Massacci and Kapetanios 2024), and sparse idiosyncratic components (Beyhum and Striaukas 2024). We next study a related subject, dynamic factor models.

8.4 Dynamic Factor Models

The last three sections mainly focused on factor analysis with static factors. Due to data structures and enrichment, however, the factors can be dynamic; see the pioneering works of Geweke (1977) and Sargent and Sims (1977). Hence, it is not surprising that the dynamic factor model (DFM) has attracted considerable attention in the last two decades across various fields such as economics, finance, marketing, and machine learning. Some review references can be found in Bai and Ng (2008b), Barhoumi et al. (2013), Stock and Watson (2016), and Doz and Fuleky (2020).

In general, dynamic factor models can be presented in two different forms, dynamic and static. The dynamic form is given as follows:

$$\begin{aligned}\text{(i) } X_t &= \mathbf{B}(L)\mathbf{f}_t + \boldsymbol{\varepsilon}_t \\ &= \mathbf{B}_0\mathbf{f}_t + \mathbf{B}_1\mathbf{f}_{t-1} + \cdots + \mathbf{B}_q\mathbf{f}_{t-q} + \boldsymbol{\varepsilon}_t \text{ and} \end{aligned} \tag{8.34}$$

$$\begin{aligned}\text{(ii) } \mathbf{f}_t &= \mathbf{C}(L)\mathbf{f}_{t-1} + \boldsymbol{\eta}_t \\ &= \mathbf{C}_1\mathbf{f}_{t-1} + \cdots + \mathbf{C}_p\mathbf{f}_{t-p} + \boldsymbol{\eta}_t, \end{aligned} \tag{8.35}$$

where $X_t = (x_{1t}, \cdots, x_{nt})' \in \mathbb{R}^n$, $\mathbf{f}_t = (f_{1t}, \cdots, f_{mt})' \in \mathbb{R}^m$ and $\boldsymbol{\varepsilon}_t = (\epsilon_{1t}, \cdots, \epsilon_{nt})' \in \mathbb{R}^n$ are defined in Sect. 8.1.1, $\mathbf{f}_{t-g} = (f_{1,t-g}, \cdots, f_{m,t-g})' \in \mathbb{R}^m$ for $g = 1, \cdots, q(p)$, $\mathbf{B}(L) = \sum_{k=0}^{q} \mathbf{B}_k L^k$, L is the backshift operator and L^k represents the backshift operator repeated k times, $\mathbf{B}_k = (\beta_{ij}^k) \in \mathbb{R}^{n\times m}$ for $i = 1, \cdots, n$, $j = 1, \cdots, m$, and $k = 0, \cdots, q$, $\mathbf{C}(L) = \sum_{l=1}^{p} \mathbf{C}_l L^{l-1}$, $\mathbf{C}_l = (c_{ij}^l) \in \mathbb{R}^{m\times m}$ for $i = 1, \cdots, m$, $j = 1, \cdots, m$, and $l = 1, \cdots, p$, and $\boldsymbol{\eta}_t = (\eta_{1t}, \cdots, \eta_{mt})' \in \mathbb{R}^m$ is the random error of the vector autoregressive model (VAR) in (8.35).

When n is fixed and $n \ll T$, we make the same assumptions of $\mathbf{f}_t$ and $\boldsymbol{\varepsilon}_t$ as in the classical factor model of Sect. 8.1.1. That is, $E(\mathbf{f}_t) = \mathbf{0}$, $\text{Cov}(\mathbf{f}_t) = \mathbf{I}_m$, $\text{Cov}(\mathbf{f}_t, \boldsymbol{\varepsilon}_t) = \mathbf{0}_{m\times n}$ and $\text{Cov}(\boldsymbol{\varepsilon}_t) = \mathbf{D} = \text{diag}(d_i) \in \mathbb{R}^{n\times n}$ for $t = 1, \cdots, T$. In addition, we assume that $E(\boldsymbol{\eta}_t) = 0$ and $\text{Cov}(\boldsymbol{\eta}_t) = \mathbf{Q}$, where $\mathbf{Q}$ is a positive definite matrix, and $\text{Cov}(\boldsymbol{\eta}_{t-v}, \boldsymbol{\varepsilon}_t) = \mathbf{0}_{m\times n}$ for $t = 1, \cdots, T$ and any $t - T \le v \le t - 1$. If both n and T are allowed to increase, one can consider penalized estimations as discussed in Sects. 8.1.2.3 and 8.2.3.

In order to estimate unknown quantities, we can consider the static form given below. Let $\mathcal{F}_t = (\mathbf{f}_t', \mathbf{f}_{t-1}', \cdots, \mathbf{f}_{t-q}')' \in \mathbb{R}^r$, where $r = m(1+q)$. In addition, let $\mathbf{B} = (\mathbf{B}_0, \mathbf{B}_1, \cdots, \mathbf{B}_q) \in \mathbb{R}^{n\times r}$. Then the static representation of (i) in (8.34) is

$$X_t = \mathbf{B}\mathcal{F}_t + \boldsymbol{\varepsilon}_t. \tag{8.36}$$

Assume that $q \ge p - 1$. Let $\mathbf{H} = (\mathbf{I}_m, \mathbf{0}_m, \cdots, \mathbf{0}_m)' \in \mathbb{R}^{r\times m}$ and $\mathbf{G}$ be an $r \times r$ matrix. The elements of $\mathbf{G}$ are defined by: (i) $\text{diag}(\mathbf{G}) = (\mathbf{C}_1, \mathbf{0}_m, \cdots, \mathbf{0}_m)' \in \mathbb{R}^{r\times r}$; (ii) the first lower-off diagonal matrices are composed of the $(r - m)$s $\mathbf{I}_m$; (iii) the first row matrices are $(\mathbf{C}_1, \cdots, \mathbf{C}_p, \mathbf{0}_m, \cdots, \mathbf{0}_m) \in \mathbb{R}^{m\times r}$, where $\mathbf{C}_1$ also appeared in (i); and (iv) the rest of the elements are 0. Then we can re-express (ii) in (8.35) as

$$\mathcal{F}_t = \mathbf{G}L\mathcal{F}_t + \mathbf{H}\boldsymbol{\eta}_t, \tag{8.37}$$

for $t = 1, \cdots, T$. This expression can be analogously obtained when $q < p - 1$.

Equations (8.36) and (8.37), together with an additional autocorrelation assumption on $\boldsymbol{\varepsilon}_t$, allow us to form a state space model. We then apply the Kalman filter (Kalman 1960) method to estimate unknown parameters in dynamic factor models when n is fixed and small; see, e.g., Stock and Watson (2016) and Doz and Fuleky (2020). As a result, maximum likelihood estimators are obtained. Some useful reference books of state space models that apply the Kalman filter can be found in Harvey (1990), Tsay (2010), Durbin and Koopman (2012), and Shumway and Stoffer (2017). In high-dimensional data where n is large, we can consider a two-step hybrid method introduced by Doz et al. (2011). To ensure that the resulting estimators are robust against the Gaussian distribution, one can consider the quasi-maximum likelihood approach; see, e.g., Doz et al. (2012), Bai and Liao (2016), and Stock and Watson (2016). In addition, Barigozzi and Luciani (2019) studied the theoretical properties of such estimators for large dynamic factor models. Empirical applications of DFM can be found in the literature such as Stock and Watson (2005), Bai and Ng (2007), and Stock and Watson (2016).

Note that the estimation of dynamic factor models depends on the values of q and p in (8.34) and (8.35), respectively. To determine the value of q, one can consider model selection methods, such as the selection criterion (IC) in (8.19). In addition, the VAR model selection criteria, such as AIC, BIC, and Hurvich and Tsai's (1993) corrected Akaike information criterion (AICc), can be used to select the number of lags p. Applying these two types of selection criteria, one can choose the best DFM. In addition, Bai and Ng (2007) proposed a method to determine the optimal number of dynamic factors. Moreover, it is of interest to study two hybrid models. One integrates the dynamic model with the FAR model in Sect. 8.3, and the other incorporates DFM with the SAR model discussed in Chap. 7. Finally, the DFM (or the state-space model) has been introduced into the field of machine learning; see, e.g., Bontempi et al. (2017), Rangapuram et al. (2018), de Stefani et al. (2019), and Andreini et al. (2020). We will discuss machine learning in Chap. 9.

8.5 Matrix Factor Models

In the above sections, we have studied factor analysis based on the vector X_t. Due to data availability and practical demands, it becomes essential to explore matrix factor analysis. To this end, Wang et al. (2019) and Chen and Fan (2023) studied matrix factor models (MFM) for high-dimensional data. The former article focused on time series data, while the latter one considered a broad scope of data structures. This section introduces MFM mainly based on the results of Chen and Fan (2023).

Let $\mathbf{X}_t$ be a $n_1 \times n_2$ matrix of observed variables for $t = 1, \cdots, T$, where n_1, n_2 and T are allowed to go to ∞. In addition, the dimension of $\mathbf{X}_t$ is comparable or greater than T. Consider the matrix factor model below:

$$\mathbf{X}_t = \mathbf{R}\mathbf{F}_t\mathbf{C}' + \mathbf{E}_t \quad (t = 1, \cdots, T), \tag{8.38}$$

where $\mathbf{F}_t$ is a $m_1 \times m_2$ latent factor matrix, $\mathbf{R}$ and $\mathbf{C}$ are a $n_1 \times m_1$ and a $n_2 \times m_2$ loading matrices, respectively. In addition, $\mathbf{E}_t$ is an error matrix that is uncorrelated with $\mathbf{F}_t$, but it can be weakly correlated across rows, columns, and observations. To estimate unknown parameters, Chen and Fan (2023) encapsulated information from the sample mean and sample covariance in definitions of the following two statistics:

$$\widehat{\mathbf{M}}_R = \frac{1}{n_1 n_2}\left[(1+\alpha)\bar{\mathbf{X}}\bar{\mathbf{X}}' + \frac{1}{T}\sum_{t=1}^{T}(\mathbf{X}_t - \bar{\mathbf{X}})(\mathbf{X}_t - \bar{\mathbf{X}})'\right] \text{ and} \tag{8.39}$$

$$\widehat{\mathbf{M}}_C = \frac{1}{n_1 n_2}\left[(1+\alpha)\bar{\mathbf{X}}'\bar{\mathbf{X}} + \frac{1}{T}\sum_{t=1}^{T}(\mathbf{X}_t - \bar{\mathbf{X}})'(\mathbf{X}_t - \bar{\mathbf{X}})\right], \tag{8.40}$$

where $\alpha \in [-1, \infty)$, $\bar{\mathbf{X}} = T^{-1}\sum_{t=1}^{T}\mathbf{X}_t$ is the sample mean, $\bar{\mathbf{X}}\bar{\mathbf{X}}'$ and $\bar{\mathbf{X}}'\bar{\mathbf{X}}$ are correspondingly to the sample row and sample column squared means, and $T^{-1}\sum_{t=1}^{T}(\mathbf{X}_t - \bar{\mathbf{X}})(\mathbf{X}_t - \bar{\mathbf{X}})'$ and $T^{-1}\sum_{t=1}^{T}(\mathbf{X}_t - \bar{\mathbf{X}})'(\mathbf{X}_t - \bar{\mathbf{X}})$ are the sample row and sample column covariance matrices, respectively. Subsequently, they proposed the α-PCA estimators of $\mathbf{R}$ and $\mathbf{C}$ as follows: $\widehat{\mathbf{R}}$ is $\sqrt{n_1}$ times the first m_1 largest eigenvectors of $\widehat{\mathbf{M}}_R$, and $\widehat{\mathbf{C}}$ is $\sqrt{n_2}$ times the first m_2 largest eigenvectors of $\widehat{\mathbf{M}}_C$. To assure model identification, the estimators need to satisfy constraints: $n_1^{-1}\widehat{\mathbf{R}}'\widehat{\mathbf{R}} = \mathbf{I}_{m_1}$ and $n_2^{-1}\widehat{\mathbf{C}}'\widehat{\mathbf{C}} = \mathbf{I}_{m_2}$. Note that the α-PCA estimators are approximately the least squares and maximum likelihood estimators; see detailed illustrations in Chen and Fan (2023).

Based on $\widehat{\mathbf{R}}$ and $\widehat{\mathbf{C}}$, we can employ (8.38) and obtain the estimator of factor matrix $\mathbf{F}_t$ below,

$$\widehat{\mathbf{F}}_t = \frac{1}{n_1 n_2}\widehat{\mathbf{R}}'\mathbf{X}_t\widehat{\mathbf{C}}. \tag{8.41}$$

Accordingly,

$$\widehat{\mathbf{X}}_t = \frac{1}{n_1 n_2}\widehat{\mathbf{R}}\widehat{\mathbf{R}}'\mathbf{X}_t\widehat{\mathbf{C}}\widehat{\mathbf{C}}'. \tag{8.42}$$

Note that here $\widehat{\mathbf{X}}_t$ is the estimator of $\mathbf{R}\mathbf{F}_t\mathbf{C}'$ in (8.38). Under appropriate assumptions, Chen and Fan (2023) demonstrated the theoretical properties of $\widehat{\mathbf{R}}$, $\widehat{\mathbf{C}}$, $\widehat{\mathbf{F}}_t$, and $\widehat{\mathbf{X}}_t$ when both m_1 and m_2 are known and fixed.

In practice, m_1 and m_2 are unknown. To this end, we can construct an objective function,

$$Q(m_1, m_2) = ||\mathbf{X}_t - \widehat{\mathbf{X}}_t|_2^2, \tag{8.43}$$

where $\widehat{\mathbf{X}}_t$ is obtained for given m_1 and m_2, and $1 \leq m_1 \leq n_1/2$ and $1 \leq m_2 \leq n_2/2$. Then, we can adapt Bai and Ng's (2002) method discussed in Sect. 8.2.4 to select $\widehat{m}_1$ and $\widehat{m}_2$. For example, by IC_1 defined in (8.19), we define the following row and column selection criteria:

$$IC_{1R}(m_1) = \log(Q(m_1, m_2)) + m_1\Big(\frac{n_1 + T}{n_1 T}\Big)\log\Big(\frac{n_1 T}{n_1 + T}\Big), \text{ for given } m_2, \text{ and} \tag{8.44}$$

$$IC_{1C}(m_2) = \log(Q(m_1, m_2)) + m_2\Big(\frac{n_2 + T}{n_2 T}\Big)\log\Big(\frac{n_2 T}{n_2 + T}\Big), \text{ for given } m_1. \tag{8.45}$$

Then, we can obtain $\widehat{m}_1$ and $\widehat{m}_2$ by iteratively searching the minimum values via Eqs. (8.44) and (8.45) until their changes are less than a pre-specified tolerance.

An alternative approach for obtaining the estimators of m_1 and m_2 is based on the eigenvalue ratio method; see, e.g., Chen and Fan (2023). Let $\widehat{\lambda}_1 \geq \cdots \geq \widehat{\lambda}_{m_1^*}$ be the first m_1^* largest eigenvalues of $\widehat{\mathbf{M}}_R$, where $m_1^* = [n_1/2]$ (or $m_1^* = [n_1/3]$). Then, the eigenvalue ratio estimator of m_1 is

$$\widehat{m}_1 = \arg \max_{1 \leq j \leq m_1^*} \frac{\widehat{\lambda}_j}{\widehat{\lambda}_{j+1}}. \tag{8.46}$$

Analogously, we can obtain $\widehat{m}_2$ based on the eigenvalues of $\widehat{\mathbf{M}}_C$. Finally, the selection of tuning parameter α has been discussed in Remark 7 of Chen and Fan (2023).

In addition to the α-PCA approach, Wang et al. (2019) employed the auto-covariance setting of $\mathbf{X}_t$ to analyze factor matrix model (8.38). This is because they mainly focus on high-dimensional time series data. For the sake of simplicity, they assumed that $\mathbf{E}_t$ is a white noise for $t = 1, \cdots, T$. On the other hand, Chen and Fan's (2023) approach is applicable as $\mathbf{E}_t$ is weakly correlated across rows, columns, and observations. To further broaden applicability, it would be of interest to extend their method from matrix to higher-order tensor, although several articles have studied high-dimensional tensor time series; see, e.g., Chen et al. (2022) and Xu et al. (2024). We will discuss tensor analysis in Chap. 10. It is also worth noting that if the factors $\mathbf{F}_t$ in matrix factor models are observed covariates, the resulting models are matrix-variate regressions (see, e.g., Ding and Cook 2018). Moreover, substituting $\mathbf{F}_t$ with the time lag of $\mathbf{X}_t$ leads to matrix-variate time series models; see, e.g., a good review article by Tsay (2024). Finally, Yu et al. (2024) extended dynamic factor models to dynamic matrix factor models with high dimensional time series.

8.6 Empirical Examples

This section introduces three examples that illustrate the application of factor models. The first example was studied by Bai and Ng (2002), who determined the number of factors in the approximate factor (AF) model. The data, obtained from the CRSP data base, comprise monthly returns of the stocks of 8436 firms from 1994:1 to 1998:12. There are 4883 firms whose stock returns are available for all of the 60 months. Then Bai and Ng (2002) employed the criteria, IC_1 and IC_2, in (8.19) to select factors, and they found that the optimal choice is two factors when the maximal number of candidate factors is 15.

The second example is for the factor-augmented regression (FAR) model, which has been analyzed by Stock and Watson (2002a,b). The observed $\mathbf{X}_t$ are 149 monthly macroeconomy variables collected from January 1959 to December 1998; see details in Stock and Watson (2002b). In addition, an 12-month-ahead variable is $y_{t+12} = \ln(IP_{t+12}/IP_t)$, where IP_t is the index of industrial production at month t. This example focuses on employing model (8.22) without including explanatory variables V_t to make 12 months ahead forecasts starting in 1970:1 and ending in 1997:12. For example, to compute the forecasts in $T = 1970{:}1$, we first use the data from $t = 1959{:}1$ to 1970:1 to form $\mathbf{X}_t$, and then fit model (8.20) to obtain PC estimates $\widehat{\mathbf{f}}_t$ via Eq. (8.16). Subsequently, the factor selection criteria on Eq. (8.19) can be applied to select the number of factors, $\widehat{m}$. Then, the resulting estimates $\widehat{\mathbf{f}}_t$, together with y_{t+12} ($t = 1959{:}1$–1969:1), are used to compute $\widehat{\boldsymbol{\alpha}}_f$ via Eq. (8.22). Finally, calculate the forecasts, $\widehat{y}_{t+12} = \widehat{\boldsymbol{\alpha}}'_f \widehat{\mathbf{f}}_t$ for $t = 1969{:}1$–$t = 1969{:}12$. This process is repeated for 1970:2–1997:12. To measure the accuracy of forecasts, the mean squared error is considered. Based on Stock and Watson's (2002a) analyses, they concluded that the AF model with $\widehat{m} = 2$ or $\widehat{m} = 3$ performs the best in comparing with other methods. Since comparing predictive accuracy has attracted considerable attention, it would be useful to study some related articles such as Diebold and Mariano (2002), Giacomini and White (2006), Diebold (2015), Coroneo and Iacone (2020), and Lan et al. (2024).

The third example taken from Chen and Fan (2023) consists of two known image data sets. The first one is the ORL database of faces, which contains a set of 40 persons' face images taken between April 1992 and April 1994. Accordingly, there are 400 images and the size of each image is 92×112 pixels. The second one is from normalized handwritten digits, which was automatically scanned from envelopes by the U.S. Postal Service (USPS). It consists of 9298 handwritten digits of "0" to "9," while Chen and Fan (2023) used a subset of the whole dataset that contains 300 images for each digit resulting in a total of 3000 images. The resolution of images is 16×16.

By applying the α-PCA method to these two datasets, Chen and Fan (2023) obtained the estimators $\widehat{\mathbf{X}}_t$, which can be viewed as the compressions of the original images. They found that $\alpha = 0$ yields the best compression results.

Chapter 9
Machine Learning and Covariance Matrices

This chapter first reviews the three types of machine learning: supervised learning, unsupervised learning, and semi-supervised learning, and briefly discusses reinforcement learning. Subsequently, we introduce three types of deep learning methods: Convolution Neural Networks (CNNs), Graph Convolutional Networks (GCNs), and Transformers, and we discuss their relationships with covariance matrices. We then introduce the concept of transfer learning and its usefulness in high-dimensional covariance analysis. An empirical example is presented to illustrate the application of deep learning.

9.1 Machine Learning

In the past decade, "Machine Learning" has not just become a popular term, it has actually impacted our daily lives in the areas of dining, transportation, clothing, housing, entertainment, education, medicine, healthcare, and others. This is mainly due to advanced technologies, demands for convenience, and market disruptions. It may not be surprising that machine learning, and the closely associated areas of big data and AI, could play such a revolutionary role in human history. The major goal of machine learning is converting data into valuable information such as predictions, classifications, and clusters. Accordingly, machine learning can be viewed as an application of AI by modeling data to assist computer systems to learn some specific tasks by themselves without being explicitly programmed.

We next review the three types of machine learning: supervised learning, unsupervised learning, and semi-supervised learning. Useful reference books about machine learning include Clarke et al. (2009), Hastie et al. (2009), Zhu and Goldberg (2009), James et al. (2013), Shalev-Shwartz and Ben-David (2014), Minh and Murino (2017), Rebala et al. (2019), Alpaydin (2020), Bernard (2021), and Jiang (2021). In addition, there are many interesting and novel articles, examples, and

W. Lan, C.-L. Tsai, *Covariance Analysis and Beyond*,
https://doi.org/10.1007/978-3-032-08796-6_9

documents that can be found in various blogs and websites such as Amazon Science Blog; https://ml.berkeley.edu/;blog.ml.cmu; https://ai.google/; Machine Learning at MIT; Machine Learning Blog Posts-IBM Developer; Blog-neptune.ai; http://ai.stanford.edu/blog, Google Cloud AI & Machine Learning Blog, and AI-Machine Learning Blog-Microsoft Tech Community.

9.2 Supervised Learning

9.2.1 Terminology

As defined in Sect. 2.2, let $Y_i = (y_{i1}, \cdots, y_{ip})'$ be the $p \times 1$ vector that denotes the i-th observation associated with p variables for $i = 1, \cdots, n$. Y_i $(i = 1, \cdots, n)$ depends on a covariate vector $X_i = (x_{i1}, \cdots, x_{iq})'$, where q is the number of covariates. Hence, Y_i is called the response vector or dependent vector, and X_i is called the predictor vector, explanatory vector, or independent vector. In the area of machine learning, the response vector is also called the label (or target) vector, and the covariate vector is called the feature (or instance) vector. When $p = 1$ or $q = 1$, the term "variable" can be used to replace "vector." For the sake of simplicity, we assume $p = 1$ in the rest of this section except for multiple category responses in Sect. 9.2.4 and denote $Y = (y_1, \cdots, y_n)' \in \mathbb{R}^n$.

Suppose that a dataset consists of n paired observations $\{(X_i, y_i) : i = 1, \cdots, n\}$. In the process of machine learning, we can split the data into three parts, a training sample, a validation sample, and a test sample with corresponding sample sizes n_1, n_2, and n_3. The purpose of the training sample is model fitting and parameter estimation, the goal for the validation sample is model selection or tuning hyperparameters, and the objective of the test sample is assessing model predictions and classifications.

As for how to randomly split the data into these three parts, there is no optimal rule. Instead, this step can be based on the sample size n, the signal to noise ratio, or the complexity of the model structure. One may adopt the 50-25-25 rule recommended by Hastie et al. (2009), or other choices such as a 70-15-15 or 80-10-10 rule, among other possibilities. It is worth noting that some researchers or practitioners may combine the training sample and validation sample into one category, or they may interchange the terms test and validation. For the sake of simplicity, we only split the data into two parts: (i) training sample for estimation and selection with the sample size n_1; and (ii) test sample for forecasts and classifications with the sample size $n_2 = n - n_1$.

In general, supervised learning has two types of tasks: regression and classification. These tasks correspond to whether the response variable (i.e., label) is continuous or discrete. In what follows, we first focus on continuous response variables and then discuss discrete cases.

9.2.2 *Fitted Errors and Prediction Errors*

Let $G = \{g_\theta(X, \boldsymbol{\alpha}) : \alpha \in \mathbb{A}$ and $\theta \in \Theta\}$ be a set of candidate regression models, where g can be either known or unknown function, $\mathbb{A}$ denotes a set of unknown regression parameters directly associated with covariates, Θ is a set of tuning (or smoothing or penalized or thresholding) parameters, and $X = (x_1, \cdots, x_q)'$. The training and testing datasets are typically denoted by X_{train} and X_{test}, respectively. To ease notation, we refer to the elements in these two datasets as X_i in this chapter, and the first element of covariates can be the constant 1. Based on the training data, $\{(X_i, y_i) : i = 1, \cdots, n_1\}$, one can obtain estimators $g_{\widehat{\boldsymbol{\theta}}}(X, \widehat{\boldsymbol{\alpha}})$ via various techniques of regression modeling, including (i) parametric regression, (ii) nonparametric regression, (iii) semiparametric regression, (iv) robust regression, and (v) penalized regression approaches, among others. For the sake of simplicity, we denote the resulting estimator by $\widehat{g}(X, \widehat{\boldsymbol{\alpha}})$. Some examples of those corresponding functions mentioned above include (i) $g(X, \boldsymbol{\alpha}) = X'\boldsymbol{\alpha}$, and $\boldsymbol{\alpha} = (\alpha_1, \cdots, \alpha_q)'$; (ii) $g(X, \boldsymbol{\alpha}) = g(X)$, where g is an unknown function; (iii) $g(X, \boldsymbol{\alpha}) = g(X'\boldsymbol{\alpha})$ or $g(X, \boldsymbol{\alpha}) = X_1'\boldsymbol{\alpha}_1 + g(X_2)$, where g is an unknown function, $X = (X_1', X_2')'$, X_1 is a $q_1 \times 1$ vector, X_2 is a $q_2 \times 1$ vector, $q_2 = q - q_1$, and $\boldsymbol{\alpha}_1$ is a $q_1 \times 1$ vector; (iv) $g(X, \boldsymbol{\alpha}) = \rho(X'\boldsymbol{\alpha})$, where ρ is an known function (see Rousseeuw and Leroy 2005); (v) $g(X, \boldsymbol{\alpha}) = X'\boldsymbol{\alpha}$ and $\sum_{j=1}^{q} |\alpha_j| \leq \tau$, where τ is a pre-specified tuning parameter. Useful reference books on those subjects include Rousseeuw (1987), Fan and Gijbels (1996), Simonoff (1996), Ruppert et al. (2003), Seber and Lee (2003), Wasserman (2004), Li and Racine (2007), and Hastie et al. (2015). It is worth noting that one may conduct model screening first before estimating parameters for high-dimensional data; see a novel paper written by Fan and Lv (2008) and its follow-up studies such as Li and Chen (2012) and Wang (2012).

To assess model fitting, one can employ various error measures such as mean squared error (MSE), deviance (Nelder and Wedderburn 1972), and generalized Pearson χ^2 statistic (McCullagh and Nelder 1989). As for which error measure should be considered, it depends on the model setting. For example, by assuming that the response Y is the sum of its conditional mean $g(X, \boldsymbol{\alpha})$ plus a random error for a given $\boldsymbol{\alpha}$ and g, we can obtain the following fitted error based on the training sample,

$$\text{MSE}_F = \frac{\sum_{i=1}^{n_1} \{y_i - \widehat{g}(X_i, \widehat{\boldsymbol{\alpha}})\}^2}{n_1}.$$

Note that the MSE-type measure is one of the most popular measures to examine model fitting. In practice, it seems natural to select the best model from a set of candidate regression models G with the smallest MSE_F. However, such a choice can cause overfitting problems. Hence, one can employ the cross-validation method (Stone 1974) to find the best model via training data; see, e.g., leave-one-out cross-validation and k-fold cross-validation in James et al. (2013).

Based on the best model selected via cross-validation, we next employ the test sample to assess its forecasting ability. To this end, we introduce the prediction error based on the test sample, $\{(X_i, y_i) : i = n_1 + 1, \cdots, n\}$,

$$\mathrm{MSE}_P = \frac{\sum_{i=n_1+1}^{n} \{y_i - \widehat{g}(X_i, \widehat{\boldsymbol{\alpha}})\}^2}{n_2}.$$

If the error computed from the best model meets our predetermined criteria, then this model can be employed for future predictions under the same environment of data collection, model structure, and assumptions. Otherwise, one can collect more data, modify model setting, employ variable selection or screening, add a penalty function, change estimation method, use ensemble models, detect outliers, or apply different model selection criteria as will be discussed in the next subsection. It is worth noting that both MSE_F and MSE_P are neither robust against outliers nor unit free. Hence, one can consider alternative error measures such as mean absolute deviation (MAD) and mean absolute percentage error (MAPE). Furthermore, if the probability distribution of response variables is known, one can consider the deviance measure mentioned above.

9.2.3 Model Selection

In Sects. 8.2 and 8.5, we have discussed factor selections. Here, we mainly focus on variable selections in the context of regression. In addition to cross-validation mentioned above, there are various selection criteria available in literature. The three most commonly used selection criteria are the Akaike information criterion (AIC), the Bayesian information criterion (BIC), and the Mallows' C_p. Useful descriptions, including selections on working high-dimensional data, can be found in Akaike (1973), Mallows (1973), Schwarz (1978), Hurvich and Tsai (1989), Li (1991), Shao (1997), Hurvich et al. (1998), McQuarrie and Tsai (1998), Burnham and Anderson (2003), Yang (2005), Chen and Chen (2008), Konishi and Kitagawa (2008), Claeskens and Hjort (2008), Wang (2009), Fan and Lv (2010), and Ding et al. (2018).

In general, a model selection criterion includes two terms: a loss function and a penalty function. These two terms are used to balance model overfitting and underfitting. Suppose the dataset follows the normal distribution, and the conditional mean response is a linear regression function, i.e., $E(y_i|X_i) = X_i'\boldsymbol{\alpha}$. Then, the resulting AIC, BIC and C_p are given below:

$$\mathrm{AIC}(k) = n_1 \log(\mathrm{MSE}_F) + 2k,$$

$$\mathrm{BIC(k)} = n_1 \log(\mathrm{MSE}_F) + \log(n_1)k,$$

and

$$C_p(k) = \text{MSE}_F/\text{MSE}_F^* - n_1 + 2k,$$

where MSE_F^* is the mean squared error that results from fitting the training data with the largest candidate model in G, k is the number of covariates in the candidate model, and $k = 1, \cdots, q$. It is worth noting that AIC may cause overfitting problems in small samples, Hurvich and Tsai (1989) proposed a correction criterion AIC_C to provide better model choices.

Based on AIC, BIC and C_p, one can find their corresponding best models. Additionally, backward stepwise selection is employed to reduce computational complexity (Hastie et al. 2009; James et al. 2013). It is worth noting that using the same dataset for model fitting and selection can cause inference problems; see Hurvich and Tsai (1990) and Efron (2014). Hence, applying test data to make inferences and predictions is a more sensible approach.

9.2.4 Classification

In practice, the response variable can be a category variable such as a binary variable. This leads to an important subject, "classification," which is the assignment of observations into groups, classes, categories, or blocks. As mentioned earlier, Section 9.2 mainly focuses on predictions. Hence, we offer only a brief review of classifications here and refer any readers interested in further study to Johnson and Wichern (2007), James et al. (2013), and Bouveyron et al. (2019).

Classification is an important topic in machine learning, and it has attracted attention across various fields. For binary responses, $y_i = 1$ or 0 for $i = 1, \cdots, n$, we can employ logistic regression to classify n observations into one of the two categories; see, e.g., Hosmer and Lemeshow (2000) and Kleinbaum and Klein (2010). In addition, the linear discriminant analysis (LDA) can be used to separate the data when the vector of covariates, X_i, follows a multivariate normal distribution with a common covariance matrix, $\mathbf{\Sigma}$. Moreover, the quadratic discriminant analysis (QDA) can be used for unequal covariance matrices, $\mathbf{\Sigma}_k$, where $k = 1, \cdots, K$, and K is the number of classes. The extension of logistic regression (i.e., multinomial logistic regression), LDA, and QDA can all be applied for more than two categories. For multiple category responses, we usually denote $Y_i = (y_{i1}, \cdots, y_{iK})'$ by a K-dimensional indicator vector with $y_{ik} = 1$ if the observation i belongs to the k-th class and $y_{ik} = 0$ otherwise, for $k = 1, \cdots, K$. In addition to those three parametric methods, a nonparametric method, the K-nearest neighbors (KNNs), can be used to allocate observations; see, e.g., Taunk et al. (2019). To compare the performance of the above four classification methods, James et al. (2013) conducted intensive simulation studies. They found that LDA and logistic regression perform satisfactorily when the decision boundaries are linear, QDA yields a better result when the boundaries are slightly nonlinear, and when the decision boundaries are complicated, KNN is the superior performer.

In addition to KNN, another widely used nonparametric classification method is the support vector machine (SVM) that was introduced by Boser et al. (1992). The aim of SVM is finding an optimally separating hyperplane that has the maximum margin from the data of each class. Note that the "support vector" refers to the observation that lies on the margin or on the incorrect side of the margin. In addition, SVM can be used not only for a linear classifier but also for nonlinear classifiers. Moreover, SVM can be robust against outliers. We can compare the performance between KNN and SVM based on the test data; see, e.g., Thanh Noi and Kappas (2018). Some useful reference books for SVM are Vapnik (1995), Cristianini and and Shawe-Taylor (2000), Steinwart and Christmann (2008), and Hastie et al. (2009). Some commonly used loss functions for assessing the performance of classifications are "0-1 loss," "hinge loss," and cross-entropy (deviance, see McCullagh and Nelder 1989). A useful reference of loss function can be found in Wang et al. (2022).

Finally, the decision tree method can be used for both regression and classification. This method is easy to understand and interpret. However, the resulting forecasts and classifications may be unstable. Consequently, a multitude of methodologies and refinements have been proposed to enhance its predictive and classifying capabilities such as bagging (Breiman 1996), random forests (Breiman 2001), and boosting (Freund and Schapire 1997). These methodologies essentially involve bootstrapping (Efron 1979; Wu 1986; Davison and Hinkley 1997) and aggregation, collectively termed as ensemble learning methods. Detailed illustrations of decision trees can be found in Breiman et al. (1984), James et al. (2013), Sheppard (2017), and Greenwell (2022).

9.3 Unsupervised Learning

In practice, the collected data may contain only covariates (i.e., features) and no response variables (i.e., labels): see, e.g., the data used in principal component analysis (PCA) and factor analysis (FA) discussed in Chap. 8. In this scenario, one can use unlabeled data to learn on itself without any supervision. This is defined as unsupervised learning. Within this context, we will discuss cluster analysis, which includes two topics: hierarchical clustering and non-hierarchical clustering. Useful reference books are Johnson and Wichern (2007), James et al. (2013), and Bouveyron et al. (2019).

9.3.1 Hierarchical Clustering Methods

Cluster analysis is used to classify observations in a dataset into several categories (or groups) so that the observations within each category are similar to each other, while observations in different categories are different from each other. To measure the similarity of observations, we commonly use measures of distance, such as the Euclidean distance, the Minkowski distance, and the normalized cosine distance. In general, there are two types of clustering algorithms: hierarchical and non-hierarchical. We next discuss the hierarchical method.

To merge clusters hierarchically, there are three commonly used dissimilarity measures: single linkage (minimum distance), complete linkage (maximum distance), and average linkage (average distance). For the sake of illustration, we consider an agglomerative clustering algorithm with the single linkage. Specifically, we begin with n clusters, and each cluster contains only one of the n observations. Accordingly, an $n \times n$ symmetric matrix is constructed by using the paired distances among the n clusters. Then identify the smallest distance between any two clusters and merge those corresponding clusters together. Subsequently, update the distance matrix based on the new cluster setting. Repeat the above merging and updating process $n - 1$ times until all observations are under a single cluster. This clustering procedure is also applicable for the complete linkage and average linkage measures, after replacing the minimum distance with maximum distance and average distance, respectively. If all three linkage measures yield the same clusters, then we should feel comfortable to make sensible interpretations. Otherwise, further steps may be needed such as checking for outliers, changing average to median or using model-based classification (see, e.g., Bouveyron et al. 2019).

9.3.2 Non-Hierarchical Clustering Methods

In contrast to the hierarchical clustering method, we now consider the most commonly used non-hierarchical clustering algorithm, K-means. It was named by MacQueen (1967), and it is an iterative centroid-based clustering method. For a given K, we iteratively assign each observation to the nearest cluster center on a numerical scale, minimizing the distances from each observation to its cluster center. The iterative process of clustering stops when no observations change clusters or the maximum number of iterations is reached. To determine the optimal number of clusters, the Silhouette coefficient (Rousseeuw 1987) or the Elbow method (Thorndike 1953) can be considered.

In addition to traditional clusterings, network classification (or community detection) has attracted much attention recently in network data. Due to the rapid advancement in technology, network data have been created across various fields, such as machine learning, business, medical and biological science, healthcare, science, engineering, and social science (Bacciu et al. 2020; Li et al. 2022). To make full use of network data, it is critical to model the nodal heterogeneity within a network via group structure. To this end, the stochastic block model (SBM) and its extensions have been widely studied and used; see, e.g., Holland et al. (1983), Bickel and Chen (2009), Bouveyron et al. (2019), and Ma et al. (2021). Specifically, SBM can be used to assess the network structure based on clustering and community detection. Some good reviews of SBM and its extensions can be found in Abbe (2018) and Lee and Wilkinson (2019).

9.4 Semi-Supervised Learning

After reviewing supervised learning and unsupervised learning, it is of interest to study semi-supervised learning. This is particularly suitable when the response variables (labels) are more expensive or difficult to collect than the covariates (or features). Suppose that the dataset consists of n paired observations $L = \{(X_i, y_i) : i = 1, \cdots, n\}$ and m additional observations of unlabeled covariates $M = \{X_i : i = n+1, \cdots, n+m\}$. Based on this dataset, it is of interest to predict the labels for the unlabeled covariates. To this end, semi-supervised learning techniques have been proposed in several areas. For example, the field of self-training (see, e.g., Amini et al. 2025) uses semi-supervised learning techniques to obtain a prediction model. Then it employs this model to make forecasts based on the subset, S, of M. We refer to S and its corresponding forecasts as the pseudo-labeled data. Subsequently, we augment the label data with the pseudo-labeled data to make forecasts. Repeat the above processes iteratively until all data in M have been used. The accuracy of self-training relies on a strong assumption, i.e., the labeled and unlabeled data follow the same probability distribution. Note that one can also split the labeled data into training and test samples. Then use the test sample to assess the accuracy of the self-training method.

In addition to self-training, mixture models have been used for semi-supervised classification and clustering (see, e.g., Zhu and Goldberg 2009; Bouveyron et al. 2019). Furthermore, Vapnik (1998) and Chapelle et al. (2006) studied semi-supervised support vector machines. Recently, Lan et al. (2022) employed the semi-supervised learning technique to impute data with a high missing rate. Finally, Kostopoulos et al. (2018) reviewed semi-supervised regressions. In sum, semi-supervised learning played a bridge role between supervised learning and unsupervised learning, which needs more work to broaden its practical applications and establish corresponding theoretical properties.

There are several methods discussed in this chapter and the previous chapter, such as LDA, PCA, and FA, all of which require estimating the unknown covariance matrix, $\mathbf{\Sigma}$. In practice, we can consider the sample estimate of $\mathbf{\Sigma}$. It is known that this estimate may not perform well as its dimension becomes large (see Bai 1999; Fan et al. 2016), since that can cause non-negligible estimation errors (Kan and Zhou 2007). Hence, we recommend employing the covariance regression approach of Zou et al. (2017) and its extensions, discussed in Chaps. 4–7, to estimate $\mathbf{\Sigma}$ when p is large. Then, use the resulting covariance estimator to construct LDA, PCA, and FA for data analyses.

In addition to the above three learning methods, reinforcement learning (Skinner 1958; Sutton and Barto 1998) has attracted great attention in the past two decades. It is a powerful learning method used for decision-making problems in dynamic environments such as robotics, game playing, and autonomous driving. To implement reinforcement learning, there are two commonly used learning algorithms, Q-learning (Watkins 1989; Watkins and Dayan 1992) and SARSA (State-Action-Reward-State-Action, Rummery and Niranjan 1994). Their major difference is that Q-learning is an off-policy algorithm, while SARSA is an on-policy algorithm; see, e.g., Zhong (2024). Based on our knowledge, there are only a few studies discussing the connection between reinforcement learning and covariance matrices. Hence, this topic can be a potential future research avenue, although we will not discuss it in this chapter.

9.5 Deep Learning

The above three sections have discussed supervised learning, unsupervised learning, and semi-supervised learning. It is of interest to note that a hot topic, deep learning, can be used within each of these three paradigms of machine learning. Deep learning (DL) is defined as the learning process that relies on multiple layers stacked sequentially. Over the past decade, deep learning has attracted considerable attention due to its powerful abilities in solving various complex problems, including detection, natural language processing, recognition, and visualization. Although deep learning techniques can be applied in three different types of machine learning, we mainly focus on supervised learning in this chapter. In the context of supervised learning, deep learning significantly enhances model performance by establishing more complex relationships between responses (or labels) and covariates (or features) through its multilayered architecture. Compared with traditional linear regression and logistic regression models, deep learning can handle nonlinear relationships and interactions between covariates so that it is more adaptable and flexible in dealing with complex and challenging tasks. However, it has some limitations, such as computational cost, data size, overfitting, interpretation, and extrapolation. Hence, it is crucial to consider these limitations when applying deep learning to real-world problems.

The concept of deep learning has a long history. Motivated by the biological neurons in the human brain, McCulloch and Pitts (1943) proposed the first computational model for neural networks. Their model mainly focuses on binary (or boolean) inputs and outputs, and each of the inputs is assigned equal weight. Subsequently, Rosenblatt (1958) introduced the "perceptron," and his model not only allows for non-binary inputs but also can assign different weights to the inputs. The above two articles, respectively, laid out the foundation and advanced the application for the artificial neural network (ANN). Figure 9.1 depicts the architecture of a simple feedforward neural network (FNN). Similar figures can be found in many articles; see, e.g., Fan et al. (2020). This is a typical type of ANN, which consists of three components: an input layer, multiple hidden layers, and an output layer.

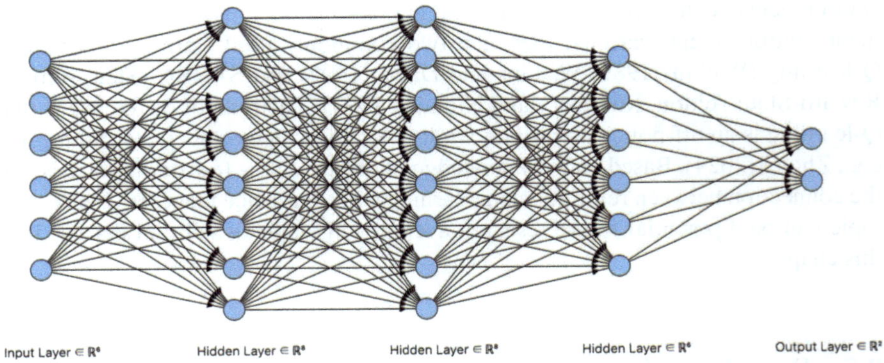

Fig. 9.1 The architecture of a simple feedforward neural network

The input data, i.e., the covariates (or features), denoted by $\mathbf{X} \in \mathbb{R}^{n \times q}$, are the initial input to the model and can be represented as $\mathbf{H}^{(0)}$. Then, at each hidden layer, we perform a series of transformations on the input data:

$$\mathbf{H}^{(l)} = \sigma(\mathbf{H}^{(l-1)}\boldsymbol{W}^{(l)} + \mathbf{b}^{(l)}), \quad l = 1, \cdots, L,$$

where $\boldsymbol{W}^{(l)} \in \mathbb{R}^{q^{(l-1)} \times q^{(l)}}$ is the weight matrix for transforming $q^{(l-1)}$-dimensional features to $q^{(l)}$-dimensional outputs at layer l, $\boldsymbol{b}^{(l)} \in \mathbb{R}^{q^{(l)}}$ represents the corresponding bias vector, and $q^{(0)} = q$. In the computation process, $\boldsymbol{W}^{(l)}$ and $\boldsymbol{b}^{(l)}$ are replicated to ensure that the same weights and biases are applied to all n samples. $\mathbf{H}^{(l-1)} \in \mathbb{R}^{n \times q^{(l-1)}}$ is the generated feature matrix from the previous layer, and L denotes the total number of hidden layers in the model. In addition, σ represents a prespecified activation function, which is typically nonlinear. The activation function plays a critical role in neural networks by learning the input features through nonlinear transformations.

In practice, there are several known activation functions being used in FNN. Here, we present seven activation functions. (i) ReLU (Rectified Linear Unit): $\sigma(z) = \max\{z, 0\}$; (ii) Leaky ReLU (Maas et al. 2013): $\sigma(z) = z$ when $z > 0$, az otherwise, where a is a small scalar; (iii) Randomized Leaky ReLU (Xu et al. 2015): $\sigma(z) = z$ when $z > 0$, αz otherwise, where $\alpha \sim U[b, c]$, and b and c are in $[0, 1)$; (iv) Tanh: $\sigma(z) = \tanh(z) = [\exp(z) - \exp(-z)]/[\exp(z) + \exp(-z)]$; (v) Sigmoid: $\sigma(z) = 1/[1 + \exp(-z)]$; (vi) Swish (Ramachandran et al. 2018): $\sigma(z) = z/(1 + \exp(-\beta z))$, where β is a tuning parameter; and (vii) Mish (Misra 2019): $\sigma(z) = z\tanh[\log\{1 + \exp(z)\}]$. A comprehensive review of the above seven, together with several more activation functions, was presented by Dubey et al. (2022). For the sake of illustration, we have depicted the most commonly used activation functions: ReLU, Sigmoid, and Tanh in Fig. 9.2.

The output layer generates final predictions or classifications. The model's parameters, denoted by $\boldsymbol{\theta}$, include the weight matrix $\boldsymbol{W}^{(l)}$ and the bias vector $\mathbf{b}^{(l)}$

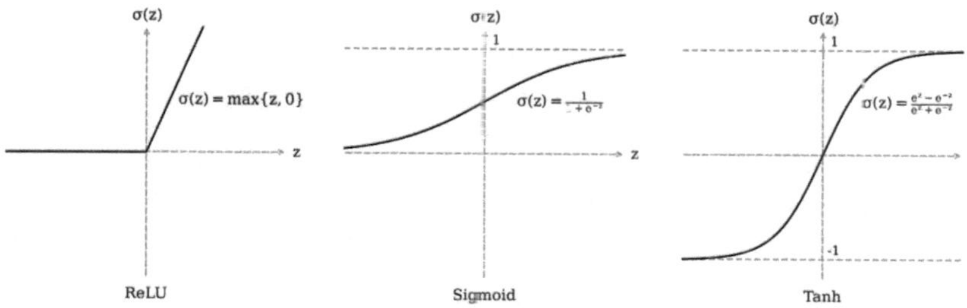

Fig. 9.2 Three commonly used activation functions in neural network: ReLU, Sigmoid, and Tanh

for $l = 1, \cdots, L$. Note that they also consist of parameters involved in the affine transformation from the final hidden layer to the output layer. All parameters are optimized for each $l = 1, \cdots, L$ by evaluating the performance of outputs. That is, for the data $\{(X_i, y_i) : i = 1, \cdots, n\}$, $\boldsymbol{\theta}$ is estimated by minimizing a prespecified loss function such as the fitted error in (9.1) given below.

$$\mathcal{L}(\boldsymbol{\theta}) = n^{-1} \sum_{i=1}^{n} \text{loss}(\widehat{y}_i(\boldsymbol{\theta}), y_i), \tag{9.1}$$

where $\widehat{y}_i(\boldsymbol{\theta})$ is the predicted output for any given $\boldsymbol{\theta}$. Note that $\widehat{y}_i(\boldsymbol{\theta})$ is obtained by mapping the L-th hidden layer to the output layer with an affine transformation function for predictions (or an affine transformation function followed by an activation function, e.g., Sigmoid, for classifications). For the sake of simplicity, we denote it by f. Accordingly, $\widehat{y}_i(\boldsymbol{\theta}) = f(\mathbf{H}^{(L)}|X_i, \boldsymbol{\theta})$.

Note that $\mathcal{L}(\cdot)$ is the empirical loss function. In the context of regression for predictions, we usually consider

$$\mathcal{L}(\boldsymbol{\theta}) = n^{-1} \sum_{i=1}^{n} (\widehat{y}_i(\boldsymbol{\theta}) - y_i)^2. \tag{9.2}$$

In the context of classifications, however, one can employ the cross-entropy measure, (i.e., the deviance in McCullagh and Nelder 1989),

$$\mathcal{L}(\boldsymbol{\theta}) = -n^{-1} \sum_{i=1}^{n} \sum_{k=1}^{K} y_{ik} \log(\widehat{y}_{ik}(\boldsymbol{\theta})), \tag{9.3}$$

to assess the goodness of fit, where K is the number of classes, and y_{ik} was defined in Sect. 9.2. Note that $\boldsymbol{\theta}$ in $\widehat{y}_i(\boldsymbol{\theta})$ and $\widehat{y}_{ik}(\boldsymbol{\theta})$ above can be replaced with $\widehat{\boldsymbol{\theta}}$ obtained by minimizing their corresponding loss functions. To implement an FNN, it would be useful to construct a selection criterion to select the number of neurons in hidden layers for reducing computation and avoiding overfitting (Haykin 2009).

The neural network's structure enjoys nonlinear characteristics so that it enables handling of complex data structures and tasks. In order to meet the demands of applications in diverse fields such as time series analysis, machine translation, image recognition, and recommendation systems, various types of deep learning models have been developed besides FNN. Here, we consider the three types of artificial neural networks (ANNs) as follows. First, Convolutional Neural Networks (CNNs) are designed for grid-like data structures, such as images, where pixels are arranged in a regular structure. CNNs utilize convolutional layers to extract and learn local features. This unique characteristic makes CNNs highly suitable for analyzing and performing visual tasks, including image classification, object detection, and facial recognition, as demonstrated in the work of Krizhevsky et al. (2017). Second, Graph Convolutional Networks (GCNs) extend the concept of convolutional operations to non-Euclidean data, particularly when data have graph structures. As a result, GCNs are empowered to handle tasks including social network analysis and recommendation systems (Kipf and Welling 2017). Third, Transformers, Recurrent Neural Networks (RNNs), and Long Short-Term Memory Networks (LSTMs), are designed to tackle sequential data. These architectures are better at modeling long-range dependencies (Elman 1990, Hochreiter and Schmidhuber 1997, Zaremba et al. 2015, Vaswani et al. 2017). According to the above discussions, the choice among CNNs, GCNs, and Transformers (or RNNs, LSTMs) depends heavily on the specific task. In the following three subsections, we will present comprehensive reviews of CNNs, GCNs, and Transformers, respectively.

It is worth noting that the above three types of neural networks and feedforward neural networks usually employ the same loss function as defined in (9.1). To minimize this loss function, one can employ the back-propagation (BP) algorithm introduced by Rumelhart et al. (1986) to adjust the weights of the connections between the neurons in different layers, to calculate the necessary gradients for the gradient descent (GD) method or its variants. This process continues iteratively until the error falls below a specified threshold value. In practical application, the aforementioned neural networks can be implemented based on "computational graphs"—a powerful tool for constructing algorithms and software packages for neural networks, such as TensorFlow (Abadi et al. 2016). In addition, after training these neural networks via training samples, one can employ the test sample and then follow Sect. 9.2 to evaluate its predictive performance. Useful references and reviews on deep learning can be found in LeCun et al. (2015), Nielsen (2015), Schmidhuber (2015), Goodfellow et al. (2016), Aggarwal (2018), Huang et al. (2020), Yuan et al. (2020), Alzubaidi et al. (2021), Bartlett et al. (2021), Fan et al. (2021), Liu (2023), Nalisnick et al. (2023), and Zheng et al. (2023). In addition, Schmidhuber (2022) presented a nice review on the history of AI and deep neural networks.

9.5.1 Convolutional Neural Networks

Traditional deep learning methods, such as feedforward neural networks (FNNs) illustrated in Fig. 9.1, face substantial challenges when handling high-dimensional and large-scale data, particularly in the context of images. The main reason is that FNNs are based on fully-connected layers so that they involve a substantial number of parameters in the modeling of high-dimensional data. For example, in the MNIST dataset of handwritten digits ($28\times28\times1$ pixel images), an FNN with a single hidden layer of 512 neurons involves 401,920 parameters to connect all input pixels to the hidden layer. The large number of parameters not only leads to computational inefficiency but also increases the risk of overfitting, especially when the quantity of training data is limited. To overcome these limitations, Convolutional Neural Networks (CNNs), introduced by LeCun et al. (1989) and LeCun et al. (1998), are able to reduce the number of parameters significantly by using localized convolutional filters and weight sharing across spatial locations, which offer a more effective solution.

In general, the main architecture of CNN comprises three key components: (i) The input layer, which utilizes a three-dimensional tensor to represent the width, height, and depth of the input data. For example, in the case of an image, the width and height correspond to the pixel dimensions, and the depth represents the number of channels (such as RGB channels for a color image); (ii) The hidden layers, which play a crucial role in feature extraction. These layers include convolutional layers (filters or kernels and receptive regions), activation layers (e.g., ReLU), pooling layers, and fully connected layers. These layers work together to learn from the local features of the input data; (iii) The output layer, which generates the final predictions or classifications based on the processed information from the hidden layers. Among these three components, the hidden layers extract information from local features and are of utmost importance. We next delve into the working mechanisms of the two fundamental components of the hidden layer: convolutional layers and pooling layers.

1. Convolutional Layer

The convolutional layer learns local features from input data by convolving small-sized filters (or convolutional kernels) over localized regions. The input feature map is represented as a three-dimensional tensor $\mathbf{X} = (x_{ijl}) \in \mathbb{R}^{d_1\times d_2\times d_3}$, where d_1 and d_2 are spatial dimensions, and d_3 is the depth (e.g., in a red-green-blue (RGB) image, the depth is 3). In the process of the convolutional operation, we need to choose the stride *a priori* for constructing the convolutional layer. A larger stride results in a smaller output feature map, while a smaller stride leads to a larger output feature map. Then, the convolutional layer applies a set of learnable filters $\mathbf{F}^l \in \mathbb{R}^{d_1^*\times d_2^*\times d_3}$ for $1 \le l \le \tilde{d}_f$ on $\mathbf{X}$, where $d_1^* \times d_2^*$ is the size of the filter and $\tilde{d}_f$ is the number of filters. Subsequently, the lmn-th element of the convolution output map $\mathbf{O}$ is defined as follows:

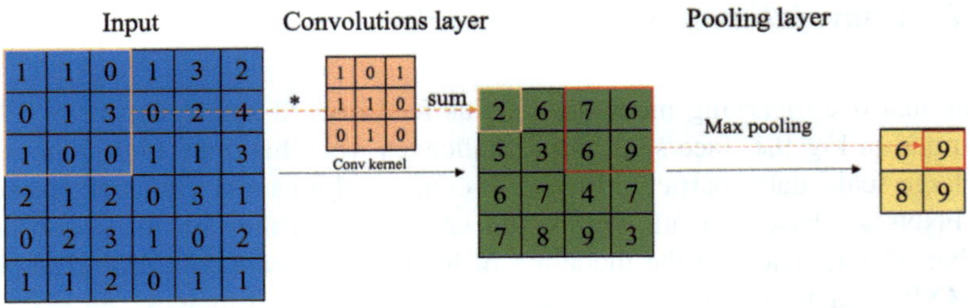

Fig. 9.3 The feature learning process of CNNs

$$\mathbf{O}^l_{mn} = \sum_{i=1}^{d_1^*} \sum_{j=1}^{d_2^*} \sum_{k=1}^{d_3} \mathbf{X}_{i+(m-1)s, j+(n-1)s, k} \mathbf{F}^l_{i,j,k}, \tag{9.4}$$

where s is the size of stride, $l = 1, \cdots, \tilde{d}_f$, $m = 1, \cdots, \tilde{d}_1$, $n = 1, \cdots, \tilde{d}_2$, $\tilde{d}_1 = (d_1 - d_1^*)/s + 1$, and $\tilde{d}_2 = (d_2 - d_2^*)/s + 1$. When $s = 1$ and $d_1^* = d_2^* = d$, the result is in the following formulation from Fan et al. (2020):

$$\mathbf{O}^l_{mn} = \sum_{i=1}^{d} \sum_{j=1}^{d} \sum_{k=1}^{d_3} \mathbf{X}_{i+(m-1), j+(n-1), k} \mathbf{F}^l_{i,j,k}. \tag{9.5}$$

Consequently, the generated feature via the convolution process captures the local feature information of the input data around that region and produces a set of feature maps that are then passed to subsequent layers for further processing. It is worth noting that the padding procedure can be considered to be adding extra pixels around the input image before applying convolution operations. This procedure prevents the loss of boundary information and maintains the spatial dimensions.

To illustrate the convolution operation, Fig. 9.3 depicts the feature learning process of the convolutional layer with a single depth ($d_3 = 1$) and one filter ($\tilde{d}_f = 1$). The input $\mathbf{X}$ is a 6×6 matrix, and the filter $\mathbf{F}$ is a 3×3 matrix defined as $\begin{bmatrix} 1 & 0 & 1 \\ 1 & 1 & 0 \\ 0 & 1 & 0 \end{bmatrix}$. By the formula in (9.5), we multiply the top-left 3×3 submatrix of the input matrix, $\begin{bmatrix} 1 & 1 & 0 \\ 0 & 1 & 3 \\ 1 & 0 & 0 \end{bmatrix}$, with the filter matrix through a dot product operation, which yields the output value 2. The filter keeps moving one stride to the right till it reaches the end. Then it starts with one stride down from the left end and repeats until the entire input map is traversed. The resulting feature map of the first depth is $\mathbf{O}^1_{mn}$, which is a 4×4 matrix.

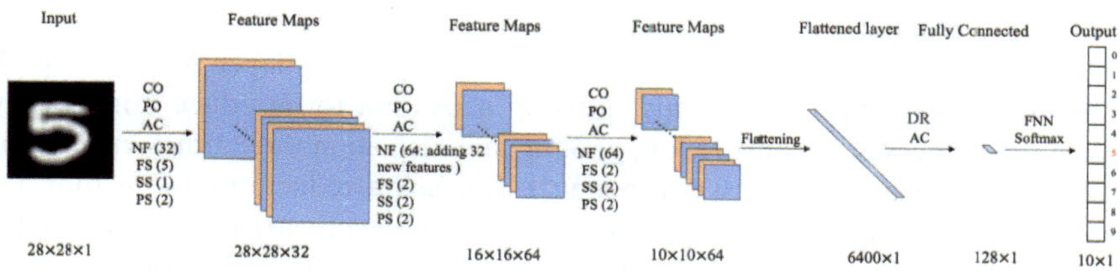

Fig. 9.4 The architecture of a CNN model, where CO, PO, AC, NF, FS, SS, PS, and DR denote convolution, pooling (e.g., Max), activation (e.g., ReLU), number of filters, filter size, stride size, padding size, and dimension reduction, respectively. In addition, WO = (WI−FS+2PS)/SS +1, where WO and WI correspond to the widths of output image and input image

2. Pooling Layer

The pooling layer is used for aggregating the information produced from the convolutional layer and the activation layer. It not only reduces the network complexity and computational cost but also retains the essential features. The most popular pooling function is max-pooling, although the average pooling or sum pooling can be considered.

Based on the max-pooling, let $\mathbf{P} \in \mathbb{R}^{\tilde{d}_f \times d_4 \times d_5}$ be the pooling layer obtained from $\tilde{\mathbf{O}} \in \mathbb{R}^{\tilde{d}_f \times \tilde{d}_1 \times \tilde{d}_2}$ with stride $\tilde{s}$, where $\tilde{\mathbf{O}}$ is an activation output from $\mathbf{O}$, $d_4 = \tilde{d}_1/q$ and $d_5 = \tilde{d}_2/q$. Then the lab-th element of $\mathbf{P}$ is given as follows:

$$\mathbf{P}^l_{ab} = \max_{1 \le i, j \le q} \tilde{\mathbf{O}}_{l, i+(a-1)\tilde{s}, j+(b-1)\tilde{s}}, \tag{9.6}$$

where $l = 1, \cdots, \tilde{d}_f$, $a = 1, \cdots, d_4$, and $b = 1, \cdots, d_5$. For the sake of illustration, the right-hand side of Fig. 9.3 displays the max-pooling without activation when $\tilde{d}_f = 1$, $\tilde{d}_1 = \tilde{d}_2 = 4$, $q = 2$, $\tilde{s} = 2$, and $d_4 = d_5 = 2$. Specifically, consider the upper and left submatrix of $\mathbf{O}^1_{4\times4}$, $\begin{bmatrix} 2 & 6 \\ 5 & 3 \end{bmatrix}$. A 2×2 max-pooling filter yields the maximum value 6. By continuously applying max-pooling filter with stride of 2 on the 4×4 matrix, $\mathbf{O}^1_{4\times4}$, the max pooling matrix is $\mathbf{P}^1_{2\times2} = \begin{bmatrix} 6 & 9 \\ 8 & 9 \end{bmatrix}$. Accordingly, this operation decreases computational complexity and preserves the most significant information.

While the aforementioned convolutional and pooling layers form the fundamental components of a CNN, more sophisticated architectures are often used in practice to enhance performance and handle complex tasks. For example, advanced models combine multiple convolution and pooling layers, together with fully connected layers and activation functions, to effectively learn and extract features from the input data. A helpful illustration of CNN can be found in Figure 2 of the seminal paper by LeCun et al. (1998), which displays the architecture of the LeNet-5 model. Note that LeNet-5 is a pioneering CNN model for handwritten character recognition. Motivated by the structure of LeCun et al. (1998), Fig. 9.4 herein depicts the process of CNN from input to output. It involves first applying a convolution layer to extract features from the input data, then using a pooling layer to downsample the feature maps, and subsequently utilizing activation to introduce nonlinearity. Repeat the processes of convolution, pooling and activation three times, and obtain flattened feature maps. Afterward, conduct dimension reductions by feeding the flattened feature maps into a fully connected layer with activations. Finally, employ the softmax function to generate the final output layer, where the softmax is the extension of sigmoid, and it is defined as $\sigma(z_j) = \exp(z_j)/\sum_{k=1}^{K}\exp(z_k)$, z_j is the j-th component of $Z = (z_1, \cdots, z_K)'$, the output of FNN obtained from the input of the fully connected layer, for $j = 1, \cdots, K$, and K is the number of classes.

In sum, the architecture of Fig. 9.4 consists of seven layers: input layer, three convolutional-pooling-activation layers, flattened layer, fully connected layer, and output layer. It is worth noting that CNNs employ convolutional filters with local receptive fields and pooling mechanisms to extract local features from input data. In comparison with FNNs, CNNs are built based on local connections rather than full connections between layers. Hence, CNNs can have sparse interactions and a small number of parameters. As a result, CNNs avoid the issue of excessive parameters and overfitting and are computationally feasible for large datasets. These merits allow CNNs to handling spatially dependent data, including widely utilized tasks in the field of computerized vision such as object detection (Redmon et al. 2015; Lin et al. 2017), image segmentation (Ronneberger et al. 2015; Kirillov et al. 2019), and facial recognition (Taigman et al. 2014; Sun et al. 2014).

9.5.2 *Graph Convolutional Networks*

In real-world application, datasets may display non-Euclidean characteristics, such as network structure that can be found in social, biological, and transportation networks (Lazer et al. 2009; Davidson et al. 2002; Yu et al. 2018). As depicted on the right-hand side of Fig. 9.5, network datasets are composed of nodes and edges. Here, nodes serve to represent entities, while edges denote the relationships between these entities (Wu et al. 2021). This differs from traditional Euclidean data, such as images or time series, which are illustrated on the left-hand side of Fig. 9.5. Owing to the irregular structure inherent in graph data, Convolutional Neural Networks (CNNs) may be unsuitable for analyzing such data. A detailed discussion about the difference between CNN and GCN can be found in Li et al. (2022).

To handle datasets with graph structures, GCNs, introduced by Kipf and Welling (2017), extended the concept of traditional convolutions to non-Euclidean domains. They achieve this by aggregating information from neighboring nodes within the observable network structure. Analogous to CNNs, the architecture of GCNs consists of three main components: (i) The input layer contains the adjacency matrix $\mathbf{A} = (a_{ij}) \in \mathbb{R}^{n\times n}$ and the feature matrix $\mathbf{X} = (X_1, \cdots, X_n)' \in \mathbb{R}^{n\times q}$, where $a_{ij} = 1$ if nodes i and j are connected for $i \neq j$, and $a_{ij} = 0$ otherwise, and $X_i \in \mathbb{R}^q$ represents the feature vector of node i for $i = 1, \cdots, n$; (ii) The hidden

Fig. 9.5 Comparison between CNNs (left-hand side) and GCNs (right-hand side)

layers include convolutional and pooling operations. These operations are used to aggregate information from the graph, enabling the network to learn meaningful representations of the graph data; and (iii) The output layer produces the final vector representation, which is then used for classification or prediction tasks.

As discussed with CNNs in Sect. 9.6.1, activation operations are typically applied after convolution operation and before the final output. Since the hidden layers play a critical role in GCNs, we next discuss the convolutional layers and pooling layers in detail.

1. Graph convolution layer

Consider an undirected graph $G = (V, E)$, where V is the set of nodes and E is the set of edges. Based on G, the adjacency matrix $\mathbf{A}$ is constructed, and $\mathbf{X} = (X_1, \cdots, X_n)'$ is its the associated feature matrix. A graph convolutional layer updates the features based on the following formula,

$$\mathbf{H}^{(l)} = \sigma\left(\tilde{\mathbf{D}}^{-\frac{1}{2}}\tilde{\mathbf{A}}\tilde{\mathbf{D}}^{-\frac{1}{2}}\mathbf{H}^{(l-1)}\mathbf{W}^{(l)}\right),$$

where $\mathbf{H}^{(l)}$ denotes the output feature matrix at layer l for $l = 1, \cdots, L$, and L is the total number of layers in the network. In addition, $\mathbf{H}^{(0)} = \mathbf{X}$ is the initial input feature matrix, $\mathbf{W}^{(l)}$ is the weight matrix, $\tilde{\mathbf{A}} = \mathbf{A} + \mathbf{I} = (\tilde{a}_{ij})$ is the adjacency matrix together with the added self-loops. Moreover, the degree matrix $\tilde{\mathbf{D}}$ is diagonal, whose diagonal element $\tilde{\mathbf{D}}_{ii}$ indicates the degree of node i in the graph (i.e., $\tilde{\mathbf{D}}_{ii} = \sum_j \tilde{a}_{ij}$), and σ is the activation function. By stacking multiple graph convolutional layers, GCNs can progressively aggregate features from their neighborhoods, thereby capturing comprehensive structural information of the entire graph.

2. Graph pooling layer

In the context of GCN, there are two major categories of tasks: (i) graph-level tasks, such as graph classifications; and (ii) node-level tasks, such as identifying individual node characteristics. In this section, we mainly focus on graph-level tasks, in which graph pooling plays an important role in downsizing the graph by integrating the information from multiple nodes so as to capture the overall structure of the graph. An illustrative example of graph pooling is presented in Fig. 9.6, which is motivated by Figure 6 of Bacciu et al. (2020). In this example, the nodes in the graph are clustered into several communities using a community

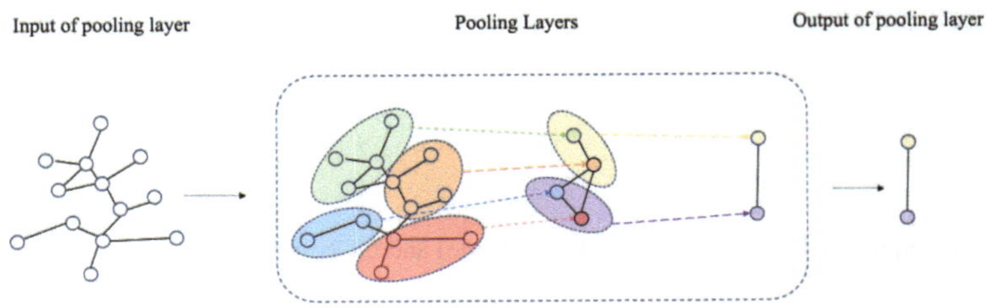

Fig. 9.6 The structure of graph pooling

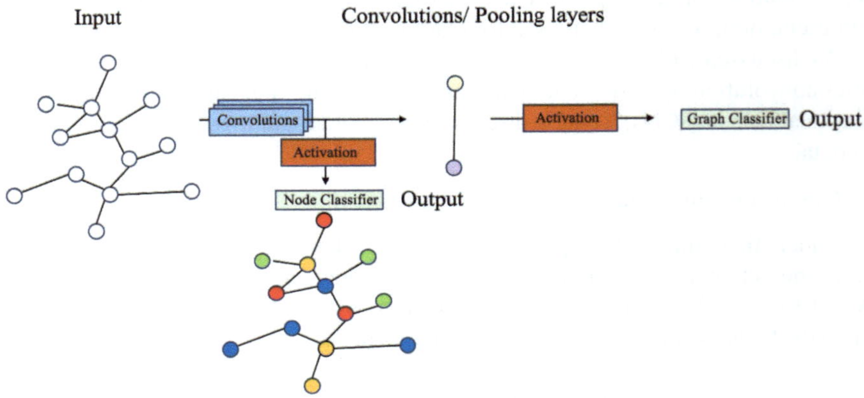

Fig. 9.7 The architecture of a GCN model

detection method like spectral clustering (Ma et al. 2019), and the nodes in each community are encircled by dashed lines. Subsequently, each community is treated as a new node, and its associated feature is calculated by summing or averaging the features of all nodes within that community. Consequently, the features of these new aggregated nodes summarize the original nodal features and are then passed to subsequent layers for further processing. For the sake of simplification, Fig. 9.6 only sketches three pooling layers, but the convolution and activation operations can be applied between pooling layers.

The above pooling method is known as node cluster pooling. Useful references and related methods can be found in Ying et al. (2018) and Ma et al. (2019). In addition to this pooling method, there are several alternative approaches. For example, the global pooling method uses the sum (or average) of the features from all nodes (Lin et al. 2014), and the downsampling pooling method utilizes the features of the top-ranked nodes based on nodal topology (Lee et al. 2019).

Having introduced the graph convolution and graph pooling layers, we are motivated by Figure 7 of Bacciu et al. (2020) and now briefly illustrate the workflow of a Graph Convolutional Network (GCN) in Fig. 9.7. This figure depicts two distinct architectures that correspond to two different types of tasks. Specifically, in a node-level task, such as node classification, we apply an activation operation to the output of the graph convolutional layers and then directly link the results to the final output layer. On the other hand, for graph-level tasks, such as graph classification, graph pooling is employed to generate a low-dimensional vector representation of the entire graph. After performing an activation operation, this representation is connected to the final output layer.

Similar to FNN, one can obtain unknown parameters in GCNs by minimizing the loss function (9.2) or (9.3). As discussed in FNN, the model is trained using back-propagation and gradient descent algorithms, including stochastic gradient descent (SGD) and adaptive moment estimation (Adam); see, e.g., Ruder (2016) and Kingma and Ba (2017). Useful references and reviews on GCN can be found in Hamilton et al. (2017), Velickovic et al. (2018), Yu et al. (2018), Xu et al. (2019), Hamilton (2020), Zhang et al. (2022), and Zhou (2020).

9.5.3 *Transformer*

As mentioned earlier in Sect. 9.6, Transformer is another well-known deep learning method that is designed to handle sequential data, such as text and speech. In order to apply the Transformer model, the data must be converted into a matrix format. For example, in the context of text data, one can convert a sentence into a sequence of n tokens (such as words, subwords, or characters), and each token is represented by a q-dimensional vector. Through this process, the data can be converted into a matrix $\mathbf{X} = (X_1, X_2, \cdots, X_n)' \in \mathbb{R}^{n \times q}$, which serves as the input for the Transformer.

Note that n mentioned above is the sequence length (the number of tokens). For example, consider the input sentence "*I went to the park to play soccer yesterday*," which will be discussed in Fig. 9.10 later. By using a tokenizer method like WordPiece (Wu et al. 2016), this sentence can be segmented into tokens such as ["I," "went," "to," "the," "park," "to," "play," "soccer," "yesterday"]. Then, we convert each token into a q-dimensional numeric vector using word embedding techniques like word2vec and GloVe. These methods help capture the semantic meaning of the tokens in a vector space (Mikolov et al. 2013; Pennington et al. 2014). The value of q, which represents the embedding dimension, varies across different algorithms. For example, it is 300 in GloVe-840B, 1024 in BERT-Large, and 12,288 in GPT-3-175B. After such conversion, we have $\mathbf{X} = (X_1, X_2, \cdots, X_n)'$ with $n = 9$, where X_i is the embedding vector of the i-th token for $i = 1, \cdots, 9$.

In the context of text translations, we denote the output of a Transformer by $Y = (y_1, y_2, \cdots, y_n)'$, which is typically a sequence data. Hence, the Transformer is well-suited to sequence-to-sequence (seq2seq) tasks. Some useful references are Sutskever et al. (2014), Vaswani et al. (2017) and Lewis et al. (2020).

The main architecture of a Transformer is an encoder–decoder structure; see Fig. 9.8. The encoder converts input data $\mathbf{X}$ into a state vector through a series of complex nonlinear transformations. The information in the state vector is then passed to the decoder to generate output data. In Fig. 9.8, the state vector is obtained

Fig. 9.8 The main architecture of a Transformer model

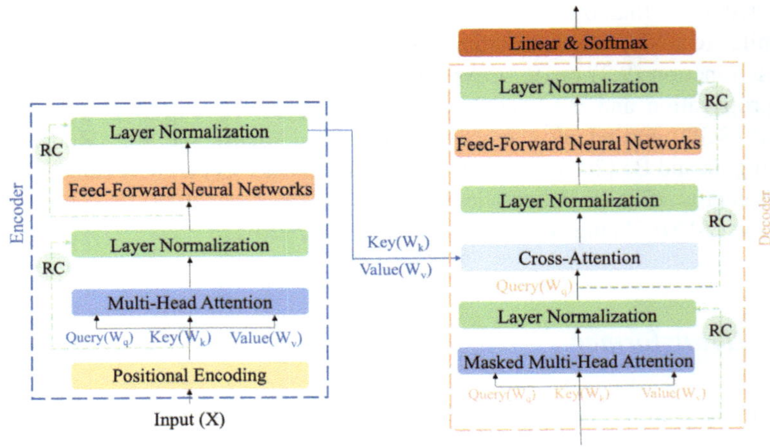

Fig. 9.9 Detailed architecture of a Transformer model

by applying complex nonlinear transformations to the input data $\mathbf{X}$ in an iterative manner. It is represented by

$$\mathbf{H}^{(l)} = \text{Encoder}(\mathbf{H}^{(l-1)}) \in \mathbb{R}^{n\times q},$$

for $l = 1, \cdots, L$, where L is the number of encoder layers in the Transformer and $\mathbf{H}^{(0)} = \mathbf{X}$.

Subsequently, the Transformer generates output $\widehat{Y} = (\widehat{y}_1, \widehat{y}_2, \cdots, \widehat{y}_n)'$ in a sequential manner. Specifically, for a given state variable $\mathbf{H}^{(L)}$ and the previously generated $\widehat{y}_1, \cdots, \widehat{y}_{i-1}$ up to position (or time) i, we generate $\widehat{y}_i$ using the following formula:

$$\widehat{y}_i = \text{Decoder}(\mathbf{H}^{(L)}, \widehat{y}_1, \cdots, \widehat{y}_{i-1}).$$

Note that the model is trained by minimizing the loss function between predictions $\widehat{Y}$ and targets Y; see (9.2) and (9.3).

The detailed architecture of the Transformer, presented in Fig. 9.9, is motivated by Vaswani et al. (2017). The Encoder layer consists of five main components: 1. positional encoding; 2. multi-head attention; 3. layer normalization (Ba et al. 2016); 4. feedforward neural networks (FNN); and 5. residual connection (labeled RC) (He et al. 2016). We next discuss them in detail.

1. Positional Encoding

In any language, the order and position of words can play a crucial role in expressing the intended meaning of a sentence. For example, in the sentence "I saw a light," the word "light" functions as a noun, referring to an object that emits brightness. On the other hand, in the sentence "She will light the candle," the word "light" serves as a verb, denoting the action of igniting something. That is, the position of a word within a sentence can be a significant indicator of both semantic meaning and grammatical role. In the Transformer architecture, positional encoding is designed to capture these positional relationships between words in a sequence. This technique generates the unique position vector of each word, allowing the model to differentiate between word positions within the semantic space.

Note that positional encoding uses sine and cosine functions with varying frequencies and phases to encode the position of each token in a sequence, ensuring that each position has a unique representation (Vaswani et al. 2017). Let i represent the index of an individual word within the sequence, k denote the embedding index, and q be the dimension of the positional encoding. Then, the position embedding is calculated according to the following rules: when k is even, $PE_{(i,k)} = \sin(i/10{,}000^{k/q})$; when k is odd, $PE_{(i,k)} = \cos(i/10{,}000^{(k-1)/q})$. Afterward, these positional embeddings are added to the corresponding word embeddings in an element-wise fashion. As a result, the model is able to process both the semantic meaning conveyed by the word embeddings and the crucial positional information represented by the positional encoding for each word in the sequence. This combination forms the input of the Transformer.

2. Multi-Head Attention

Because the word of "Attention" appears in both the Encoder and the Decoder in Fig. 9.9, this subsection will discuss all three places that attention appears: multi-head attention in the Encoder, cross-attention in the Decoder, and masked multi-head attention in the Decoder. We begin with multi-head attention in the Encoder. As proposed in Vaswani et al. (2017), the Transformer generates a series of attention matrices at each layer, essentially capturing the "relevance" or "attention" of different tokens within an input sequence. Specifically, the ij-th elements of the attention matrix $\mathbf{A}^{(l)} = (a_{ij}^{(l)}) \in \mathbb{R}^{n\times n}$, for the l-th encoder layer, are defined as follows:

$$a_{ij}^{(l)} = \frac{\exp\left((\mathbf{h}_i^{(l-1)}\mathbf{W}_q^{(l)})(\mathbf{h}_j^{(l-1)}\mathbf{W}_k^{(l)})'/\sqrt{d_k}\right)}{\sum_{s=1}^{n}\exp\left((\mathbf{h}_s^{(l-1)}\mathbf{W}_q^{(l)})(\mathbf{h}_s^{(l-1)}\mathbf{W}_k^{(l)})'/\sqrt{d_k}\right)}, \tag{9.7}$$

where $\mathbf{h}_i^{(l-1)} \in \mathbb{R}^q$ and $\mathbf{h}_j^{(l-1)} \in \mathbb{R}^q$ are the i-th and j-th elements of $\mathbf{H}^{(l-1)} = (\mathbf{h}_1^{(l-1)}, \cdots, \mathbf{h}_n^{(l-1)})'$, $\mathbf{H}^{(l-1)} \in \mathbb{R}^{n\times q}$ is the output of the $(l-1)$-th encoder layer, and $\mathbf{H}^0 = \mathbf{X}$ is the initial encoder input. Based on $\mathbf{A}^{(l)}$ and $\mathbf{H}^{(l-1)}$, the output of the attention operation is computed by Attention(Q, K, V) $= \mathbf{A}^{(l)}\left(\mathbf{H}^{(l-1)}\mathbf{W}_v\right)$.

In addition, the $\mathbf{W}_q^{(l)} \in \mathbb{R}^{q\times d_k}$, $\mathbf{W}_k^{(l)} \in \mathbb{R}^{q\times d_k}$, and $\mathbf{W}_v^{(l)} \in \mathbb{R}^{q\times d_v}$ in (9.7) are learnable projection matrices that play crucial roles in the attention mechanism by mapping the input representations into the query, key, and value spaces, and Q, K and V represent the query, key, and value, respectively. Moreover, d_k and d_v control the projection size, and $d_k = d_v = q$ is commonly considered; see Vaswani et al. (2017). Additionally, the scaling factor $1/\sqrt{d_k}$ prevents numerical instability between high-dimensional queries and keys. The above attention process allows the model to focus on relevant parts of the input data by calculating attention weights based on these projected representations.

To illustrate the workflow of the query-key-value mechanism, we take the sentence "*I went to the park to play soccer yesterday*" as an example. When the model processes the token '*play*', the vectors of query, key, and value are generated in the following manner. Assume that $X_{\text{play}} \in \mathbb{R}^q$ is the word embedding vector corresponding to the word "*play*." Then, we evaluate the query vector as $\mathbf{q}_{\text{play}} = \mathbf{W}_q' X_{\text{play}}$. This query vector characterizes the contextual information that related to the token "*play*," such as the activity (play soccer) and the location (the park). To generate the key vectors, let $X_i \in \mathbb{R}^q$ denote the word embedding of the i-th token for $i = 1, \cdots, 9$. Then the resulting key vector is $\mathbf{k}_i = \mathbf{W}_k' X_i$. The key vector acts as an identifier for each token. That in turn allows the model to assess the degree of relevance of each token with respect to the query vector. We finally generate the value vectors. For each token $i = 1, \cdots, 9$, the value vector $\mathbf{v}_i$ is given by $\mathbf{v}_i = \mathbf{W}_v' X_i$. The value vector carries the actual semantic content of each token. These values are utilized to generate the output based on the matching between the query and the keys, which is determined by the attention scores.

So far, we have mainly focused on a single attention matrix. However, a single attention matrix may restrict the model's capacity to effectively and stably capture the relationships that can be salient within a large sample of rich text. To overcome this limitation, the multi-head attention mechanism is proposed. This mechanism divides the attention computation process into multiple parallel heads. Each individual head utilizes distinct parameter matrices, $\mathbf{W}_q$, $\mathbf{W}_k$, and $\mathbf{W}_v$, to project the input sequence into separate subspaces. The structure of the multi-head attention mechanism has several advantages. Firstly, it enables efficient parallel processing so that the computations for each head can be executed simultaneously without incurring additional computational complexity. Secondly, integrating the results from multiple heads significantly enhances the stability and robustness of the model. This is achieved by aggregating diverse representations of the input sequence, which allows the model to capture a more comprehensive range of relationships.

The elements of the attention matrix $\mathbf{A}_m^{(l)} = (a_{m,ij}^{(l)}) \in \mathbb{R}^{n\times n}$, for the m-th head in the l-th layer, are defined below:

$$a_{m,ij}^{(l)} = \frac{\exp\left((\mathbf{h}_i^{(l-1)}\mathbf{W}_{q,m}^{(l)})(\mathbf{h}_j^{(l-1)}\mathbf{W}_{k,m}^{(l)})'/\sqrt{d_k}\right)}{\sum_{s=1}^{n}\exp\left((\mathbf{h}_i^{(l-1)}\mathbf{W}_{q,m}^{(l)})(\mathbf{h}_s^{(l-1)}\mathbf{W}_{k,m}^{(l)})'/\sqrt{d_k}\right)}, \tag{9.8}$$

for $l = 1, \cdots, L$ and $m = 1, 2, \cdots, M$, where L is the total number of layers, M is the total number of attention heads, $\mathbf{W}_{q,m}^{(l)} \in \mathbb{R}^{q \times d_k}$ and $\mathbf{W}_{k,m}^{(l)} \in \mathbb{R}^{q \times d_k}$ are the query and key parameter matrices associated with the m-th head and l-th layer, respectively, and $d_k = q/M$ for each head. The outputs of the multi-head are obtained by concatenating the results of all heads as follows:

$$\text{MultiHead(Q, K, V)} = \text{Concat}\Big(\mathbf{A}_1^{(l)}\mathbf{H}^{(l-1)}\mathbf{W}_{v,1}, \cdots, \mathbf{A}_M^{(l)}\mathbf{H}^{(l-1)}\mathbf{W}_{v,M}\Big)\mathbf{W}_O^{(l)}, \tag{9.9}$$

where $\mathbf{W}_{v,m}^{(l)} \in \mathbb{R}^{q \times d_v}$ is the value matrix associated with the l-th layer and the m-th head, and $\mathbf{W}_O^{(l)} \in \mathbb{R}^{Md_v \times q}$ is a weight matrix for generating the final representation. Note that the Concat$(\cdot)$ operator is used to stack the outputs of the M heads. For example, suppose that the input has a dimension of $n \times q$, where $q = 64$ and the number of attention heads is $M = 8$. Then, each attention head generates an output of size $n \times d_v$, where $d_v = q/M = 8$. By stacking the outputs of all 8 attention heads together along the feature dimension, we obtain a combined representation of size $n \times 64$ (Vaswani et al. 2017). After concatenating the outputs of all M attention heads, we integrate them with the weight matrix $\mathbf{W}_O^{(l)} \in \mathbb{R}^{Md_v \times q}$ to generate the final representation of the multi-head attention mechanism.

After completing the discussion of attention in the Encoder on the left-hand side of Fig. 9.9, we next discuss the attention mechanisms in the Decoder on the right-hand side of Fig. 9.9. They are cross-attention (also known as encoder–decoder attention) and masked multi-head attention. In the cross-attention mechanism, the query is generated from the output of the previous Decoder layer. Its effect can be represented by $\mathbf{H}_{\text{Decoder}}^{(l-1)}\mathbf{W}_q$. However, the key and value are from the output of the Encoder, and their effects are calculated by $\mathbf{H}_{\text{Encoder}}\mathbf{W}_k$ and $\mathbf{H}_{\text{Encoder}}\mathbf{W}_v$, respectively. This configuration integrates the two fundamental components of the Transformer architecture that enhance the decoder abilities.

We next introduce the masked multi-head attention. In language generation tasks, it is important to ensure that words can be predicted in a sequential manner. Hence, it is essential to prevent the model from seeing future information during the prediction process, as this would lead to unrealistic and incorrect outputs. To this end, the Decoder utilizes masked multi-head attention. Specifically, this mechanism ensures that, at each prediction step, the model predictions only rely on previously generated tokens without involving future ones. Accordingly, during the training process, a mask is applied by setting the attention weights of future tokens to zero.

3. Layer Normalization

Layer normalization plays an important role in stabilizing the training process by re-centering and re-scaling the inputs. For the q-dimensional vector $X = (x_1, x_2, \cdots, x_q)'$, the process of layer normalization is given as follows: $\text{LayerNorm}(X) = \gamma \odot \frac{X-\mu}{\sigma} + \beta$, where $\mu = \frac{1}{q}\sum_{i=1}^{q} x_i$ and $\sigma = \sqrt{\frac{1}{q}\sum_{i=1}^{q}(x_i - \mu)^2}$. Note that the γ and β are learnable parameters, and $\odot$ denotes element-wise multiplication (Xu et al. 2019), which is the Hadamard product defined in Sect. 1.6.3.

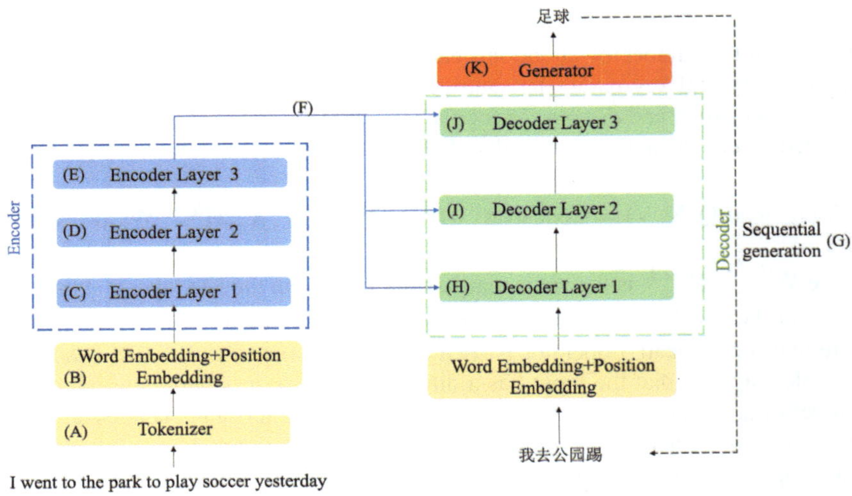

Fig. 9.10 A Transformer model is used to translate the English sentence *"I went to the park to play soccer yesterday"* into the Chinese sentence *"wo qu gong yuan ti zu qiu le zuo tian"*

4. Feedforward Neural Network (FNN)

The FNN component can be naturally applied to sequence data. Its structure has been presented in Fig. 9.1.

5. Residual Connection (RC)

During neural network training, residual connection (He et al. 2016) is designed to mitigate the vanishing gradient problem and model degradation by introducing short paths. Specifically, for each sublayer in FNN (or multi-head attention), the input $\mathbf{H}$ is added to the output of the sublayer ($\mathbf{H}$), resulting in $\mathbf{H}+$ sublayer($\mathbf{H}$). Accordingly, the residual connection enhances the speed and reliability of parameter updates during the FNN training process, leading to faster and more stable convergence. This, together with some activation operators and the layer normalization, serves as the input to the next layer. In Fig. 9.9, this operation is labeled as "RC".

Having discussed each of the components appearing in the Encoder (including the attention components of the Decoder), we now turn to the Decoder on the right-hand side of Fig. 9.9. The Decoder layer basically consists of four main components: 1. masked multi-head attention; 2. layer normalization; 3. cross-attention; and 4. feedforward Neural Network (FNN). Each of these components has been introduced in the context of the Encoder, so we refer the reader to the discussion above.

After introducing all components of the Transformer depicted in Fig. 9.9, we next present an analytical example to illustrate a machine translation task via the Transformer; see Fig. 9.10. In step (A), each word in the input sentence *"I went to the park to play soccer yesterday"* is tokenized. Subsequently, in step (B), these tokens are converted into word embeddings. These word embeddings are then combined with positional embeddings to form the input to the model. In steps C, D, and E, the inputs are fed into the Encoder. Here, the multi-head attention mechanism is used to capture the contextual relationships between the words. The final output of the Encoder is step (F), which serves as the relevant context for the Decoder to start its translation process. In step (G), the Decoder generates the translation in the target language. To this end, step (G) makes use of both cross-attention and masked attention to integrate the output from the Encoder and the previous decoding state. Within each decoder layer (steps H, I, and J), the model further refines the translation process by combining the encoder context and previously generated tokens from the Decoder. Finally, in step (K), the generator produces the translated word "zuqiu," which means "soccer" in Chinese. This process continues iteratively until the entire sentence has been successfully translated.

As previously discussed, the Encoder–Decoder structure depicted in Figs. 9.9 and 9.10 is widely employed in machine-based language translation. In addition, two other architectures can be considered: the Encoder-only and Decoder-only types. Encoder-only models are well-suited for tasks that require text understanding, such as sentence classification and entity recognition. Decoder-only models, on the other hand, are utilized for generative tasks like text generation. In large language models (LLMs), the GPT (Generative Pre-trained Transformer) series (see Radford et al. 2018), DeepSeek models (see Liu et al. 2024), and Gemini 3 Pro introduced by Google in 2025 serve as prominent examples of the Decoder-only architecture. In contrast, the BERT (Bidirectional Encoder Representations from Transformers) series (Devlin et al. 2019) act like powerful Encoder-only models. All of these models are designed to generate text by leveraging patterns and knowledge acquired during a pre-training phase on vast datasets. More importantly, they have demonstrated their impressive ability to execute large language models and have thereby profoundly influenced the lives of human beings. A historical review of BERT and GPT can be found in Zhao et al. (2023).

9.6 Covariance Matrices and Deep Learning

9.6.1 Covariance Matrix Estimation via Deep Learning

Deep learning methods are typically focusing on the conditional mean $E(\mathbf{Y}|\mathbf{X})$ of the response $\mathbf{Y}$ for given covariates $\mathbf{X}$. Recently, however, several studies have utilized deep learning to estimate the conditional covariance matrix $\text{Cov}(\mathbf{Y}|\mathbf{X})$. Hence, this subsection presents three types of methods for estimating covariance matrices. The first method employs autoencoders to estimate a covariance matrix within the context of high-dimensional factor models, while the second approach applies deep learning methods to model the elements of the Cholesky decomposition of the inverse of a covariance matrix. Finally, the third method uses Graph

Convolutional Networks (GCNs) to estimate covariance matrices by incorporating both feature and network information. We next discuss them in detail.

1. Deep learning for covariance matrix estimation via factor models
In high-dimensional data, estimating the covariance matrix becomes a challenging task. In this case, the factor model studied in Chap. 8 offers an effective approach for covariance matrix modeling. As described in Chap. 8, the factor model typically assumes the following form (see Anderson 2003):

$$x_{it} = \alpha_{it} + \boldsymbol{\beta}_i' \mathbf{f}_t + \epsilon_{it}, \tag{9.10}$$

where x_{it} represents the observation of unit i at time t for $i = 1, \cdots, n$ and $t = 1, \cdots, T$. In addition, α_{it} is the intercept, $\mathbf{f}_t = (f_{1t}, \cdots, f_{mt})' \in \mathbb{R}^m$ denotes the m-dimensional latent factors, $\boldsymbol{\beta}_i = (\beta_{i1}, \cdots, \beta_{im})' \in \mathbb{R}^m$ comprises the unknown factor loadings, and ϵ_{it} satisfies the condition $E(\epsilon_{it}|\mathbf{f}_t) = 0$ for all $t = 1, \cdots, T$.

Under (9.10), the covariance matrix of $X_t = (x_{1t}, x_{2t}, \cdots, x_{nt})'$ can be expressed below:

$$\boldsymbol{\Sigma} = \mathbf{B}\boldsymbol{\Sigma}_f \mathbf{B}' + \boldsymbol{\Sigma}_{\boldsymbol{\varepsilon}}, \tag{9.11}$$

where $\mathbf{B} = (\boldsymbol{\beta}_1, \cdots, \boldsymbol{\beta}_n)' \in \mathbb{R}^{n\times m}$ is the factor loading matrix, $\boldsymbol{\Sigma}_f = \mathrm{Cov}(\mathbf{f}_t)$ represents the covariance matrix of the latent factors, and $\boldsymbol{\Sigma}_{\boldsymbol{\varepsilon}} = \mathrm{Cov}(\boldsymbol{\varepsilon}_t)$ is the covariance matrix of the error vector $\boldsymbol{\varepsilon}_t = (\epsilon_{1t}, \cdots, \epsilon_{nt})'$. Note that Chap. 8 assumes that $\boldsymbol{\Sigma}_f = \mathbf{I}_m$, but this section relaxes this assumption so that the latent factors can be correlated, allowing for more complex relationships between them and the observed variables. In addition, we adopt the notation X_t and (n, T) from (9.10), but it is only used in the context of factor models. In the other contexts, n denotes the sample size and p represents the dimension of $\boldsymbol{\Sigma}$.

Based on (9.11), to estimate the covariance matrix $\boldsymbol{\Sigma}$, one needs to estimate $\mathbf{B}$, $\boldsymbol{\Sigma}_f$, and $\boldsymbol{\Sigma}_{\boldsymbol{\varepsilon}}$. A common approach to estimate $\mathbf{B}$ and $\boldsymbol{\Sigma}_f$ is the principal component analysis (PCA), in which $\boldsymbol{\Sigma}_{\boldsymbol{\varepsilon}}$ is typically assumed to be a diagonal matrix and estimated from the residuals. Since the number of parameters involved in $\mathbf{f}_t$ and $\mathbf{B}$ are of order $O(n)$, the parameters that need to be estimated are very large, especially as n grows.

To overcome this challenge, some deep learning methods have attracted attention. For example, Huynh and Lenhard (2022) proposed modeling the factor $\mathbf{f}_t$ as a function of the observations X_t. Specifically, they assumed $\mathbf{f}_t = g(X_t)$, where $g : \mathbb{R}^n \to \mathbb{R}^m$ is a mapping function, and m is the dimension of $\mathbf{f}_t$. Based on the factor model described in (9.10), the corresponding loss function, denoted by $\mathcal{L}$, is formulated as follows:

$$\mathcal{L}(\mathbf{B}, g, \boldsymbol{\alpha}_t) = \frac{1}{T}\sum_{t=1}^{T} \|X_t - \mathbf{B}g(X_t) - \boldsymbol{\alpha}_t\|^2, \tag{9.12}$$

where $\boldsymbol{\alpha}_t = (\alpha_{1t}, \alpha_{2t}, \cdots, \alpha_{nt})'$. To estimate the unknown function g, Huynh and Lenhard (2022) proposed an asymmetric autoencoder with L encoding layers.

Analogous to the Transformer framework discussed in Sect. 9.5.3, an autoencoder typically consists of two main parts (see Hinton and Salakhutdinov 2006), namely the encoder and the decoder. Firstly, an encoder $f : \mathbb{R}^n \to \mathbb{R}^m (m \ll n)$ that compresses input data into a lower-dimensional representation. Accordingly, the encoder f is used to estimate g, and X_t acts as the input of the network. In addition, $\mathbf{a}^{(l)}$ represents the output in the l-th hidden layer of the encoder for $l = 1, \cdots, L$, which is computed as follows: $\mathbf{a}^{(l)} = \sigma\left(\mathbf{W}^{(l)}\mathbf{a}^{(l-1)} + \mathbf{b}^{(l)}\right)$; see Sect. 9.5 for more details. Here, $\mathbf{a}^{(0)} = X_t$, $\mathbf{W}^{(l)}$, and $\mathbf{b}^{(l)}$ are unknown parameters, and σ is an activation function. Accordingly, the final layer of the encoder yields the estimated factors.

Note that activation functions generally transform linear functions into nonlinear ones, and the output of these functions is often restricted to a particular interval, such as $[0, 1]$ or $[0, +\infty)$. However, because the latent factors $\mathbf{f}_t$ can be any value within the range $(-\infty, +\infty)$, the activation function is omitted from the last encoding layer. Consequently, the estimated latent factors $\mathbf{f}_t$ are given by $\widehat{\mathbf{f}_t} = \mathbf{a}^{(L)} = \mathbf{W}^{(L)}\mathbf{a}^{(L-1)} + \mathbf{b}^{(L)} = \mathbf{W}^{(L)} \cdot \sigma\left(\mathbf{W}^{(L-1)} \cdot \ldots \sigma\left(\mathbf{W}^{(1)}X_t + \mathbf{b}^{(1)}\right) \cdots + \mathbf{b}^{(L-1)}\right) + \mathbf{b}^{(L)}$. For the sake of simplicity, we define $\boldsymbol{\phi}_l = \mathbf{W}^{(l)}\mathbf{a}^{(l-1)} + \mathbf{b}^{(l)}$. Then, the estimated latent factors are denoted by $\widehat{\mathbf{f}_t} = \boldsymbol{\phi}_L \circ \sigma \circ \boldsymbol{\phi}_{L-1} \circ \sigma \circ \boldsymbol{\phi}_{L-2} \circ \cdots \circ \sigma \circ \boldsymbol{\phi}_1$.

Secondly, the decoder involves a single layer $h : \mathbb{R}^m \to \mathbb{R}^n$, that employs the outputs of the encoder, $\widehat{\mathbf{f}_t}$, to generate their predictions in the output layer. Specifically, $\widehat{X}_t = \mathbf{W}\widehat{\mathbf{f}_t} + \mathbf{b}$, where $\mathbf{W}$ is the weight matrix and $\mathbf{b}$ is the bias vector. In the context of the factor model, $\mathbf{W} = \mathbf{B}$ and $\mathbf{b} = \boldsymbol{\alpha}_t$. In other words, $\mathbf{B}$ and $\boldsymbol{\alpha}_t = \boldsymbol{\alpha}$ correspond to the parameters of the fully connected layer of the decoder. For illustrative purposes and for the sake of simplicity, Fig. 9.11 presents the architecture of an autoencoder with $L = 2$. Note that this architecture is adapted from Figure 1 of Huynh and Lenhard (2022), in which $L = 3$ as considered.

We next estimate those unknown parameters involved in the process of the autoencoder. The parameters in the encoder are $\mathbf{W}^{(l)}$ and $\mathbf{b}^{(l)}$ for $l = 1, \cdots, L$, and the parameters in the decoder are $\mathbf{B}$ and $\boldsymbol{\alpha}_t$ for $t = 1, \cdots, T$. For the sake of convenience, we group all parameters together and refer to them as $\boldsymbol{\theta}$. Since the number of parameters is relatively large, a ridge penalty is added into the loss function (9.12) with $\mathbf{f}_t = \widehat{\mathbf{f}_t}$ to avoid overfitting. The resulting penalty function is

$$\mathcal{L}(\boldsymbol{\theta}) = \frac{1}{T}\sum_{t=1}^{T} \|X_t - \mathbf{B}(\boldsymbol{\phi}_L \circ \sigma \circ \boldsymbol{\phi}_{L-1} \circ \sigma \circ \boldsymbol{\phi}_{L-2} \circ \cdots \circ \sigma \circ \boldsymbol{\phi}_1) - \boldsymbol{\alpha}_t\|^2 + \lambda\|\boldsymbol{\theta}\|^2, \tag{9.13}$$

where λ is a regularization parameter. By minimizing the loss function (9.13), we are able to estimate the factor loading matrix $\mathbf{B}$ and the factors $\mathbf{f}_t$. Accordingly, the covariance matrix of X_t, $\boldsymbol{\Sigma}$, in (9.11) can be estimated using the results of $\widehat{\mathbf{B}}$,

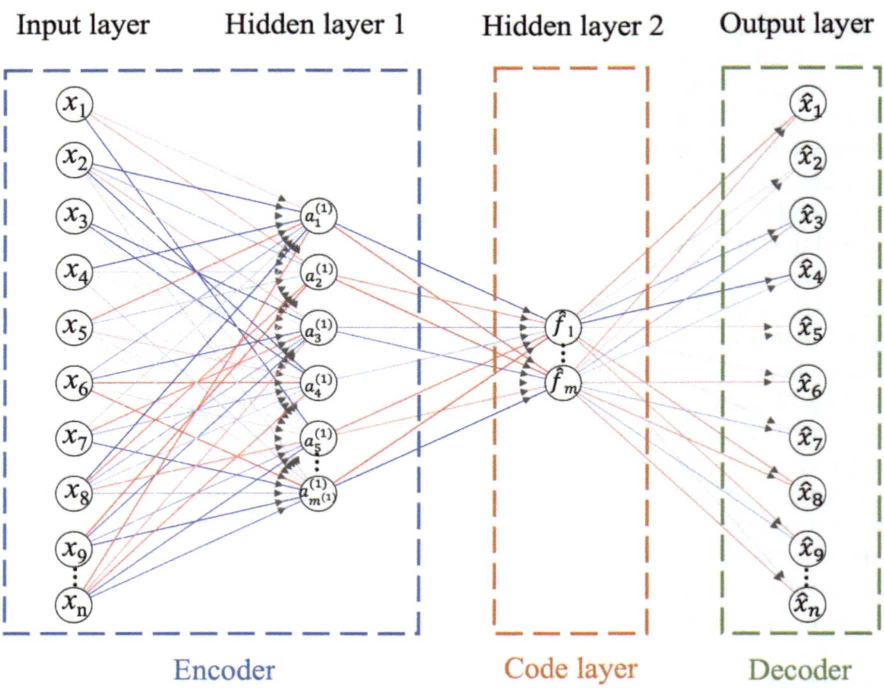

Fig. 9.11 The encoder estimates latent factors through multiple layers, each applying $\mathbf{a}^{(l)} = \sigma\left(\mathbf{W}^{(l)}\mathbf{a}^{(l-1)} + \mathbf{b}^{(l)}\right)$ for $l = 1, 2$. The code layer reduces the input dimension from n to m. The decoder is a linear layer that connects the code layer to the decode layer

$\widehat{\boldsymbol{\Sigma}}_f = T^{-1}\sum_{t=1}^{T}(\widehat{\mathbf{f}}_t - \bar{\mathbf{f}})(\widehat{\mathbf{f}}_t - \bar{\mathbf{f}})'$, and $\widehat{\boldsymbol{\Sigma}}_{\boldsymbol{\varepsilon}} = T^{-1}\sum_{t=1}^{T}\widehat{\boldsymbol{\varepsilon}}_t\widehat{\boldsymbol{\varepsilon}}_t'$, where $\bar{\mathbf{f}}$ is the sample mean of $\widehat{\mathbf{f}}_t$ and $\widehat{\boldsymbol{\varepsilon}}_t = X_t - \widehat{\mathbf{B}}\widehat{\mathbf{f}}_t - \widehat{\boldsymbol{\alpha}}_t$ are the estimators of error terms for $t = 1, \cdots, T$.

In the above discussion, the factor loadings $\boldsymbol{\beta}_i$ $(i = 1, \cdots, n)$ are assumed to be time-invariant. In real practice, however, $\boldsymbol{\beta}_i$ can be time-varying; see Su and Wang (2017). To relax this limitation, Wang and Singh (2024) modeled $\boldsymbol{\beta}_i$ as a function of the input features $Z_t \in \mathbb{R}^q$ and estimated them using Kolmogorov–Arnold Networks (KANs); see Liu et al. (2025). This approach is designed to account for time-varying factor loadings so that the model is suitable in real-world applications.

2. Deep learning for covariance matrix estimation via the Cholesky decomposition

In addition to factor-based models discussed above, some studies have concentrated on estimating covariance matrices via Cholesky decompositions (see Williams 1996; Dorta et al. 2018; Barratt and Boyd 2023). These methods basically utilize deep learning techniques to estimate the elements of the lower triangular matrix, which is obtained using the Cholesky decomposition on the inverse of the covariance matrix.

Let $Y \in \mathbb{R}^p$ be a multivariate normal random response vector with mean $\mathbf{0}$ and covariance matrix $\boldsymbol{\Sigma}$. In addition, let $U \in \mathbb{R}^q$ be a covariate vector. To estimate $\boldsymbol{\Sigma}$ with a sample size of only one (i.e., $n = 1$), Williams (1996) employed neural networks, together with the covariate vector U, to estimate the elements of the Cholesky decomposition of the inverse covariance matrix $\boldsymbol{\Sigma}^{-1}$. To this end, we apply the Cholesky decomposition introduced in Sect. 1.6.4 and obtain that

$$\boldsymbol{\Sigma}^{-1} = \mathbf{G}\mathbf{G}',$$

where $\mathbf{G} = (g_{j_1 j_2}) \in \mathbb{R}^{p \times p}$ is a lower triangular matrix with strictly positive diagonal elements. Then, define the transformation of $\mathbf{G}$ as follows: $T(\mathbf{G}) = (\tilde{g}_{j_1 j_2}) \in \mathbb{R}^{p \times p}$, where

$$\tilde{g}_{j_1 j_2} = \begin{cases} \log(g_{j_1 j_1}) & \text{for } j_1 = j_2 \\ g_{j_1 j_2} & \text{for } j_1 \neq j_2 \end{cases}.$$

Subsequently, consider

$$H^{(L)} = \text{vech}(T(\mathbf{G})), \tag{9.14}$$

where vech($\mathbf{M}$) is an operator that vectorizes the low triangular part (including the diagonal) of any square matrix $\mathbf{M}$ into a column vector.

Note that $H^{(L)} \in \mathbb{R}^{p(p+1)/2}$ is the final output of an FNN with L layers that was introduced in Sect. 9.5, where the l-th layer is

$$H^{(l)} = \sigma(\mathbf{W}^{(l)} H^{(l-1)} + \mathbf{b}^{(l)}),$$

for $l = 1, \cdots, L$. In addition, $H^{(l)} \in \mathbb{R}^{r^{(l)}}$, $\mathbf{W}^{(l)} \in \mathbb{R}^{r^{(l)} \times r^{(l-1)}}$, and $\mathbf{b}^{(l)} \in \mathbb{R}^{r^{(l)}}$ are the output vector, weight matrix, and bias vector of the l-th layer, respectively. Moreover, the input vector is $H^{(0)} = U$ and $r^0 = q$. The detailed architecture of this procedure is depicted in Fig. 9.12.

Let $\boldsymbol{\theta}$ be an unknown parameter vector, where each element represents a parameter in $\mathbf{W}^{(l)}$ and $\mathbf{b}^{(l)}$, for $l = 1, \cdots, L$. To estimate it, we consider the maximum likelihood approach by maximizing the following log-likelihood after omitting irrelevant constants,

$$\ell_{FNN}(\boldsymbol{\theta}) = \log(|\mathbf{G}(\boldsymbol{\theta})|) - \frac{1}{2}\|\mathbf{G}(\boldsymbol{\theta})'Y\|_2^2, \tag{9.15}$$

where $\mathbf{G}(\boldsymbol{\theta}) := \mathbf{G} = T^{-1}(\text{vech}^{-1}(H^{(L)}))$. In addition, $T^{-1}(\cdot)$ and $\text{vech}^{-1}(\cdot)$ are the inverse operators of $T(\cdot)$ and vech$(\cdot)$, respectively. Accordingly, we obtain the estimated $\boldsymbol{\theta}$, that is $\widehat{\boldsymbol{\theta}}_{FNN} = \text{argmax}_{\boldsymbol{\theta}} \ell_{FNN}(\boldsymbol{\theta})$. This leads to the estimator of $\mathbf{G}$, $\widehat{\mathbf{G}}_{FNN} = \mathbf{H}(\widehat{\boldsymbol{\theta}}_{FNN})$, and the estimator of $\boldsymbol{\Sigma}$, $\widehat{\boldsymbol{\Sigma}} = (\widehat{\mathbf{G}}_{FNN}\widehat{\mathbf{G}}'_{FNN})^{-1}$.

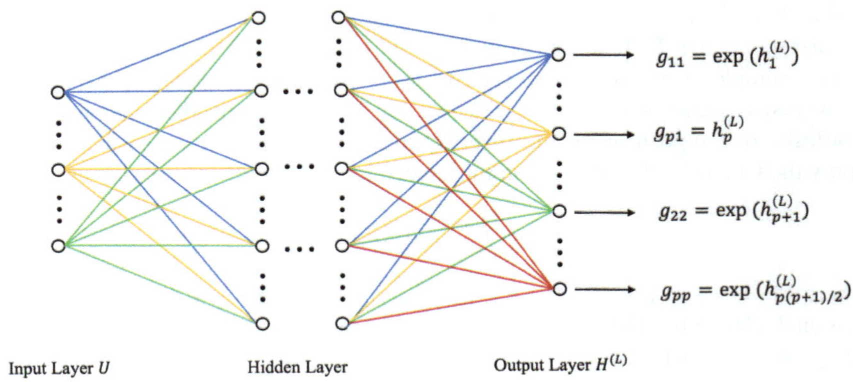

Fig. 9.12 The architecture of an FNN for covariance matrix estimation, where $H^{(L)} = (h_1^{(L)}, \cdots, h_{p(p+1)/2}^{(L)})' \in \mathbb{R}^{p(p+1)/2}$ is the final output of the FNN with L layers

As shown in Fig. 9.12, when the dimension p is large, the matrix $\mathbf{G}$ typically involves a vast number of unknown parameters. This not only results in high computational costs but also raises concerns about potential overfitting. To overcome this challenge and simplify the FNN process, Dorta et al. (2018) proposed a strategy for image data. Specifically, they imposed a fixed sparsity pattern on the matrix $\mathbf{G}$ and used FNN to estimate only the non-zero elements of the matrix. In other words, for $j_1 \geq j_2$, they assume that $g_{j_1 j_2} \neq 0$ if a patch of size s centered at j_1 includes the pixel j_2. As a result, when pixels j_1 and j_2 are neighbors in the image plane, $g_{j_1 j_2}$ is non-zero; otherwise, $g_{j_1 j_2} = 0$. Under this strategy, the number of the non-zero elements in $\mathbf{G}$ is reduced from $p(p+1)/2$ to $p((s^2-1)/2+1)$, where $s \ll p$. Consequently, in a high-dimensional scenario, the $\mathbf{G}$ can be estimated efficiently using Eq. (9.15).

In addition to the challenge due to high-dimension data, Barratt and Boyd (2023) noted that the optimization in (9.15) is non-convex. This non-convexity raises a significant challenge for finding the globally optimal estimator. Hence, Barratt and Boyd (2023) proposed an alternative approach to model the elements of the matrix $\mathbf{G}$, and obtain estimators with a convex regularizer.

Specifically, they modeled $\mathbf{G}$ as follows:

$$\text{diag}(\mathbf{G}) = \mathbf{A}U + B \text{ and } \text{offdiag}(\mathbf{G}) = \mathbf{C}U + D,$$

where diag($\cdot$) is the vector of diagonal entries, and offdiag($\cdot$) extracts the strictly lower-triangular entries in a prespecified fixed order. In addition, $\mathbf{A} \in \mathbb{R}^{p \times q}$, $B \in \mathbb{R}^p$, $\mathbf{C} \in \mathbb{R}^{\frac{p(p-1)}{2} \times q}$, and $D \in \mathbb{R}^{\frac{p(p-1)}{2}}$. The covariate vector U is scaled such that $\|U\|_\infty \leq 1$.

To estimate the unknown parameters $\mathbf{A}$, B, $\mathbf{C}$, and D, Barratt and Boyd (2023) introduced the following objective function,

$$\ell(\mathbf{A}, B, \mathbf{C}, D) = \log(|\mathbf{G}|) - \frac{1}{2}\|\mathbf{G}'\mathbf{Y}\|_2^2 - R(\mathbf{A}, B, \mathbf{C}, D), \tag{9.16}$$
$$s.t. \quad \text{diag}(\mathbf{G}) = \mathbf{A}U + B, \quad \text{offdiag}(\mathbf{G}) = \mathbf{C}U + D, \quad \|\mathbf{A}\|_{\text{row},1} \leq B - \epsilon,$$

where $\epsilon > 0$. Note that $R(\cdot)$ is a convex regularizer, $\|\mathbf{A}\|_{\text{row},1} \in \mathbb{R}_+^p$ is the vector of ℓ_1 norms of the rows of $\mathbf{A}$, and the last constraint guarantees that $\text{diag}(\mathbf{G}) \geq \epsilon$ for any $\|U\|_\infty \leq 1$. The function $R(\cdot)$ can be either the trace inverse regularization or dual norm regularization, introduced in Barratt and Boyd (2023). Incorporating such a convex regularizer ensures that Eq. (9.16) is a convex optimization problem. Accordingly, the estimators of $\mathbf{A}, B, \mathbf{C}, D$ are obtained as follows: $\widehat{\mathbf{A}}, \widehat{B}, \widehat{\mathbf{C}}, \widehat{D} = \text{argmax}_{\mathbf{A},B,\mathbf{C},D}\ell(\mathbf{A}, B, \mathbf{C}, D)$. As a result, $\text{diag}(\widehat{\mathbf{G}}) = \widehat{\mathbf{A}}U + \widehat{B}$ and $\text{offdiag}(\widehat{\mathbf{G}}) = \widehat{\mathbf{C}}U + \widehat{D}$. Subsequently, the estimator of the covariance matrix is $\widehat{\mathbf{\Sigma}} = (\widehat{\mathbf{G}}\widehat{\mathbf{G}}')^{-1}$. Analogous to the approaches in Williams (1996) and Dorta et al. (2018), we can straightforwardly employ neural networks to model $\text{diag}(\mathbf{G})$ and $\text{offdiag}(\mathbf{G})$ for finding the estimator of $\mathbf{\Sigma}$.

3. Covariance matrix estimation via GCN

We have applied two deep learning methods to estimate $\mathbf{\Sigma}$. Motivated by the approach of Wang et al. (2023), we utilize graph convolutional networks (GCNs) to effectively combine both covariate and network information together to estimate covariance matrices. To this end, we assume that $Y \in \mathbb{R}^p$ is a multivariate normal random vector with mean $\mathbf{0}$ and covariance matrix $\mathbf{\Sigma}$. In addition, let $\mathbf{Z} = (Z_1, \cdots, Z_p)' \in \mathbb{R}^{p\times q}$ be the covariate matrix, where the $Z_j \in \mathbb{R}^q$ are q-dimensional covariates for $j = 1, \cdots, p$. We then consider the adjacency matrix $\mathbf{A} = (a_{j_1j_2}) \in \mathbb{R}^{p\times p}$ that encodes the network information among p units, where $a_{j_1j_2} = 1$ if there exists an edge from j_1 to j_2 and $a_{j_1j_2} = 0$ otherwise. For the sake of completeness, we assume $a_{jj} = 0$.

It is worth noting that the adjacency matrix can be directed ($\mathbf{A} \neq \mathbf{A}'$) or undirected ($\mathbf{A} = \mathbf{A}'$). We then model the covariance matrix as

$$\mathbf{\Sigma} = \lambda\mathbf{I_p} + \mathbf{H}^{(L)}\mathbf{H}^{(L)'},$$

where $\lambda > 0$ is a weight parameter, and $\mathbf{H}^{(L)} \in \mathbb{R}^{p\times r^{(L)}}$ is the final output of a GCN model with L layers; see Sect. 9.5.2. This output can be obtained via the following iterative process:

$$\mathbf{H}^{(l)} = \sigma\left(\tilde{\mathbf{D}}^{-\frac{1}{2}}\tilde{\mathbf{A}}\tilde{\mathbf{D}}^{-\frac{1}{2}}\mathbf{H}^{(l-1)}\mathbf{W}^{(l)}\right), \tag{9.17}$$

for $l = 1, \cdots, L$, where $\tilde{\mathbf{A}} = (\tilde{a}_{j_1j_2}) := \mathbf{A} + \mathbf{I}$ and $\tilde{\mathbf{D}} = \text{diag}\{\tilde{d}_1, \cdots, \tilde{d}_p\}$ with $\tilde{d}_{j_1} = \sum_{j_2}\tilde{a}_{j_1j_2}$ for $j_1, j_2 = 1, \cdots, p$. In addition, the matrix $\mathbf{W}^{(l)} \in \mathbb{R}^{r^{(l-1)}\times r^{(l)}}$ is the learnable parameter of the l-th layer, the matrix $\mathbf{H}^{(l)} \in \mathbb{R}^{p\times r^{(l)}}$ is the output in the l-th layer with $\mathbf{H}^{(0)} = \mathbf{Z}$, and $\sigma(\cdot)$ denotes an activation function for $l = 1, \cdots, L$.

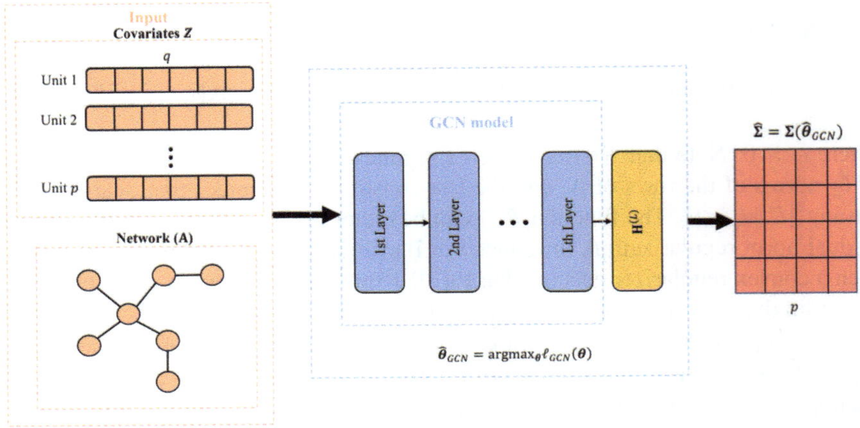

Fig. 9.13 The architecture of a GCN model for covariance matrix estimation

After the L iterating processes in (9.17), the final output $\mathbf{H}^{(L)}$ enables us to effectively capture both the covariate information and the information about the L-hop neighbors. Here, the $(l+1)$-hop neighbors refer to a set of paths connecting any two units j_1 and j_2 through l intermediate units. The architecture of a GCN model for covariance matrix estimation is shown in Fig. 9.13, which is adapted from Figure 1 of Wang et al. (2023).

We next employ the maximum likelihood method to estimate all unknown parameters involved in the GCN model. The log-likelihood function is given below:

$$\ell_{GCN}(\boldsymbol{\theta}) = -\frac{p}{2}\log(2\pi) - \frac{1}{2}\log|\boldsymbol{\Sigma}(\boldsymbol{\theta})| - \frac{1}{2}Y'\boldsymbol{\Sigma}^{-1}(\boldsymbol{\theta})Y, \tag{9.18}$$

where $\boldsymbol{\Sigma}(\boldsymbol{\theta}) = \lambda\mathbf{I_p} + \mathbf{H}^{(L)}\mathbf{H}^{(L)'}$, and $\boldsymbol{\theta}$ is the unknown parameter vector that results from $\mathbf{W}^{(l)}$, for $l = 1, \cdots, L$, and the weight parameter λ. Then, the estimator of $\boldsymbol{\theta}$ is $\widehat{\boldsymbol{\theta}}_{GCN} = \text{argmax}_{\boldsymbol{\theta}}\ell_{GCN}(\boldsymbol{\theta})$. Accordingly, the estimator of $\boldsymbol{\Sigma}$ is $\widehat{\boldsymbol{\Sigma}} = \boldsymbol{\Sigma}(\widehat{\boldsymbol{\theta}}_{GCN})$. Note that the maximum likelihood estimator relies on the normality assumption of Y. Without imposing this assumption, the $\boldsymbol{\theta}$ can be estimated by minimizing the least-squares loss function $\|YY' - \boldsymbol{\Sigma}(\boldsymbol{\theta})\|$.

9.6.2 *Improving Deep Learning using Covariance Matrices*

In the previous section, we have applied deep learning methods to estimate the covariance matrix. In contrast, some researchers have incorporated covariance matrices to enhance deep learning models. Hence, this section introduces the three pertinent methods, which are given below. 1. Global covariance pooling, which uses the covariance matrix to capture the second-order feature interactions; 2.

Covariance neural networks, which utilize the covariance matrix for graph filters, thereby enhancing feature representation in convolutional operations; and (3) Cross-covariance attention, which employs the covariance matrix to reduce computational complexity in Transformer models.

1. Global covariance pooling

Let $\mathbf{X} \in \mathbb{R}^{d_1 \times d_2 \times d_3}$ be the output of the final convolution layer in CNNs, where d_1, d_2, and d_3 are the width, height, and depth (dimension) of the feature. Traditional pooling methods such as mean, max, and sum pooling only consider the first-order representation of features. For instance, the feature tensor $\mathbf{X}$ is reshaped into a matrix $\bar{\mathbf{X}} \in \mathbb{R}^{d_1 d_2 \times d_3}$, where $\bar{X}_i$ represents the i-th row of $\bar{\mathbf{X}}$ for $i = 1, \cdots, d_1 d_2$. Accordingly, the sum pooling is $x_s^{\text{1-order}} = \sum_{i=1}^{d_1 d_2} \bar{X}_i$, and the mean pooling is $x_m^{\text{1-order}} = (d_1 d_2)^{-1} \sum_{i=1}^{d_1 d_2} \bar{X}_i$. These approaches, however, only account for first-order representation without capturing the interactions between features.

In recent work, by utilizing the covariance matrix of features, Lin et al. (2015) and Li et al. (2017) proposed global covariance pooling methods to characterize second-order representation with the form of $x_c^{\text{2-order}} = \bar{\mathbf{X}}'\mathbf{J}\bar{\mathbf{X}}$, where $\mathbf{J} = \frac{1}{d_1 d_2}\left(\mathbf{I} - \frac{1}{d_1 d_2}\mathbf{1}\mathbf{1}'\right)$ and $\mathbf{I}$ is the $(d_1 d_2) \times (d_1 d_2)$-dimensional identity matrix. It follows that the second-order representation $x_c^{\text{2-order}}$ is based on the sample covariance matrix of $\bar{\mathbf{X}}$.

It is noteworthy that the sample covariance matrix is usually not stable (see Bai 1999; Fan et al. 2016). To improve model performance by mitigating biases and stabilizing covariance estimators, various normalization methods have been proposed. For example, by assuming that $\bar{\mathbf{X}}$ has been centralized, Ionescu et al. (2015) proposed the matrix logarithm normalization $\log(\bar{\mathbf{X}}'\bar{\mathbf{X}} + \tau\mathbf{I})$, where τ is a tuning parameter to prevent small or zero eigenvalues in the sample covariance matrix. The logarithm-transformed matrix operator was previously discussed in Sect. 3.3.2. In addition, Li et al. (2017) and Wang et al. (2021) introduced matrix power normalization, where the sample covariance matrix is replaced by its power $(\bar{\mathbf{X}}'\mathbf{J}\bar{\mathbf{X}})^{\alpha}$, and $\alpha > 0$ is a prespecified constant. Note that the most widely used method is the matrix square-root normalization by setting $\alpha = 1/2$ (Lin and Maji 2017; Wang et al. 2017; Li et al. 2018; Wang et al. 2020). In addition, the effectiveness of global covariance pooling methods has been demonstrated by Wang et al. (2020) from the optimization perspective, and they found that global covariance pooling has three notable merits compared with traditional first-order pooling methods: (a) significantly sped up convergence of CNNs; (b) improved robustness against image corruption and perturbations; and (c) provided better generalizations across various vision tasks such as object detection.

2. CoVariance Neural Networks

The graph convolutional networks introduced in Sect. 9.5.2 relies on an observed network (or graph) structure. In some real applications, the network structure is partly observed or unobservable, which limits the practical utility of GCNs. To overcome this challenge, Sihag et al. (2022) introduced a covariance convolutional network architecture, coVariance Neural Network (VNN), which utilized the sample covariance matrix as a graph representation for constructing a covariance (graph) filter. This new architecture is inspired by the similarity between principal component analysis (PCA) and the graph Fourier transform (GFT), as both rely on eigen-decomposition.

In graph Fourier transform for network data, consider the Laplacian matrix $\mathcal{L} = \mathbf{D} - \mathbf{A}$, where $\mathbf{A} = (a_{ij})$ is the adjacency matrix and $\mathbf{D} = \text{diag}\{D_{11}, \cdots, D_{nn}\}$ is the associated degree matrix with $D_{ii} = \sum_j a_{ij}$. The Laplacian matrix provides a global perspective of the graph's structure (Chung 1997). We then decompose the Laplacian matrix $\mathcal{L}$ as $\mathcal{L} = \mathbf{V}\mathbf{\Lambda}_1\mathbf{V}'$, where $\mathbf{V}$ is the eigenvector matrix and $\mathbf{\Lambda}_1$ is its associated eigenvalue matrix. As a result, the graph Fourier transform is $\mathbf{X}^* = \mathbf{V}'\mathbf{X}$, where $\mathbf{X} = (X_1, \cdots, X_n)' \in \mathbb{R}^{n \times q}$ is the feature matrix. Similarly, in PCA, we decompose the covariance matrix $\mathbf{\Sigma} = \text{Cov}(\mathbf{X}_i)$ into $\mathbf{\Sigma} = \mathbf{U}\mathbf{\Lambda}_2\mathbf{U}'$, where $\mathbf{U}$ is the eigenvector matrix and $\mathbf{\Lambda}_2$ is its eigenvalue matrix. Accordingly, the PCA transform is $\tilde{\mathbf{X}} = \mathbf{U}'\mathbf{X}$.

Motivated by the similarity discussed above, Sihag et al. (2022) introduced the coVariance Fourier Transform (VFT) and coVariance Filters, which perform convolution operations on the sample covariance matrix. Specifically, the VFT projects the data into the eigen-space of the sample covariance matrix $\widehat{\mathbf{\Sigma}}_s$ using the eigenvector matrix $\mathbf{U}$ that is $\tilde{\mathbf{X}} = \mathbf{U}'\mathbf{X}$ as defined above. Accordingly, a coVariance filter is formulated as a polynomial of the sample covariance matrix $\widehat{\mathbf{\Sigma}}_s$ that is $H(\widehat{\mathbf{\Sigma}}_s) = \sum_{k=0}^m h_k \widehat{\mathbf{\Sigma}}_s^k$, where h_k is a real coefficient for $k = 0, \cdots, m$, and m is the degree of the polynomial used to characterize the complexity of the filter. In contrast, a traditional graph filter is a polynomial of the graph Laplacian $\mathcal{L}$ that is $H(\mathcal{L}) = \sum_{k=0}^m h_k \mathcal{L}^k$.

Note that the coVariance filter $H(\widehat{\mathbf{\Sigma}}_s)$ plays a crucial role in the fundamental building block of the VNN architecture, named the coVariance perceptron. It is defined as $\Phi(\mathbf{x}; \widehat{\mathbf{\Sigma}}_s, H) \triangleq \sigma\big(H(\widehat{\mathbf{\Sigma}}_s)\mathbf{x}\big)$, where σ is an activation function. In the VNN architecture, each layer sequentially performs two operations: (i) A covariance filter $H(\widehat{\mathbf{\Sigma}}_s) = \sum_{k=0}^m h_k \widehat{\mathbf{\Sigma}}_s^k$ is applied to integrate information from different signal scales (features); and (ii) Nonlinear activation functions $\sigma(\cdot)$ are utilized. Note that multiple coVariance perceptrons can be iteratively stacked to construct the deep-layered VNN architecture. As demonstrated by Sihag et al. (2022), VNN is more stable and robust compared to PCA.

3. Cross-covariance attention

In Sect. 9.5.3, we have briefly discussed the Transformer architecture and attention mechanism. Despite its remarkable success in various tasks, the Transformer faces challenges when dealing with long sequence datasets due to its quadratic-computation complexity. As defined in (9.7), for an input matrix $\mathbf{X} \in \mathbb{R}^{n \times q}$, the attention mechanism requires calculating the queries, keys, and values (i.e., $\mathbf{Q} = \mathbf{X}\mathbf{W}_q \in \mathbb{R}^{n \times d_k}$, $\mathbf{K} = \mathbf{X}\mathbf{W}_k \in \mathbb{R}^{n \times d_k}$, and $\mathbf{V} = \mathbf{X}\mathbf{W}_v \in \mathbb{R}^{n \times d_v}$). The resulting computational complexity is $O\{n \cdot q \cdot (2d_k + d_v)\}$. Subsequently, the attention mechanism for computing the Gram matrix, $\mathbf{Q}\mathbf{K}' = \mathbf{X}\mathbf{W}_q\mathbf{W}_k'\mathbf{X}' \in \mathbb{R}^{n \times n}$, results in the complexity $O(n^2 \cdot d_k)$. In addition, the output, Attention$(\mathbf{Q}, \mathbf{K}, \mathbf{V}) =$ Softmax$\left(\frac{\mathbf{QK}'}{\sqrt{d_k}}\right)\mathbf{V}$, has the computational complexity $O(n^2 \cdot d_v)$. In sum, the total computational complexity of the attention mechanism is $O\{n \cdot q \cdot (2d_k + d_v)\} + O(n^2 \cdot d_k) + O(n^2 \cdot d_v) \approx O(n^2 \cdot d_k)$ as $d_v = d_k = q$ and $n \gg q$. In addition to the computational cost, the memory required to store the attention matrix is $O(n^2)$. As n increases, hence, the growing computational cost and memory usage make the Transformer impractical, limiting its scalability.

To address this issue, El-Nouby et al. (2021) proposed the cross-covariance attention (XC-Attention) mechanism, XC-Attention$(\mathbf{Q}, \mathbf{K}, \mathbf{V}) = \mathbf{V} \cdot A_{XC}(\mathbf{K}, \mathbf{Q})$, where $A_{XC}(\mathbf{K}, \mathbf{Q}) =$ Softmax$\left(\frac{\mathbf{K}'\mathbf{Q}}{\tau}\right)$ and τ is a learnable parameter. This is different from the traditional attention that relies on the Gram matrix $\mathbf{QK}' = \mathbf{XW}_q\mathbf{W}_k'\mathbf{X}'$, while XC-Attention builds on $\mathbf{K}'\mathbf{Q} = \mathbf{W}_k'\mathbf{X}'\mathbf{XW}_q$ that is associated with the unnormalized sample covariance matrix $\widehat{\boldsymbol{\Sigma}} = n^{-1}\mathbf{X}'\mathbf{X}$. By operating on the sample covariance matrix $\widehat{\boldsymbol{\Sigma}} \in \mathbb{R}^{q \times q}$ rather than the Gram matrix $\mathbf{G} = \mathbf{XX}' \in \mathbb{R}^{n \times n}$, the computational burden shifts from token-level to feature-level. Given that the feature dimension d_k is typically much smaller than that of the number of tokens n (i.e., $n \gg d_k$), this approach reduces the computational complexity from $O(n^2 \cdot d_k)$ to $O(n \cdot d_k^2)$. As a result, it enables us to effectively alleviate the computational and memory constraints when dealing with a long sequence dataset.

It is worth noting that the sample covariance matrix is still used regardless of whether it captures second-order interactions, enhances feature representation, or reduces computational complexity. As previously mentioned, the sample covariance estimator can become problematic as its dimension becomes large (see Bai 1999; Fan et al. 2016). Therefore, we suggest utilizing the methodologies presented in Chaps. 4–8 to estimate the covariance matrix $\boldsymbol{\Sigma}$. Then, use the resulting covariance estimator to construct covariance pooling and covariance neural networks.

9.7 Transfer Learning and Covariance Matrices

9.7.1 Transfer Learning

In Sect. 9.6.1, we employed the three deep learning methods to estimate covariance matrices. However, these methods are only based on a single dataset. An alternative approach is integrating multiple data sources, and this can be a powerful tool to improve the learning accuracy with only the target dataset by leveraging information from different yet related source datasets. For the sake of illustration, Fig. 9.14 depicts the learning processes of traditional machine learning (ML) and transfer learning (TL) in panels (a) and (b), respectively; see, e.g., Pan and Yang (2010). It indicates that traditional ML techniques learn each task independently, while TL techniques transfer knowledge from some related tasks to a target task. A typical application of TL involves improving accuracy for car recognition by integrating

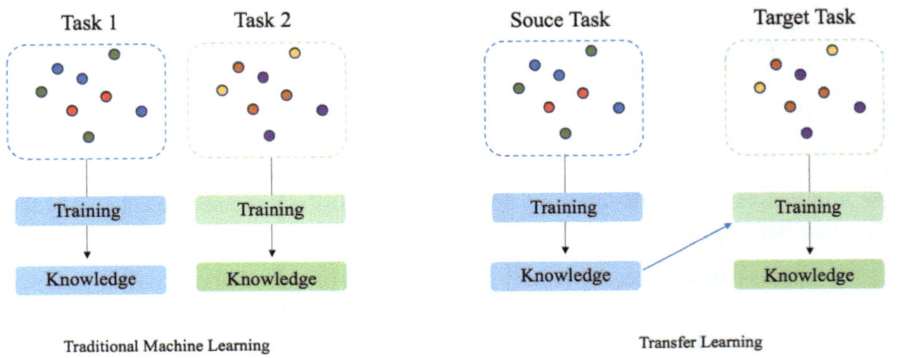

Fig. 9.14 The difference in learning processes between traditional machine learning and transfer learning

both car-labeled data and some truck-labeled data (Weiss et al. 2016). In this example, the car dataset is our target, whereas the truck-labeled dataset is source.

In the last two decades, transfer learning (TL) techniques have been extensively applied to diverse fields. In the context of computer vision, they have been widely used for image classification (Shaha and Pawar 2018) and object detection (Bu et al. 2021). In natural language processing, TL techniques also play a crucial role (Daumé 2007). In addition, TL has witnessed significant development in various biomedical applications. For example, it has predicted protein localization (Mei et al. 2011), assisted in biological imaging diagnosis (Shin et al. 2016), predicted drug sensitivity (Turki et al. 2017), and conducted integrative analysis of multi-omics data (Sun and Hu 2016). For exploring insights into transfer learning techniques, see Yang et al. (2020), Torrey and Shavlik (2010), Zhuang et al. (2021), and references therein. Despite significant achievements of TL in algorithm developments and real-life applications, only a few of these techniques contribute to establishing statistical theory in covariance matrix estimations. In the next two subsections, we will study TL in estimating the high-dimensional covariance matrix and its inverse, respectively, with some theoretical justifications.

9.7.2 Transfer Learning in Covariance Matrix Estimation

In estimating covariance matrices, it is expected that integrating knowledge from related source datasets can enhance the accuracy of target covariance matrix estimation. Suppose that we have an i.i.d. target sample, $X_i^{(0)} \in \mathbb{R}^q$ for $i = 1, \cdots, n_0$, where each $X_i^{(0)}$ follows a distribution with the covariance matrix $\mathbf{\Sigma}^{(0)}$. Additionally, we have source samples $X_i^{(k)} \in \mathbb{R}^q$, for $i = 1, \cdots, n_k$ and $k = 1, \cdots, K$. These source samples are i.i.d. from source distributions with covariance matrices $\mathbf{\Sigma}^{(k)}$. In addition, source samples are independent of the target sample. Our goal is to combine the source samples with the target sample to estimate the target covariance matrix $\mathbf{\Sigma}^{(0)}$.

To address this task, Cai and Kim (2024) recently introduced a transfer learning method for estimating the covariance matrix of a bandable structure (Qiu and Chen 2012; Qiu and Chen 2015; Bien et al. 2016; He and Chen 2016). Note that the bandable structure of $\mathbf{\Sigma}$ is defined in (2.16) of Sect. 2.2. In order to quantify the similarity between the target and the source of bandable covariance matrices, Cai and Kim (2024) defined the disparity matrix $\mathbf{\Delta}^{(k)} = \mathbf{\Sigma}^{(0)} - \mathbf{\Sigma}^{(k)}$, where $k = 1, \cdots, K$ and K is the total number of source samples. In addition, they utilized the ℓ_1-norm $\|\mathbf{\Delta}^{(k)}\|_1$ to measure the distance between the target and the source covariance matrices, where $\|\cdot\|_1$ represents the matrix ℓ_1-norm defined in Sect. 1.6.2. Accordingly, a value of the ℓ_1-norm decreases, the source sample more closely resembles the target sample. Based on this metric, the informative set of source samples with a specified disparity threshold $\delta \geq 0$ is denoted by

$$\mathcal{A}_1 = \left\{1 \leq k \leq K : \|\mathbf{\Delta}^{(k)}\|_1 \leq \delta\right\}.$$

Consequently, as long as δ is relatively small, the datasets within $\mathcal{A}_1$ can provide valuable information for improving the estimation of the target covariance matrix.

Without loss of generality, we consider $K = 1$. Under this setting, Cai and Kim (2024) proposed a blockwise tridiagonal estimator for the target covariance matrix $\mathbf{\Sigma}^{(0)}$ and then established the minimax optimal rate of convergence for the estimator under the spectral norm. However, the estimator depends not only on the decay rates in the classes of bandable covariance matrices induced by $\mathbf{\Sigma}^{(0)}$ and $\mathbf{\Sigma}^{(1)}$, respectively, but also on the disparity threshold δ. To address this issue, they further introduced a novel data-driven algorithm and established the adaptive rate of convergence.

In addition to bandable covariance matrices, some known structured covariance matrices, such as sparse and spiked covariance matrices, have been extensively studied in the literature. Hence, extending transfer learning techniques to estimating these structured covariance matrices in high-dimensional data can provide a promising direction for future research. Another interesting direction in the transfer learning framework is the estimation of the precision matrix (i.e., the inverse of the covariance matrix) for Gaussian graphical models (GGMs), which will be discussed in the next subsection.

9.7.3 *Transfer Learning in Gaussian Graphical Models*

Gaussian graphical models (GGMs) have been widely used to model the dependence structure among a set of random variables in various applications (Varoquaux et al. 2010; Zhao et al. 2014; Drton and Maathuis 2017; Glymour et al. 2019). In GGMs, the parameter of interest is the precision matrix, $\mathbf{\Theta} = \mathbf{\Sigma}^{-1}$, which uniquely determines the conditional dependence structure of its corresponding graph. However, due to high dimensionality and relatively small sample sizes, estimating a precision matrix based on a single sample often results in high uncertainty and low accuracy in link detection. This motivates researchers to consider transfer learning in high-dimensional GGMs. Assume that the target sample $X_i^{(0)} \in \mathbb{R}^q$ is i.i.d. generated from $N(\mu, \mathbf{\Sigma}^{(0)})$ for $i = 1, \cdots, n_0$. In addition, assume that the observations $X_i^{(k)} \in \mathbb{R}^q$ in source samples are i.i.d. generated from $N(\mu^{(k)}, \mathbf{\Sigma}^{(k)})$ for $i = 1, \cdots, n_k$ and $k = 1, \cdots, K$. Let $\mathbf{\Theta}^{(0)} = \{\mathbf{\Sigma}^{(0)}\}^{-1}$ and $\mathbf{\Theta}^{(k)} = \{\mathbf{\Sigma}^{(k)}\}^{-1}$ be the precision matrices calculated from the target dataset and the k-th source dataset, respectively.

To quantify the similarity between the target model and source models, Li et al. (2023) employed the Kullback–Leibler (KL) divergence to measure the difference between the k-th source model and the target model. That is,

$$\mathcal{D}_{\mathrm{KL}}(N_{\mathbf{\Sigma}^{(k)}}||N_{\mathbf{\Sigma}^{(0)}}) = \frac{1}{2}\mathrm{tr}(\mathbf{\Delta}^{(k)}) - \frac{1}{2}\log\det(\mathbf{I} + \mathbf{\Delta}^{(k)}), \tag{9.19}$$

where $\mathbf{\Delta}^{(k)} = \mathbf{\Theta}^{(0)}\mathbf{\Sigma}^{(k)} - \mathbf{I}$ is the k-th divergence matrix, and $N_{\mathbf{\Sigma}^{(k)}}$ and $N_{\mathbf{\Sigma}^{(0)}}$ denote the normal distributions with zero means and covariance matrices $\mathbf{\Sigma}^{(k)}$ and $\mathbf{\Sigma}^{(0)}$, respectively.

By the definition of $\mathbf{\Theta}^{(0)} = \{\mathbf{\Sigma}^{(0)}\}^{-1}$, it is natural to consider the moment equation,

$$\mathbf{I} = \mathbf{\Theta}^{(0)}\mathbf{\Sigma}^{(0)}. \tag{9.20}$$

To integrate source data information, from $k = 1, \cdots, K$, we re-express the moment equation (9.20) as

$$\mathbf{I} = \mathbf{\Theta}^{(k)}\mathbf{\Sigma}^{(k)} = \mathbf{\Theta}^{(0)}\mathbf{\Sigma}^{(k)} - \mathbf{\Delta}^{(k)}, \tag{9.21}$$

for any $k = 1, \cdots, K$, where $\mathbf{\Delta}^{(k)}$ is the k-th divergence matrix defined in (9.19). It is worth noting that $\mathbf{\Delta}^{(k)}$ characterizes the bias in the moment equation (9.21). To simultaneously utilize all the source datasets, denote the weighted averages of the covariance and divergence matrices by $\mathbf{\Sigma}^{\mathcal{K}} = \sum_{k=1}^{K}\alpha_k\mathbf{\Sigma}^{(k)}$ and $\mathbf{\Delta}^{\mathcal{K}} = \sum_{k=1}^{K}\alpha_k\mathbf{\Delta}^{(k)}$, respectively, where $\alpha_k = n_k/n$ for $n = \sum_{k=1}^{K} n_k$. Hence, by (9.21), we obtain

$$\mathbf{\Theta}^{(0)}\mathbf{\Sigma}^{\mathcal{K}} - \mathbf{\Delta}^{\mathcal{K}} - \mathbf{I} = 0. \tag{9.22}$$

This moment equation will be used for estimating $\mathbf{\Theta}^{(0)}$ via the constraint l_1 minimization method (CLIME) of Cai et al. (2011).

Before presenting the algorithm of Li et al. (2023), we first introduce some useful notation and definitions. For any matrix $\mathbf{A}$, denote A_j the j-th column of $\mathbf{A}$. In addition, for the target sample, we randomly split the data into two disjoint folds

$\mathcal{I}$ and $\mathcal{I}^c$ such that $|\mathcal{I}| = cn_0$ for some constant $0 < c < 1$. Moreover, let $\bar{X} = \frac{1}{|\mathcal{I}|}\sum_{i\in\mathcal{I}} X_i$ and $\widetilde{X} = \frac{1}{|\mathcal{I}^c|}\sum_{i\in\mathcal{I}^c} X_i$. Then, compute $\widehat{\boldsymbol{\Sigma}} = \frac{1}{|\mathcal{I}|}\sum_{i\in\mathcal{I}} X_i X_i' - \bar{X}\bar{X}'$ and $\widetilde{\boldsymbol{\Sigma}} = \frac{1}{|\mathcal{I}^c|}\sum_{i\in\mathcal{I}^c} X_i X_i' - \widetilde{X}\widetilde{X}'$, which are the estimation and aggregation covariance matrices, respectively. Finally, for the source samples, denote $\bar{X}^{\mathcal{K}} = n^{-1}\sum_{k=1}^{K}\sum_{i=1}^{n_k} X_i^{(k)}$ and $\widehat{\boldsymbol{\Sigma}}^{\mathcal{K}} = n^{-1}\sum_{k=1}^{K}\sum_{i=1}^{n_k}[X_i^{(k)}(X_i^{(k)})' - \bar{X}^{\mathcal{K}}(\bar{X}^{\mathcal{K}})']$.

Based on the target sample and the method of Cai et al. (2011), we next compute the single-task CLIME estimator,

$$\widehat{\boldsymbol{\Theta}}^{(CL)} = \operatorname{argmin}_{\boldsymbol{\Theta}\in\mathbb{R}^{q\times q}} \|\boldsymbol{\Theta}\|_1 \quad \text{subject to} \quad \|(\widehat{\boldsymbol{\Sigma}} + |\mathcal{I}|^{-1}\mathbf{I})\boldsymbol{\Theta} - \mathbf{I}\|_\infty \le \lambda_{\text{CL}}, \tag{9.23}$$

where $\|\cdot\|_1$ and $\|\cdot\|_\infty$ represent the matrix ℓ_1-norm and ℓ_∞-norm defined in Sect. 1.6.2, respectively, and $\lambda_{\text{CL}} > 0$ is a tuning parameter. Note that $\widehat{\boldsymbol{\Theta}}^{(CL)}$ is obtained by solving moment equation (9.20) after replacing its $\boldsymbol{\Sigma}^{(0)}$ with $\widehat{\boldsymbol{\Sigma}}$ under some sparsity constraint, and adding the diagonal matrix $|\mathcal{I}|^{-1}\mathbf{I}$ ensures that $\widehat{\boldsymbol{\Sigma}} + |\mathcal{I}|^{-1}\mathbf{I}$ is positive definite. Since $\widehat{\boldsymbol{\Theta}}^{(CL)}$ is evaluated using only target data, it is expected that utilizing the information from both the target study and the source study will improve the efficiency of $\widehat{\boldsymbol{\Theta}}^{(CL)}$.

Using the estimator $\widehat{\boldsymbol{\Theta}}^{(CL)}$ obtained from the target study, Li et al. (2023) proposed a Trans-CLIME method for estimating high-dimensional GGMs. In addition, they established the minimax optimal rate of convergence for estimating the GGMs under the Frobenius norm. Moreover, they demonstrated that the rates achieved by transfer learning are faster than the corresponding minimax rates in a single-task setting. As for empirical application, they applied their proposed Trans-CLIME method to infer gene networks in a target area of brain tissue by leveraging the gene expressions from multiple other areas of brain tissue. The use of Trans-CLIME led to a significant reduction in prediction errors. For the sake of illustration, we summarize their estimation method in Algorithm 9.1.

In this algorithm, the first two steps produce an initial estimator $\widehat{\boldsymbol{\Theta}}$ of the target graph $\boldsymbol{\Theta}^{(0)}$. Specifically, based on the moment equation (9.22), the optimization in Step 1 can be viewed as an adaptive thresholding of an initial estimator $\boldsymbol{\Delta}^{\mathcal{K}}$, $(\widehat{\boldsymbol{\Theta}}^{(CL)})'\widehat{\boldsymbol{\Sigma}}^{\mathcal{K}} - \mathbf{I}$. Note that the CLIME estimator $\widehat{\boldsymbol{\Theta}}^{(CL)}$ is not necessarily symmetric, the transpose is used. Then, using the result of the approximate sparse parameter $\boldsymbol{\Delta}^{\mathcal{K}}$ in Step 1 and the moment equation (9.22), Step 2 obtains an estimator of the target graph $\boldsymbol{\Theta}^{(0)}$. When the similarities between source datasets and the target dataset are sufficiently high, Step 2 yields a desirable transfer learning estimator of the target graph. However, if the similarities are low, the resulting estimator $\widehat{\boldsymbol{\Theta}}$ may not be as good as the single-task estimator $\widehat{\boldsymbol{\Theta}}^{(CL)}$. To address this issue, the third step in Algorithm 9.1 performs a selection or aggregation, based on $\widehat{\boldsymbol{\Theta}}^{(CL)}$ and $\widehat{\boldsymbol{\Theta}}$, to generate the final estimator. This procedure yields a weight vector, $\widehat{\mathbf{v}}_j$, that is either $(0, 1)'$ or $(1, 0)'$ motivated by the following moment equation,

$$\widetilde{\boldsymbol{\Sigma}}(\widehat{\boldsymbol{\Theta}}_j^{(CL)}, \widehat{\boldsymbol{\Theta}}_j)\widehat{\mathbf{v}}_j - \mathbf{e}_j \approx 0.$$

Here, $\mathbf{e}_j$ is a q-dimensional column vector with all elements being zero except for the j-th element being 1. Accordingly, this allows us to make a selection between $\widehat{\mathbf{\Theta}}_j^{(CL)}$ and $\widehat{\mathbf{\Theta}}_j$. It is important to note that the sample-splitting step ensures that both $\widehat{\mathbf{\Theta}}^{(CL)}$ and $\widehat{\mathbf{\Theta}}$ are independent of $\widetilde{\mathbf{\Sigma}}$ calculated from the samples used for aggregation.

Algorithm 9.1: Trans-CLIME algorithm

Input: Target sample $(X_1^{(0)}, \cdots, X_{n_0}^{(0)})$ and source sample $\{X_1^{(k)}, \cdots, X_{n_k}^{(k)}\}_{k=1}^K$.
Output: $\widehat{\mathbf{\Theta}}^{(0)}$.
Step 1. Based on (9.21), compute $\widehat{\mathbf{\Delta}}^{\mathcal{K}}$ via

$$\widehat{\mathbf{\Delta}}^{\mathcal{K}} = \text{argmin}_{\mathbf{\Delta}\in\mathbb{R}^{q\times q}} \|\mathbf{\Delta}\|_1 \quad \text{subject to} \quad \|\mathbf{\Delta} - \{(\widehat{\mathbf{\Theta}}^{(CL)})'\widehat{\mathbf{\Sigma}}^{\mathcal{K}} - \mathbf{I}\}\|_\infty \le \lambda_{\mathbf{\Delta}},$$

where $\lambda_{\mathbf{\Delta}}$ is the tuning parameter.
Step 2. Based on $\widehat{\mathbf{\Delta}}^{\mathcal{K}}$ and (9.22), compute $\widehat{\mathbf{\Theta}}$ given below.

$$\widehat{\mathbf{\Theta}} = \text{argmin}_{\mathbf{\Theta}\in\mathbb{R}^{q\times q}} \|\mathbf{\Theta}\|_1 \quad \text{subject to} \quad \|(\widehat{\mathbf{\Sigma}}^{\mathcal{K}} + n^{-1}\mathbf{I})\mathbf{\Theta} - (\widehat{\mathbf{\Delta}}^{\mathcal{K}} + \mathbf{I})'\|_\infty \le \lambda_{\mathbf{\Theta}},$$

where $\lambda_{\mathbf{\Theta}}$ is the tuning parameter.
Step 3. For $j = 1, \cdots, q$, compute $\widehat{\mathbf{\Theta}}_j^{(0)}$ via

$$\widehat{\mathbf{\Theta}}_j^{(0)} = (\widehat{\mathbf{\Theta}}_j^{(CL)}, \widehat{\mathbf{\Theta}}_j)\widehat{\mathbf{v}}_j,$$

where

$$\widehat{\mathbf{v}}_j = \text{argmin}_{\mathbf{v}\in\{(0,1)',(1,0)'\}} \mathbf{v}'\widehat{W}(j)\mathbf{v} - 2\mathbf{v}'(\widehat{\mathbf{\Theta}}_{jj}^{(CL)}, \widehat{\mathbf{\Theta}}_{jj})$$

and

$$\widehat{W}(j) = \begin{pmatrix} (\widehat{\mathbf{\Theta}}_j^{(CL)})'\widetilde{\mathbf{\Sigma}}\widehat{\mathbf{\Theta}}_j^{(CL)} & (\widehat{\mathbf{\Theta}}_j^{(CL)})'\widetilde{\mathbf{\Sigma}}\widehat{\mathbf{\Theta}}_j \\ (\widehat{\mathbf{\Theta}}_j^{(CL)})'\widetilde{\mathbf{\Sigma}}\widehat{\mathbf{\Theta}}_j & \widehat{\mathbf{\Theta}}_j'\widetilde{\mathbf{\Sigma}}\widehat{\mathbf{\Theta}}_j \end{pmatrix}.$$

In addition to traditional GGMs , the TL framework has been expanded to semi-parametric graphical models (He et al. 2022). However, these approaches are restricted to vector-value data. Hence, Ren et al. (2024) proposed a transfer learning framework for tensor GGMs, which are applicable to tensor data. Detailed discussion of tensor analysis is given in the next chapter.

9.8 An Empirical Example

To illustrate the performance of covariance matrix estimators, we consider the three deep learning methods presented in Sect. 9.6.1, which are the Autoencoder (AE) method, the Feedforward Neural Network (FNN) method, and the Graph Convolutional Network (GCN) method. In order to make comparison, we collect the daily return of $p = 278$ stocks included in the China Securities Index 300 (CSI 300 index) from January 3, 2017, to December 31, 2024. The data consist of $n = 1943$ observations, and, for each observation $i = 1, \cdots, n$, the response variables $Y_i \in \mathbb{R}^p$ represent the returns of the 278 stocks. Note that in Sect. 9.6.1 (I), we denote the responses, the dimension of responses, and the sample size by X_i, n, and T, respectively, in the factor model used for the AE method. Correspondingly, p and n represent the dimension of the responses Y_i and the sample size defined above for the FNN and GCN methods. For the sake of simplicity, however, we slightly abuse notation and use p and n rather than n and T to represent the dimension of responses and the sample size, respectively, for the AE method.

Note that the FNN method requires an input vector that is associated with the covariance matrix of Y_i. To this end, we collect the three factors involved in the Fama–French three-factor model (Fama and French 1993) as the input vector. We denote it by $U_i = (\text{MARKET}_i, \text{SMB}_i, \text{HML}_i)' \in \mathbb{R}^3$. Here, MARKET_i represents the "market excess return," which measures the difference between of the broad A-share market return and the risk-free rate. In addition, SMB_i and HML_i, respectively, represent the "Small minus Big" for measuring the return spread between small-cap and large-cap stocks and the "High minus Low" for capturing the return spread between high book-to-market and low book-to-market stocks. Motivated by the Fama–French three-factor model, we also set $m = 3$ used in the Autoencoder (AE) method.

To implement GCN, we further collect the covariate matrices $\mathbf{Z}_i \in \mathbb{R}^{p\times q}$, which represent $q = 42$ firms' fundamentals. Each column of $\mathbf{Z}_i$ has been normalized to have zero mean and unit variance. These fundamentals include total assets, operating revenue, debt-to-asset ratio, and earnings per share, among others. A detailed description of these firms' fundamentals is presented in Table 9.1. Since the GCN method requires a network structure of the p stocks, we also create an adjacency matrix $\mathbf{A} = (a_{j_1 j_2}) \in \mathbb{R}^{p\times p}$ based on the firms' supply-chain relationships (Shi et al. 2022). Here, $a_{j_1 j_2} = 1$ if there exists a supply chain from firm j_1 (supplier) to firm j_2 (customer) for $j_1 \neq j_2$, otherwise $a_{j_1 j_2} = 0$. For the sake of completeness, self-connections are excluded, that is, $a_{jj} = 0$ for all $j = 1, \cdots, p$. It is important to note that the supply-chain network is directed and static. In addition, the estimated network density of this supply-chain network is $\widehat{\pi} = \{p(p-1)\}^{-1} \sum_{j_1 \neq j_2} a_{j_1 j_2} = 0.950 \times 10^{-3}$, which suggests that the supply-chain network is highly sparse.

To compare three covariance matrix estimations, we evaluate their performance based on Markowitz portfolio optimization (Markowitz 1952). To this end, we employ the commonly used rolling window approach (see, e.g., DeMiguel et al.

Table 9.1 The definition and description of the $q = 42$ firms' fundamentals

Firm fundamentals	Definition and description
V1	Total Assets
V2	Total Liabilities
V3	Paid-in Capital (or Share Capital)
V4	Total Owners' Equity (or Shareholders' Equity)
V5	Operating Profit
V6	Total Profit (Profit Before Tax)
V7	Net Profit
V8	Net Profit Attributable to Owners of the Parent Company
V9	Total Operating Revenue
V10	Total Operating Costs
V11	Accounts Payable Payment Period
V12	Book-to-Market Ratio (B/M, Shareholders' Equity/Market Capitalization)
V13	Net Asset Value per Share (Total Equity/Total Shares Outstanding)
V14	Growth in Net Operating Cash Flow
V15	Debt-to-Asset Ratio (Total Liabilities/Total Assets)
V16	Debt-to-Equity Ratio (Total Liabilities/Shareholders' Equity)
V17	Diluted Earnings per Share (Net Profit/Weighted Average Shares)
V18	Net Profit Margin (DuPont Analysis)
V19	EBITDA
V20	Earnings-to-Price Ratio (EPS/Stock Price)
V21	Equity-to-Fixed Assets Ratio (Shareholders' Equity/Fixed Assets)
V22	Equity Multiplier (Total Assets/Shareholders' Equity)
V23	Parent Company Equity-to-Total Liabilities Ratio (Parent Company's Equity/Parent Company's Total Liabilities)
V24	Enterprise Value (Market Cap+Debt-Cash & Cash Equivalents)
V25	Enterprise Value-to-EBITDA (Enterprise Value/EBITDA)
V26	Enterprise Value Excluding Cash (Market Cap+Debt)
V27	Operating Cost Ratio (Total Costs/Total Revenue)
V28	Cost-to-Profit Ratio (Total Costs/Net Profit)
V29	Total Operating Expenses (Selling+Admin+Financial Expenses)
V30	Fixed Asset Ratio (Fixed Assets/Total Assets)
V31	Core Operational Profit
V32	Intangible Asset Ratio (Intangible Assets/Total Assets)
V33	Liabilities per Share (Total Liabilities/Shares)
V34	Non-Interest-Bearing Current Liabilities
V35	Operating-to-Total Profit Ratio (Operating Profit/Total Profit)
V36	Total Operating Revenue per Share (Revenue/Total Shares)
V37	Price-to-Book (P/B, inverse of B/M)
V38	Price-to-Earnings Ratio (Stock Price/EPS)
V39	Operating Profit Margin (Operating Profit/Revenue)
V40	Retained Earnings (Cumulative Net Income-Cumulative Dividends Declared)
V41	Diluted Return on Equity (Net Profit/Average Shareholders' Equity)
V42	Tangible Assets per Share (Tangible Equity/Shares)

2009 and Zou et al. 2017) with a window size $T^* = 70$ to estimate covariance matrices, obtain the optimal portfolio weights, and then assess portfolio performance. To accomplish this task, for each window $t = T^*, \cdots, n-1$ with observations $k = t-69, \cdots, t$, we first estimate the covariance matrices of X_k and Y_k by the AE and FNN (or GCN) methods, respectively, given below.

Autoencoder (AE) For each window t, using 70 observations ranged from $t-69$ to t, AE employs a neural encoder–decoder structure to obtain the estimate of the covariance matrix $\widehat{\boldsymbol{\Sigma}}_{AE,t}$. Specifically, the code layer comprises a second hidden layer ($L = 2$) of FNN with the hyperbolic tangent (tanh) activation function, and the decoder involves a single linear layer. The FNN outputs, together with Eqs. (9.12) and (9.11) with $m = 3$, yield $\widehat{\boldsymbol{\Sigma}}_{AE,t}$. Note that this FNN model was trained for 500 epochs using the Adam optimizer (Kingma and Ba 2017) with learning rate = 1×10^{-5}. During the training process, it minimizes the MSE between the reconstructed $\widehat{X}_t$ and the observed returns X_t, as defined in Eq. (9.12).

Feedforward Neural Network (FNN) For each k ($k = t-69, \cdots, t$), we consider a three-layer FNN model with the hyperbolic tangent (tanh) activation function in each layer. Note that the dimension of the final output is set to $p(p-1)/2 \times 1$, which corresponds to the lower triangular elements of the Cholesky decomposition of $\boldsymbol{\Sigma}^{-1}$. In addition, the Adam optimizer (Kingma and Ba 2017) was used to train the model with a learning rate of 1×10^{-4} for 500 epochs. After getting the output from FNN, we then employ Eqs. (9.15) and (9.14) to obtain an estimate of the covariance matrix $\widehat{\boldsymbol{\Sigma}}_{FNN,k}$. Subsequently, for each t, we average $\widehat{\boldsymbol{\Sigma}}_{FNN,k}$ over the observations $k = t-69, \cdots, t$ to obtain the estimator of the covariance matrix $\widehat{\boldsymbol{\Sigma}}^{(t)}_{FNN} = \sum_{k=t-T^*+1}^{t} \widehat{\boldsymbol{\Sigma}}_{FNN,k}/T^*$.

Graph Convolutional Network (GCN) Analogous to the method of Wang et al. (2023), we consider a three-layer model for each single observation k, where k ranges from $t-69$ to t. In each layer of this model, the tanh activation function is utilized, and for each observation k, a ($p \times 16$)-dimensional final output is generated. As used in AE and FNN, we also adopt the Adam optimizer with a learning rate of 1×10^{-4} to train the model for 500 epochs. The output, in conjunction with Eqs. (9.17)–(9.18), yields the estimate of the covariance matrix, $\widehat{\boldsymbol{\Sigma}}_{GCN,k}$. Subsequently, for each window t, we calculate the averaging covariance matrix, $\widehat{\boldsymbol{\Sigma}}^{(t)}_{GCN} = \sum_{k=t-T^*+1}^{t} \widehat{\boldsymbol{\Sigma}}_{GCN,k}/T^*$ for window t.

Using the estimates of the covariance matrices obtained from AE/FNN/GCN methods (i.e., the $\widehat{\boldsymbol{\Sigma}}^{(t)}$s), we next construct the optimal portfolio weights by minimizing the variance of the portfolio,

$$\widehat{\mathbf{w}}_t = \text{argmin}_{\mathbf{w}_t} \mathbf{w}_t' \widehat{\boldsymbol{\Sigma}}^{(t)} \mathbf{w}_t, \quad \text{s.t.} \quad \mathbf{w}_t' \mathbf{1} = 1, \tag{9.24}$$

for $t = T^*, \cdots, n-1$, where $\mathbf{1} = (1, \cdots, 1)' \in \mathbb{R}^p$. The closed-form of weight estimate is $\widehat{\mathbf{w}}_t = \widehat{\boldsymbol{\Sigma}}^{(t)-1}\mathbf{1}/(\mathbf{1}'\widehat{\boldsymbol{\Sigma}}^{(t)-1}\mathbf{1})$. In addition, we consider the optimal portfolio with the short-sale constraint (Jagannathan and and Ma 2003). By solving

Table 9.2 The comparison of the out-of-sample portfolios calculated by the AE, FNN, GCN, and EW methods under the two optimization schemes: (1) global minimum-variance portfolios, and (2) minimum-variance portfolios with short-sale constraint ($w_i \geq 0$)

Model	Portfolios	Mean	SD	SR
AE	Unconstrained	0.0012	0.0560	0.0235
	Constrainted	0.0012	0.0546	0.0269
FNN	Unconstrained	0.0008	0.0600	0.0206
	Constrainted	0.0009	0.0590	0.0214
GCN	Unconstrained	0.0013	0.0403	0.0310
	Constrainted	0.0014	0.0409	0.0316
EW	–	0.0003	0.0124	0.0206

the above problem (9.24) with an additional constraint that $\mathbf{w}_t \geq \mathbf{0}$, we can obtain the constrained portfolio weight $\widehat{\mathbf{w}}_t^{cons}$. Based on $\widehat{\mathbf{w}}_t$ or $\widehat{\mathbf{w}}_t^{cons}$, we calculate the out-of-sample portfolio return evaluated at $i = t + 1$ for $t = T^*, \cdots, n - 1$ and denote it by $\widehat{r}_{t+1}$.

Using the portfolio returns $\widehat{r}_{t+1} = \widehat{\mathbf{w}}_t' Y_{t+1}$, we assess the performance of the portfolios via the following three measures: (i) the sample mean (Mean); (ii) the sample standard deviation (SD); and (iii) the Sharpe ratio (SR), which is calculated using the formula SR=(Mean-the average risk-free rate)/SD; see Sharpe (1966). For the sake of comparison, we also include the equally weighted (EW) portfolio (i.e., $\mathbf{w}_t = (1, \cdots, 1)'/p$) as a benchmark.

Table 9.2 compares four methods and two optimization schemes. It shows that the Graph Convolutional Network (GCN) method with a short-sale constraint yields the highest Sharpe ratio (0.0316); hence, it performs the best. This finding is sensible since the GCN approach utilizes additional interrelationships among stocks via graph structures. In addition, constrained portfolios outperform non-constrained portfolios across all three models. This finding is consistent with that of DeMiguel et al. (2009) and Zou et al. (2017). Moreover, all three neural network methods outperform the equal-weighted benchmark. Hence, we conclude that the optimal portfolio weights resulting from deep learning covariance estimations, especially through graph-based learning, are superior to naive diversification.

Chapter 10
Tensor and Covariance Matrices

This chapter briefly reviews tensor operations and then focuses on tensor covariances; it consists of four sections. Section 10.1 presents two tensor operations that are tensor products and decompositions. To analyze linear relationships between two tensors (i.e., multi-arrays), Sect. 10.2 introduces tensor covariance, tensor correlation, and tensor canonical correlation. Subsequently, Sect. 10.3 studies three types of tensor covariance models, which are the separable covariance model, the core shrinkage covariance model, and the tensor factor model. To illustrate the empirical application of tensor-based covariance matrices, Sect. 10.4 presents two real examples including speech recognition and object tracking.

10.1 Tensor Operations

10.1.1 Tensor

The term "tensor" has been widely used in physics and mathematics for more than two centuries (Ricci and Levi-Civita 1900). In the last two decades, tensor analysis has attracted considerable attention due to its ability to manage high-dimensional data and provide insightful interpretations. Tensors have been applied and have made an impact across many fields; see, e.g., facial recognition (Khan et al. 2017), imaging analysis (Zhou et al. 2013; Li et al. 2018; Mokhtari et al. 2019; Deng et al. 2023), longitudinal data analysis (Hoff 2015), network analysis (Nakatsuji et al. 2017; Gitelman 2022), neural networks (Novikov et al. 2015), object or target tracking (Li and Sun 2011; Zhou et al. 2018; Govaers 2019), recommender systems (Bi et al. 2018), speaker recognitions (Saito et al. 2011; Yu et al. 2013; Saito et al. 2019), and time series modeling (Chen et al. 2022; Wang et al. 2022; Barigozzi et al. 2023). A few useful references are available for studying tensors; see, e.g., Cichocki et al. (2015), Rabanser et al. (2017), Bi et al. (2021), Sun et al. (2021),

W. Lan, C.-L. Tsai, *Covariance Analysis and Beyond*,
https://doi.org/10.1007/978-3-032-08796-6_10

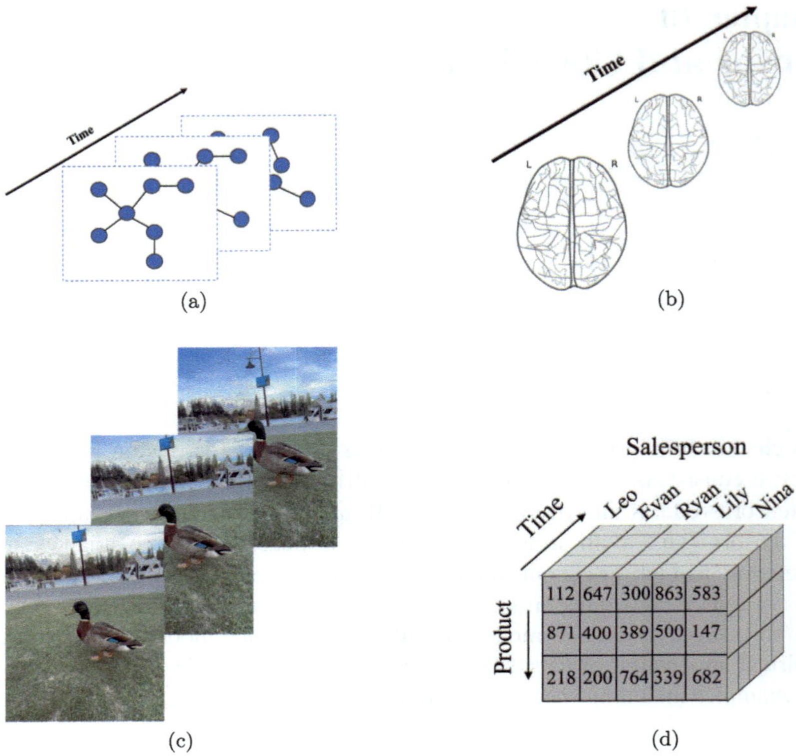

Fig. 10.1 Applications of tensor data. (**a**) Dynamic social network data. (**b**) fMRI data. (**c**) Image data. (**d**) Sales data

and Bi et al. (2022). In this chapter, we mainly adapt a seminal paper of Kolda and Bader (2009) to introduce tensors.

In Chap. 2, we studied one-dimensional vectors and two-dimensional matrices. With the rapid development and widespread application of information technology in the last two decades, much real-world data have been collected in more than two dimensions, including dynamic network data such as import–export volume data, fMRI data, image data, and sales data. For the sake of illustration, the visualizations of such multi-dimensional data are succinctly presented in Fig. 10.1.

For multi-dimensional data, vectors or matrices are usually unable to capture their intricate patterns, dependencies, and interactions. To overcome this limitation, it is natural to use tensors to represent data in multi-dimensional arrays. To this end, consider an Nth-order tensor (or N-dimensional array) $\mathcal{X} \in \mathbb{R}^{I_1 \times I_2 \times \cdots \times I_N}$, where N is the number of tensor orders of $\mathcal{X}$, and I_k is the marginal dimension of the k-th mode for $k = 1, \cdots, N$. Accordingly, a vector is a first-order tensor, and a matrix is a second-order tensor. In addition, denote the $(i_1, i_2, \cdots, i_N)$-th element of $\mathcal{X}$ by

$x_{i_1 i_2 \cdots i_N}$, where the indices $i_1 = 1, \cdots, I_1$, $i_2 = 1, \cdots, I_2$, and $i_N = 1, \cdots, I_N$. The L_2-norm of $\mathcal{X}$ is $||\mathcal{X}|| = \{\sum_{i_1=1}^{I_1} \sum_{i_2=1}^{I_2} \cdots \sum_{i_N=1}^{I_N} x_{i_1 i_2 \cdots i_N}^2\}^{1/2}$.

For a given N-dimensional array $\mathcal{X}$, we next define the two types of subarrays, i.e., fiber and slice. A fiber is a subarray in which all but one indicator are fixed. Hence, the mode-k fiber is defined as $X_{i_1 i_2 \cdots i_{k-1} : i_{k+1} \cdots i_N} \in \mathbb{R}^{I_k}$. For the purpose of illustration, we consider a third-order tensor $\mathcal{X} \in \mathbb{R}^{I_1 \times I_2 \times I_3}$. Accordingly, $\mathcal{X}$ induces the column, row, and tube fibers, they are denoted by $X_{:i_2 i_3} = (X_{i_1 i_2 i_3})_{i_1} \in \mathbb{R}^{I_1}$, $X_{i_1 : i_3} = (X_{i_1 i_2 i_3})_{i_2} \in \mathbb{R}^{I_2}$, and $X_{i_1 i_2 :} = (X_{i_1 i_2 i_3})_{i_3} \in \mathbb{R}^{I_3}$ for the indices $i_1 = 1, \cdots, I_1$, $i_2 = 1, \cdots, I_2$, and $i_3 = 1, \cdots, I_3$, respectively. Similarly, a slice is a subarray in which all but two indicators are fixed. Hence, the mode-kl slice is defined as $\mathbf{X}_{i_1 i_2 \cdots i_{k-1} : i_{k+1} \cdots i_{l-1} : i_{l+1} \cdots i_N} \in \mathbb{R}^{I_k \times I_l}$ for $k < l$ and $N \geq 2$. The mode-kl slices of a third-order tensor are the horizontal, lateral, and frontal slices, which are denoted by $\mathbf{X}_{i_1 ::} = (X_{i_1 i_2 i_3})_{i_2 i_3} \in \mathbb{R}^{I_2 \times I_3}$, $\mathbf{X}_{: i_2 :} = (X_{i_1 i_2 i_3})_{i_1 i_3} \in \mathbb{R}^{I_1 \times I_3}$, and $\mathbf{X}_{:: i_3} = (X_{i_1 i_2 i_3})_{i_1 i_2} \in \mathbb{R}^{I_1 \times I_2}$, respectively. After introducing the fiber and slice, we next present the mode-k matricization (or unfolding) that reshapes the elements of an N-th order tensor $\mathcal{X}$ into a matrix denoted by $\mathbf{X}_{(k)} \in \mathbb{R}^{I_k \times (I_1 \cdots I_{k-1} I_{k+1} \cdots I_N)}$. Specifically, we transform the element $x_{i_1 i_2 \cdots i_N}$ of $\mathcal{X}$ into the element $x_{i_k j}$ of $\mathbf{X}_{(k)}$ for $j = 1 + \sum_{l=1, l \neq k}^{N} (i_l - 1) K_l$, where $K_l = \Pi_{r=1, r \neq k}^{l-1} I_r$. Based on this matricization, we can define the k-rank of $\mathcal{X}$, denoted by $rank_k(\mathcal{X})$, which is the column rank of $\mathbf{X}_{(k)}$. Let $G_k = rank_k(\mathcal{X})$ for $k = 1, \cdots, N$. We then refer to $\mathcal{X}$ as rank-$(G_1, G_2, \cdots, G_N)$ tensor.

To better understand the above definitions, we take a third-order tensor $\mathcal{X} \in \mathbb{R}^{2 \times 3 \times 2}$ as an example. The elements of this tensor are: $x_{111} = 1$, $x_{121} = 5$, $x_{131} = 9$, $x_{211} = 3$, $x_{221} = 7$, $x_{231} = 11$, $x_{112} = 2$, $x_{122} = 6$, $x_{132} = 10$, $x_{212} = 4$, $x_{222} = 8$, and $x_{232} = 12$. Based on this tensor, we can identify its various fiber components. There are six column fibers, which are $\{X_{:11} = (1, 3)'$, $X_{:12} = (2, 4)'$, $X_{:21} = (5, 7)'$, $X_{:22} = (6, 8)'$, $X_{:31} = (9, 11)'$, $X_{:32} = (10, 12)'\}$. In addition, there are four row fibers, namely $\{X_{1:1} = (1, 5, 9)'$, $X_{1:2} = (2, 6, 10)'$, $X_{2:1} = (3, 7, 11)'$, $X_{2:2} = (4, 8, 12)'\}$. Moreover, there are six tube fibers, denoted by $\{X_{11:} = (1, 2)'$, $X_{12:} = (5, 6)'$, $X_{13:} = (9, 10)'$, $X_{21:} = (3, 4)'$, $X_{22:} = (7, 8)'$, $X_{23:} = (11, 12)'\}$. It should be noted that all fibers are expressed as column vectors. In addition to fibers, this tensor has two horizontal slices, three lateral slices, and two frontal slices. The details of these slices are as follows:

$$\mathbf{X}_{1::} = \begin{pmatrix} 1 & 2 \\ 5 & 6 \\ 9 & 10 \end{pmatrix}, \quad \mathbf{X}_{2::} = \begin{pmatrix} 3 & 4 \\ 7 & 8 \\ 11 & 12 \end{pmatrix};$$

$$\mathbf{X}_{:1:} = \begin{pmatrix} 1 & 2 \\ 3 & 4 \end{pmatrix}, \quad \mathbf{X}_{:2:} = \begin{pmatrix} 5 & 6 \\ 7 & 8 \end{pmatrix}, \quad \mathbf{X}_{:3:} = \begin{pmatrix} 9 & 10 \\ 11 & 12 \end{pmatrix};$$

$$\text{and } \mathbf{X}_{::1} = \begin{pmatrix} 1 & 5 & 9 \\ 3 & 7 & 11 \end{pmatrix} \quad \mathbf{X}_{::2} = \begin{pmatrix} 2 & 6 & 10 \\ 4 & 8 & 12 \end{pmatrix}.$$

Finally, the three mode-k matricizations are given below.

$$\mathbf{X}_{(1)} = \begin{pmatrix} 1 & 5 & 9 & 2 & 6 & 10 \\ 3 & 7 & 11 & 4 & 8 & 12 \end{pmatrix}, \quad \mathbf{X}_{(2)} = \begin{pmatrix} 1 & 3 & 2 & 4 \\ 5 & 7 & 6 & 8 \\ 9 & 11 & 10 & 12 \end{pmatrix};$$

$$\text{and } \mathbf{X}_{(3)} = \begin{pmatrix} 1 & 3 & 5 & 7 & 9 & 11 \\ 2 & 4 & 6 & 8 & 10 & 12 \end{pmatrix}.$$

10.1.2 Tensor Products

After introducing tensors, we next explore tensor operations, which are critical for understanding how two tensors interact with each other. For this purpose, we introduce three fundamental tensor products, which are generalizations of commonly used operations from vectors and matrices to multidimensional arrays. The first two products basically extend the inner and outer products of two vectors discussed in Sect. 1.6.3, to two N-dimensional arrays. Let $\mathcal{X} = (x_{i_1 i_2 \cdots i_N}) \in \mathbb{R}^{I_1 \times I_2 \times \cdots \times I_N}$ and $\mathcal{Y} = (y_{i_1 i_2 \cdots i_N}) \in \mathbb{R}^{I_1 \times I_2 \times \cdots \times I_N}$, where $i_k = 1, \cdots, I_k$ for $k = 1, \cdots, N$. The inner product of these two tensors is defined as $\mathcal{X} \cdot \mathcal{Y} = \sum_{i_1=1}^{I_1} \sum_{i_2=1}^{I_2} \cdots \sum_{i_N=1}^{I_N} x_{i_1 i_2 \cdots i_N} y_{i_1 i_2 \cdots i_N}$. Notably, $\mathcal{X} \cdot \mathcal{X}$ is equivalent to the squared norm of $\mathcal{X}$, i.e., $\|\mathcal{X}\|^2$. Furthermore, let $\mathcal{Z} = (z_{j_1 j_2 \cdots j_N}) \in \mathbb{R}^{J_1 \times J_2 \times \cdots \times J_N}$. The outer product of $\mathcal{X}$ and $\mathcal{Z}$ is denoted by $\mathcal{X} \otimes_{outer} \mathcal{Z} = (x_{i_1 i_2 \cdots i_N} z_{j_1 j_2 \cdots j_N}) \in \mathbb{R}^{I_1 \times I_2 \times \cdots \times I_N \times J_1 \times J_2 \times \cdots \times J_N}$, where $x_{i_1 i_2 \cdots i_N} z_{j_1 j_2 \cdots j_N}$ is its $(i_1, i_2, \cdots, i_N, j_1, j_2, \cdots, j_N)$-th element. Note that the tensor outer product, $\otimes_{outer}$ is different from the Kronecker product, $\otimes$, defined in Sect. 1.6.3. For example, consider $\mathcal{X} = (1, 2)'$ and $\mathcal{Z} = (3, 4)'$. Then, $\mathcal{X} \otimes_{outer} \mathcal{Z} = \begin{pmatrix} 3 & 4 \\ 6 & 8 \end{pmatrix}$, while $\mathcal{X} \otimes \mathcal{Z} = (3, 4, 6, 8)' = vec(\mathcal{Z} \otimes_{outer} \mathcal{X})$. With a slight abuse of notation, for the rest of this chapter, we use $\otimes$ rather than $\otimes_{outer}$ to denote the tensor outer product.

We now introduce the k-mode product. Let $\mathcal{X} \in \mathbb{R}^{I_1 \times I_2 \times \cdots \times I_N}$ and $\mathbf{D} \in \mathbb{R}^{J \times I_k}$. The k-mode product of $\mathcal{X}$ and $\mathbf{D}$ is defined as $\mathcal{X} \times_k \mathbf{D} = (\sum_{i_k=1}^{I_k} x_{i_1 \cdots i_{k-1} i_k i_{k+1} \cdots i_N} d_{j i_k}) \in \mathbb{R}^{I_1 \times I_2 \times \cdots \times I_{k-1} \times J \times I_{k+1} \times \cdots \times I_N}$, where $\sum_{i_k=1}^{I_k} x_{i_1 \cdots i_{k-1} i_k i_{k+1} \cdots i_N} d_{j i_k}$ is the $(i_1, \cdots, i_{k-1}, j, i_{k+1}, \cdots, i_N)$-th element of the tensor $\mathcal{X} \times_k \mathbf{D}$. To better understand this operation, we consider the two frontal slices of the tensor $\mathcal{X} \in \mathbb{R}^{2 \times 3 \times 2}$ and the 2×2 matrix $\mathbf{D}$, which are given below:

$$\mathbf{X}_{::1} = \begin{pmatrix} 1 & 5 & 9 \\ 3 & 7 & 11 \end{pmatrix}, \quad \mathbf{X}_{::2} = \begin{pmatrix} 2 & 6 & 10 \\ 4 & 8 & 12 \end{pmatrix}, \quad \text{and } \mathbf{D} = \begin{pmatrix} 13 & 14 \\ 15 & 16 \end{pmatrix}.$$

Then, the resulting first-mode product $\mathcal{Z} = \mathcal{X} \times_1 \mathbf{D} \in \mathbb{R}^{2 \times 3 \times 2}$ is

$$\mathbf{Z}_{::1} = \begin{pmatrix} 55 & 163 & 271 \\ 63 & 187 & 311 \end{pmatrix} \text{ and } \mathbf{Z}_{::2} = \begin{pmatrix} 82 & 190 & 298 \\ 94 & 218 & 342 \end{pmatrix},$$

where $\mathbf{Z}_{::1}$ and $\mathbf{Z}_{::2}$ are the two frontal slices of the tensor $\mathcal{Z}$.

10.1.3 Tensor Decompositions

Having introduced tensor products, we now extend the concept of matrix decomposition to tensors and introduce two commonly-used tensor decomposition methods, i.e., CANDECOMP/PARAFAC (CP) decomposition and Tucker decomposition. These decomposition methods allow us to effectively utilize and extract information from tensors.

We first introduce the CP decomposition method, named by Kiers (2000) from related articles. The CP decomposition presents a tensor as a linear combination of rank-1 tensors. Specifically, for an N-dimensional array $\mathcal{X} \in \mathbb{R}^{I_1 \times I_2 \times \cdots \times I_N}$, the rank-$G$ CP decomposition of $\mathcal{X}$ is given as follows:

$$\mathcal{X} = [\![\lambda; \widetilde{\mathbf{A}}_1, \cdots, \widetilde{\mathbf{A}}_N]\!] := \sum_{r=1}^{G} \lambda_r A_1^{(r)} \otimes \cdots \otimes A_N^{(r)},$$

where G is a positive integer, $\widetilde{\mathbf{A}}_k = (\mathbf{A}_k^{(1)}, \cdots, \mathbf{A}_k^{(G)}) \in \mathbb{R}^{I_k \times G}$, and $\mathbf{A}_k^{(r)} = (a_{k1}^{(r)}, \cdots, a_{kI_k}^{(r)})' \in \mathbb{R}^{I_k}$ are normalized to length one, for $k = 1, \cdots, N$ and $r = 1, \cdots, G$. In addition, the weight vector is $\boldsymbol{\lambda} = (\lambda_1, \cdots, \lambda_G)' \in \mathbb{R}^G$, and $\otimes$ denotes the vector outer product. Note that $\boldsymbol{\lambda} = \mathbf{1}_G$ results in the classical definition of CP decomposition. After taking the decomposition, the elements of $\mathcal{X}$ can be expressed as $x_{i_1 i_2 \cdots i_N} = \sum_{r=1}^{G} \lambda_r \prod_{k=1}^{N} a_{k i_k}^{(r)}$, where $i_k = 1, \cdots, I_k$. A sufficient condition for the uniqueness of the CP decomposition is given in Sidiropoulos and Bro (2000). As for the applications of the CP decomposition, they can be found in Sect. 3.5 of Kolda and Bader (2009), Zhou et al. (2013), Liu et al. (2015), and Son et al. (2022).

We next introduce the Tucker decomposition, which was proposed by Tucker (1966). The Tucker decomposition is a method that decomposes a tensor into a core tensor multiplied by a matrix along each mode. Let $\mathcal{C} \in \mathbb{R}^{J_1 \times J_2 \times \cdots \times J_N}$ be the core tensor, and for $k = 1, \cdots, N$, let $\widetilde{\mathbf{B}}_k = (b_{i_k j_k}) = (\mathbf{B}_{k1}, \cdots, \mathbf{B}_{kJ_k}) \in \mathbb{R}^{I_k \times J_k}$. Then, the Tucker decomposition of a tensor $\mathcal{X}$ is defined below.

$$\begin{aligned}\mathcal{X} = [\![\mathcal{C}; \widetilde{\mathbf{B}}_1, \cdots, \widetilde{\mathbf{B}}_N]\!] : &= \mathcal{C} \times_1 \widetilde{\mathbf{B}}_1 \cdots \times_N \widetilde{\mathbf{B}}_N \\ &= \sum_{j_1=1}^{J_1} \sum_{j_2=1}^{J_2} \cdots \sum_{j_N=1}^{J_N} c_{j_1 j_2 \cdots j_N} \mathbf{B}_{1j_1} \otimes \mathbf{B}_{2j_2} \cdots \otimes \mathbf{B}_{Nj_N},\end{aligned}$$

where $j_k = 1, \cdots, J_k$ and $J_k \leq I_k$ for $k = 1, \cdots, N$. Based on the above decomposition, the entries of the tensor $\mathcal{X}$ are given as follows:

$$x_{i_1 i_2 \cdots i_N} = \sum_{j_1=1}^{J_1} \sum_{j_2=1}^{J_2} \cdots \sum_{j_N=1}^{J_N} c_{j_1 j_2 \cdots j_N} b_{i_1 j_1} b_{i_2 j_2} \cdots b_{i_N j_N},$$

for $i_k = 1, \cdots, I_k$ and $k = 1, \cdots, N$. Typically, the matrices $\widetilde{\mathbf{B}}_k$ are orthogonal, and the tensor $\mathcal{C}$ is referred to as the core tensor. It is worth noting that the Tucker decomposition is usually not unique.

It can be demonstrated that the CP decomposition is a special case of the Tucker decomposition. This can be shown by assuming that $J_1 = J_2 = \cdots = J_N = G$ and $\mathcal{C}$ is a super-diagonal tensor, i.e., $c_{j_1 j_2 \cdots j_N} \neq 0$ when $j_1 = j_2 = \cdots = j_N$, and $c_{j_1 j_2 \cdots j_N} = 0$ otherwise. In addition, the Tucker decomposition can be regarded as an extension of the standard two-way principal components analysis (PCA). Consider a third-order tensor $\mathcal{X} \in \mathbb{R}^{I_1 \times I_2 \times I_3}$ that satisfies $\mathcal{X} = \mathcal{C} \times_1 \widetilde{\mathbf{B}}_1 \times_2 \mathbf{I}_{I_2} \times_3 \mathbf{I}_{I_3}$, where $\mathcal{C} \in \mathbb{R}^{I_1 \times I_2 \times I_3}$, $\widetilde{\mathbf{B}}_1 \in \mathbb{R}^{I_1 \times I_1}$, and $\mathbf{I}_m$ represents an $m \times m$ identity matrix. Detailed illustrations can be found in Tucker (1966), Timmerman and Kiers (2000), and Section 4 of Kolda and Bader (2009). Note that the higher-order singular value decomposition (HOSVD) of a tensor is a particular type of orthogonal Tucker decomposition; see, e.g., De Lathauwer et al. (2000a), Rajwade et al. (2012), and Vannieuwenhoven et al. (2012). The applications of the Tucker decomposition can be found in Kolda and Bader (2009), Cichocki et al. (2015), Liu et al. (2020), Bhatt et al. (2021), and Hou et al. (2021).

In practice, both the CP and Tucker decompositions are useful for extracting valuable information from data through dimensionality reduction. In the case of the third-order tensor, Ceulemans and Kiers (2006) studied the selection between the CP and Tucker decompositions. The algorithms for computing these two types of decompositions can be found in the works of Kolda and Bader (2009), Rabanser et al. (2017), Kaya and Robert (2019), and Hale and Prater-Bennette (2021). After reviewing tensors and their operations, we will present some covariance related subjects in the next two sections.

10.2 Covariance and Correlation in Tensor Analysis

10.2.1 Tensor Covariance and Tensor Correlation

In the previous section, we have introduced tensor products and tensor decompositions, which allow us to understand tensors' interactions and simplifications. We next focus on tensor covariance and tensor correlation, which are crucial for quantifying the relationships between responses. In particular, they are valuable for analyzing complex datasets where traditional covariance and correlation measures are insufficient.

To define tensor covariance and correlation, we begin with matrix-variate data, which are a third-order tensor. In the last two decades, matrix-variate data have appeared in various fields of study such as speech recognition (Hoff et al. 2023), finance market (Zhang et al. 2023; Chen et al. 2024a), image analysis (Zhang and Schneider 2010; Leng and Pan 2018), international trade (Chen et al. 2024a), environmental monitoring (Mardia and Goodall 1993), signal processing (Werner et al. 2008), gene expression (Yin and Li 2012), and radar detection (Greenewald et al. 2016).

To analyze matrix-variate data, let $\mathbf{Y}_i = (y_{i,j_1 j_2}) \in \mathbb{R}^{p\times q}$ be i.i.d. random response matrices with mean $E(y_{i,j_1 j_2}) = 0$. For $i = 1, \cdots, n$, the elements of covariance and correlation of matrix-variate data are defined as $\sigma_{j_1 j_2 j_1' j_2'} = \text{cov}(y_{i,j_1 j_2}, y_{i,j_1' j_2'})$ and $\rho_{j_1 j_2 j_1' j_2'} = \text{corr}(y_{i,j_1 j_2}, y_{i,j_1' j_2'})$, respectively. According to Deng et al. (2023), we denote the tensor covariance and tensor correlation by $\boldsymbol{\Sigma} = (\sigma_{j_1 j_2 j_1' j_2'}) \in \mathbb{R}^{p\times q\times p\times q}$ and $\mathbf{R} = (\rho_{j_1 j_2 j_1' j_2'}) \in \mathbb{R}^{p\times q\times p\times q}$, respectively, where $j_1, j_1' = 1, \cdots, p$ and $j_2, j_2' = 1, \cdots, q$. An alternative and common approach for defining tensor covariance and tensor correlation is based on the vectorization data (Dutilleul 1999; Manceur and Dutilleul 2013; Hoff 2011). Specifically, $\boldsymbol{\Sigma} = \text{Cov}\{\text{vec}(\mathbf{Y}_i)\} \in \mathbb{R}^{pq\times pq}$ and $\mathbf{R} = \text{Corr}\{\text{vec}(\mathbf{Y}_i)\} \in \mathbb{R}^{pq\times pq}$, where $\text{vec}(\cdot)$ is the vectorization operator. In the remainder of this chapter, covariance and correlation refer to the vectorization of tensor data and not to the matrix variate data except for the general definition below.

The definition of covariance and correlation for matrix-variate data can be straightforwardly extended to that of tensor data. Consider the i.i.d. Nth-order random tensors $\mathcal{Y}_i = (y_{i,j_1 j_2\cdots j_N}) \in \mathbb{R}^{I_1\times I_2\times\cdots\times I_N}$. Then, the tensor covariance and correlation can be defined as $\boldsymbol{\Sigma} = (\sigma_{j_1 j_2\cdots j_N j_1' j_2'\cdots j_N'}) \in \mathbb{R}^{I_1\times I_2\times\cdots\times I_N\times I_1\times I_2\times\cdots\times I_N}$ and $\mathbf{R} = (\rho_{j_1 j_2\cdots j_N j_1' j_2'\cdots j_N'}) \in \mathbb{R}^{I_1\times I_2\times\cdots\times I_N\times I_1\times I_2\times\cdots\times I_N}$, respectively, where $\sigma_{j_1 j_2\cdots j_N j_1' j_2'\cdots j_N'} = \text{cov}(y_{i,j_1 j_2\cdots j_N}, y_{i,j_1' j_2'\cdots j_N'})$ and $\rho_{j_1 j_2\cdots j_N j_1' j_2'\cdots j_N'} = \text{corr}(y_{i,j_1 j_2\cdots j_N}, y_{i,j_1' j_2'\cdots j_N'})$. In vectorized responses, however, we have $\boldsymbol{\Sigma} = \text{Cov}\{\text{vec}(\mathcal{Y}_i)\} \in \mathbb{R}^{I_1 I_2\cdots I_N\times I_1 I_2\cdots I_N}$ and $\mathbf{R} = \text{Corr}\{\text{vec}(\mathcal{Y}_i)\} \in \mathbb{R}^{I_1 I_2\cdots I_N\times I_1 I_2\cdots I_N}$. Note that the $j_1 + [\{\sum_{k=2}^{N}(j_k - 1)\prod_{l=1}^{k-1} I_l\}]$-th element of $\text{vec}(\mathcal{Y}_i)$ is $y_{i,j_1 j_2\cdots j_N}$.

Remark The covariance matrix $\boldsymbol{\Sigma}$ (or correlation matrix $\mathbf{R}$) plays an important role in multivariate analysis, such as principal component analysis and discriminant analysis. As discussed in Sect. 2.2.3, when the number of samples is relatively small compared to the number of variables, the sample covariance matrix becomes unstable or inconsistent. It is expected that tensor data further exacerbate this phenomenon. This is because the dimensionality of tensor data often grows exponentially with the tensor order. Hence, obtaining reliable covariance estimates is critical in high-dimensional tensors, and Sect. 10.3 provides a few possible approaches to address this challenging task.

10.2.2 Tensor Canonical Correlation

After introducing tensor correlation, we next analyze tensor canonical correlation (Kim et al. 2007, TCCA), which is an extension of canonical correlation analysis (Hotelling 1936, CCA) in vector data. Tensor canonical correlation analysis mainly focuses on identifying the most correlated linear combinations of tensors from different data sources. In practice, there are multiple types of tensor-structured data in multi-modality data analysis (e.g., the different types of brain imaging tensors in neuroscience data). Then, tensor canonical correlation can help extract some insightful findings that might remain hidden when using only tensor correlation.

It is known that canonical correlation analysis (CCA) has been widely used for many decades across various fields of study including business, economics, engineering, psychometrics, and science. In the context of two-dimensional arrays, Lee and Choi (2007) proposed a two-dimensional CCA (2D-CCA) to analyze 2D images. Subsequently, Kim et al. (2007) introduced TCCA in multi-dimensional data, which could simultaneously detect and classify multiple actions in videos. Additionally, Luo et al. (2015) developed TCCA for analyzing canonical correlations of multiple views and successfully extracted high-order correlation information from different perspectives. However, neither of these two articles analyzed convergence properties for their proposed approaches. Recently, Chen et al. (2021) extended 2D-CCA and obtained a TCCA method with theoretical justification. In the rest of this section, we adopt their TCCA approach.

Before delving into tensor canonical correlation analysis (TCCA), we first provide a comprehensive review of canonical correlation analysis (CCA). Without loss of generality, we assume that all the random variables discussed below have been centered with mean zero. Let $X \in \mathbb{R}^p$, $Y \in \mathbb{R}^q$, $A \in \mathbb{R}^p$, and $B \in \mathbb{R}^q$. Then the classical CCA is

$$\max_{A\in\mathbb{R}^p, B\in\mathbb{R}^q} \operatorname{corr}(A'X, B'Y) = \max_{A\in\mathbb{R}^p, B\in\mathbb{R}^q} \frac{\operatorname{cov}(A'X, B'Y)}{\sqrt{\operatorname{var}(A'X)\operatorname{var}(B'Y)}},$$

which is equivalent to the following form,

$$\max_{A,B} A'\mathbf{\Sigma}_{XY}B \ \ s.t. \ \ A'\mathbf{\Sigma}_{XX}A = B'\mathbf{\Sigma}_{YY}B = 1, \tag{10.1}$$

where $\mathbf{\Sigma}_{XX} = E(XX')$, $\mathbf{\Sigma}_{XY} = E(XY')$, and $\mathbf{\Sigma}_{YY} = E(YY')$. The solutions of A and B in (10.1) are their corresponding eigenvectors with maximum eigenvalues of matrices $\mathbf{\Sigma}_{XX}^{-1}\mathbf{\Sigma}_{XY}\mathbf{\Sigma}_{YY}^{-1}\mathbf{\Sigma}_{YX}$ and $\mathbf{\Sigma}_{YY}^{-1}\mathbf{\Sigma}_{YX}\mathbf{\Sigma}_{XX}^{-1}\mathbf{\Sigma}_{XY}$, respectively, where $\mathbf{\Sigma}_{YX} = \mathbf{\Sigma}'_{XY}$; see Johnson and Wichern (2007). In practice, one can estimate the canonical correlation coefficients by replacing $\mathbf{\Sigma}_{XX}$, $\mathbf{\Sigma}_{XY}$, and $\mathbf{\Sigma}_{YY}$ with their corresponding sample estimators.

To simplify the introduction of TCCA, we follow the approach of Chen et al. (2021) and assume that $\mathcal{X}$ and $\mathcal{Y} \in \mathbb{R}^{I_1\times I_2\times\cdots\times I_N}$ with the same mode and shape. In addition, assume that $\mathcal{A} = A_1 \otimes \cdots \otimes A_N \in \mathbb{R}^{I_1 \times I_2 \times \cdots \times I_N}$ and $\mathcal{B} = B_1 \otimes \cdots \otimes B_N \in \mathbb{R}^{I_1 \times I_2 \times \cdots \times I_N}$ are rank-1 CP decompositions, where $A_k \in \mathbb{R}^{I_k}$ and $B_k \in \mathbb{R}^{I_k}$ for $k = 1, \cdots, N$. Then Chen et al. (2021) proposed TCCA given below.

$$\max_{\mathcal{A},\mathcal{B}} \operatorname{corr}(\mathcal{A} \cdot \mathcal{X}, \mathcal{B} \cdot \mathcal{Y})$$

where $\cdot$ represents the tensor inner product defined in Sect. 10.1.2. Suppose that $\mathcal{X}_i$ and $\mathcal{Y}_i$ are i.i.d. samples drawn from the random tensors $\mathcal{X}$ and $\mathcal{Y}$, respectively, for $i = 1, \cdots, n$. Then the sample tensor canonical correlation is

$$\max_{\mathcal{A},\mathcal{B}} \frac{\sum_{i=1}^{n} (\mathcal{A} \cdot \mathcal{X}_i)(\mathcal{B} \cdot \mathcal{Y}_i)}{\sqrt{\sum_{i=1}^{n} (\mathcal{A} \cdot \mathcal{X}_i)^2 \sum_{i=1}^{n} (\mathcal{B} \cdot \mathcal{Y}_i)^2}}.$$

The modified version of the higher-order power method (De Lathauwer et al. 2000b, HOPM) developed by Chen et al. (2021) can be used to compute the canonical components. In addition, they presented the three applications of TCCA using data for gene expression, air pollution, and electricity demands. More TCCA applications on face recognition, hand-gesture recognition, and biometric structure prediction can be found in Lee and Choi (2007), Kim et al. (2007), and Luo et al. (2015), respectively. In addition to TCCA, the multilinear CCA (MCCA), an alternative extension of CCA, has been discussed in Lu et al. (2013).

10.3 Tensor Covariance Models

In the previous section, we have introduced tensor covariance, tensor correlation, and tensor canonical correlation. Tensor correlation can be derived from tensor covariance, and tensor canonical correlation is also closely related to tensor covariance. Therefore, this section mainly focuses on modeling tensor covariance matrices based on the separable covariance model, core shrinkage covariance model, and tensor factor model discussed below.

10.3.1 Separable Covariance Model

Let $\mathbf{Y}_i \in \mathbb{R}^{p \times q}$ be i.i.d. random response matrices with mean $\mathbf{0}$ and covariance matrix $\mathbf{\Sigma} = (\sigma_{j_1 j_2}) := \operatorname{Cov}\{\operatorname{vec}(\mathbf{Y}_i)\} \in \mathcal{S}_{pq}^{+}$ for $i = 1, \cdots, n$, where $\mathcal{S}_{pq}^{+} = \{\mathbf{\Sigma} \in \mathbb{R}^{pq \times pq} : \mathbf{\Sigma} \text{ is a positive definite matrix}\}$. A natural estimator of $\mathbf{\Sigma}$ is the sample covariance matrix $\widehat{\mathbf{\Sigma}}_s = n^{-1} \sum_{i=1}^{n} \operatorname{vec}(\mathbf{Y}_i)\operatorname{vec}'(\mathbf{Y}_i)$. However, the sample covariance matrix $\widehat{\mathbf{\Sigma}}_s$ performs poorly as $pq/n \to \pi \in (0, \infty]$ (Wachter 1978; Johnstone 2001; Johnstone and Lu 2009). To address the challenges of high dimensionality, we consider a separability assumption on the covariance matrix $\boldsymbol{\Sigma}$. That is,

$$\boldsymbol{\Sigma} = \boldsymbol{\Sigma}_2 \otimes \boldsymbol{\Sigma}_1 \in \mathcal{S}_{pq}^{+}, \tag{10.2}$$

where $\boldsymbol{\Sigma}_1 \in \mathcal{S}_p^{+}$ and $\boldsymbol{\Sigma}_2 \in \mathcal{S}_q^{+}$ are row covariance and column covariance matrices, respectively. Useful references are Dawid (1981), Dutilleul (1999), Werner et al. (2008), Hoff (2011), Leng and Tang (2012), Tsiligkaridis and Hero (2013), Roś et al. (2016), Soloveychik and Trushin (2016), Derksen and Makam (2021), and Drton et al. (2021).

The separable structure in (10.2) can provide a simple yet meaningful interpretations; see Zhang et al. (2023). In addition, model (10.2) effectively reduces the number of parameters from $pq(pq+1)/2$ to $p(p+1)/2 + q(q+1)/2$. However, the number of parameters can be large when either p or q tends to infinity. Thus, additional assumptions are required to further reduce covariance parameters. It is worth noting that several structured covariance matrices have been adapted for the case of high-dimensional covariance estimation. These include the bandable structure (Bickel and Levina 2008b), the tapering structure (Furrer and Bengtsson 2007; Cai et al. 2010), and the sparse structure (Fan et al. 2016); see a comprehensive summary of these methods in Chap. 2. This motivates us to study parameter estimation by imposing each of those three structures on $\boldsymbol{\Sigma}_1$ and $\boldsymbol{\Sigma}_2$, separately. The remainder of this subsection is organized as follows: Sect. 10.3.1.1 presents the bandable and tapering covariance models, while Sect. 10.3.1.2 analyzes the sparse covariance model.

10.3.1.1 Bandable and Tapering Covariance Models

We first introduce the bandable covariance model by assuming that both $\boldsymbol{\Sigma}_1$ and $\boldsymbol{\Sigma}_2$ are bandable; see Zhang et al. (2023). Let $\mathbf{A} = (a_{j_1 j_2}) \in \mathbb{R}^{d \times d}$, where d represents either p or q. We then define the class of bandable matrices as

$$\mathcal{B}_d(k) = \left\{ \mathbf{A} \in \mathbb{R}^{d \times d} : a_{j_1 j_2} = 0 \text{ for any } |j_1 - j_2| > k,\ 1 \le j_1, j_2 \le d \right\}.$$

The estimator of a bandable covariance matrix can be obtained by optimizing the following loss function:

$$\begin{aligned} \ell_1(\boldsymbol{\Sigma}_1, \boldsymbol{\Sigma}_2) &= \|\widehat{\boldsymbol{\Sigma}}_{\mathbf{s}} - \boldsymbol{\Sigma}_2 \otimes \boldsymbol{\Sigma}_1\|_F^2, \\ s.t. \quad \boldsymbol{\Sigma}_1 &\in \mathcal{B}_p(k_1) \quad \text{and} \quad \boldsymbol{\Sigma}_2 \in \mathcal{B}_q(k_2), \end{aligned} \tag{10.3}$$

where k_1 and k_2 are tuning parameters that control the regularization level of banding. Accordingly, the banded covariance estimators of $\boldsymbol{\Sigma}_1$ and $\boldsymbol{\Sigma}_2$ are

$\left(\widehat{\boldsymbol{\Sigma}}_1^{ba}(k_1), \widehat{\boldsymbol{\Sigma}}_2^{ba}(k_2)\right) = \operatorname{argmin}_{\boldsymbol{\Sigma}_1,\boldsymbol{\Sigma}_2}\ell_1(\boldsymbol{\Sigma}_1, \boldsymbol{\Sigma}_2)$, where the superscript ba denotes a bandable estimator.

Since the objective function (10.3) has two constraints, estimating $\boldsymbol{\Sigma}_1$ and $\boldsymbol{\Sigma}_2$ is quite challenging. To effectively solve this optimization problem, Zhang et al. (2023) transformed (10.3) into an equivalent unconstrained objective function,

$$\ell_2(\boldsymbol{\Sigma}_1, \boldsymbol{\Sigma}_2) = \|\tilde{\boldsymbol{\Sigma}}^{ba}(k_1, k_2) - \boldsymbol{\Sigma}_2 \otimes \boldsymbol{\Sigma}_1\|_F^2, \tag{10.4}$$

where $\tilde{\boldsymbol{\Sigma}}^{ba}(k_1, k_2) = \widehat{\boldsymbol{\Sigma}}_s * \{B_{k_2}(\mathbf{1}_q\mathbf{1}_q') \otimes B_{k_1}(\mathbf{1}_p\mathbf{1}_p')\}$ is a doubly banded matrix of $\widehat{\boldsymbol{\Sigma}}_s$ with bandwidths k_1 and k_2. In addition, $B_{k_1}(\mathbf{1}_p\mathbf{1}_p') = [\mathbf{1}(|j_1 - j_2| \le k_1)]$ and $B_{k_2}(\mathbf{1}_q\mathbf{1}_q') = [\mathbf{1}(|j_1 - j_2| \le k_2)]$ are two constraint matrices, $\mathbf{1}_p$ and $\mathbf{1}_q$ are $p \times 1$ and $q \times 1$ vectors of 1s, $\mathbf{1}(\cdot)$ is an indicator function, and $*$ represents the Hadamard product. As a result, the banded covariance estimator of $\boldsymbol{\Sigma}$ is $\widehat{\boldsymbol{\Sigma}}^{ba}(k_1, k_2) = \widehat{\boldsymbol{\Sigma}}_2^{ba}(k_2) \otimes \widehat{\boldsymbol{\Sigma}}_1^{ba}(k_1)$, and the tuning parameters (k_1, k_2) can be chosen via the resampling procedure proposed by Bickel and Levina (2008b).

We next introduce the tapering covariance model, which assumes that both $\boldsymbol{\Sigma}_1$ and $\boldsymbol{\Sigma}_2$ are tapered. To this end, for any $k \ge 0$, Zhang et al. (2023) defined the following tapering matrix $T_k(\mathbf{A})$ for any $d \times d$ matrix $\mathbf{A} = (a_{j_1 j_2})$,

$$(T_k(\mathbf{A}))_{j_1, j_2} = \begin{cases} a_{j_1 j_2}, & \text{when } |j_1 - j_2| \le [k/2]; \\ (2 - \frac{|j_1 - j_2|}{[k/2]})a_{j_1 j_2}, & \text{when } [k/2] < |j_1 - j_2| \le k; \\ 0, & \text{otherwise.} \end{cases} \tag{10.5}$$

where $j_1, j_2 = 1, \cdots, d$. Analogous to (10.4), Zhang et al. (2023) introduced the following unconstrained objective function to obtain tapering estimators of covariance matrices,

$$\ell_3(\boldsymbol{\Sigma}_1, \boldsymbol{\Sigma}_2) = \|\tilde{\boldsymbol{\Sigma}}^{ta}(k_1, k_2) - \boldsymbol{\Sigma}_2 \otimes \boldsymbol{\Sigma}_1\|_F^2,$$

where the superscript ta indicates a tapered estimator, $\tilde{\boldsymbol{\Sigma}}^{ta}(k_1, k_2) = \widehat{\boldsymbol{\Sigma}}_s * \{T_{k_2}(\mathbf{1}_q\mathbf{1}_q') \otimes T_{k_1}(\mathbf{1}_p\mathbf{1}_p')\}$, and $T_{k_1}(\mathbf{1}_p\mathbf{1}_p')$ and $T_{k_2}(\mathbf{1}_q\mathbf{1}_q')$ are two tapering matrices. With a slight abuse of notation, we still use the notation (k_1, k_2) to represent the tuning parameters that control the regularization level of tapering. Then, the tapering estimators of $\boldsymbol{\Sigma}_1$ and $\boldsymbol{\Sigma}_2$ are $\left(\widehat{\boldsymbol{\Sigma}}_1^{ta}(k_1), \widehat{\boldsymbol{\Sigma}}_2^{ta}(k_2)\right) = \operatorname{argmin}_{\boldsymbol{\Sigma}_1,\boldsymbol{\Sigma}_2}\ell_3(\boldsymbol{\Sigma}_1, \boldsymbol{\Sigma}_2)$. Consequently, the tapering estimator of $\boldsymbol{\Sigma}$ is $\widehat{\boldsymbol{\Sigma}}^{ta}(k_1, k_2) = \widehat{\boldsymbol{\Sigma}}_2^{ta}(k_2) \otimes \widehat{\boldsymbol{\Sigma}}_1^{ta}(k_1)$, and the tuning parameters (k_1, k_2) can be selected via the resampling method.

In addition to obtaining the bandable and tapering estimators of $\boldsymbol{\Sigma}$, Zhang et al. (2023) showed the upper bounds of their convergence rates. To present these results, they adopted and adapted the definitions from Bickel and Levina (2008b) and Cai et al. (2010), respectively, and defined them as

$$\tilde{U}(\alpha,\varepsilon)=\Big\{\mathbf{A}:\max_{1\le j_1\le d}\sum_{j_2=1}^{d}\big(|a_{j_1j_2}|:|j_1-j_2|>k\big)\le C_1k^{-\alpha}\text{ for all }k\ge 1,$$
$$\text{and }0<\varepsilon\le\gamma_{\min}(\mathbf{A})\le\gamma_{\max}(\mathbf{A})\le\tfrac{1}{\varepsilon}<\infty\Big\},$$

and

$$\tilde{M}(\alpha,\varepsilon)=\Big\{\mathbf{A}:|a_{j_1j_2}|\le C_2|j_1-j_2|^{-\alpha-1}\text{ for }j_1\ne j_2$$
$$\text{and }0<\varepsilon\le\gamma_{\min}(\mathbf{A})\le\gamma_{\max}(\mathbf{A})\le\frac{1}{\varepsilon}\Big\},$$

where $\mathbf{A}$ is a $d\times d$ covariance matrix, $\varepsilon<1$, and $C_1>0$, $C_2>0$ and $\alpha>0$ are fixed constants. Here, $\gamma_{\min}(\mathbf{A})$ and $\gamma_{\max}(\mathbf{A})$ are the smallest and largest eigenvalues of $\mathbf{A}$. Note that the set $\tilde{U}(\alpha,\varepsilon)$ is a class of approximately bandable covariance matrices with a certain uniformity property across any rows $1\le j_1\le d$. A similar definition can be found in (2.16) of Sect. 2.5. In addition, the set $\tilde{M}(\alpha,\varepsilon)$ is a class of tapering covariance matrices.

To introduce theoretical results for the estimators of both bandable and tapering covariance matrices, define, for $u\in\{1,2\}$,

$$\tilde{\alpha}_u=\begin{cases}2\alpha_u & \text{when }\mathbf{\Sigma}_1\in\tilde{U}(\alpha_1,\varepsilon)\text{ and }\mathbf{\Sigma}_2\in\tilde{U}(\alpha_2,\varepsilon)\\ 2\alpha_u+1 & \text{when }\mathbf{\Sigma}_1\in\tilde{M}(\alpha_1,\varepsilon)\text{ and }\mathbf{\Sigma}_2\in\tilde{M}(\alpha_2,\varepsilon),\end{cases}$$

and denote $\mathbf{1}_{\eta,d}(k)=\mathbf{1}(\eta=ba,k<d-1)+\mathbf{1}(\eta=ta,k<2d-2)$ for $\eta\in\{ba,ta\}$ and $d\in\{p,q\}$. Accordingly, $\widehat{\mathbf{\Sigma}}^{\eta}(k_1,k_2)$ represents either the bandable or tapering covariance matrix according to $\eta\in\{ba,ta\}$. Then, Zhang et al. (2023) demonstrated the following theorem.

Theorem 10.1 *Let $\mathbf{Y}_i\in\mathbb{R}^{p\times q}$ be i.i.d. random response matrices with the true covariance matrix $\mathbf{\Sigma}=\mathbf{\Sigma}_2\otimes\mathbf{\Sigma}_1$. Assume that the elements of $\mathbf{Y}_i$ satisfy the finite fourth-moment condition. Then, for a universal finite constant $C_0>0$, we have that*

$$\mathrm{E}\left(\frac{\|\widehat{\mathbf{\Sigma}}^{\eta}(k_1,k_2)-\mathbf{\Sigma}\|_F^2}{pq}\right)\le C_0\left(\frac{k_1k_2}{n}+\mathbf{1}_{\eta,p}(k_1)k_1^{-\tilde{\alpha}_1}+\mathbf{1}_{\eta,q}(k_2)k_2^{-\tilde{\alpha}_2}\right).$$

The above theorem provides an upper bound for $\frac{\|\widehat{\mathbf{\Sigma}}^{\eta}(k_1,k_2)-\mathbf{\Sigma}\|_F^2}{pq}$. Under some additional conditions, Zhang et al. (2023) also obtained the lower bound and demonstrated that there exist sufficient conditions for which the convergence rates of the proposed banded and tapering covariance estimators are minimax optimal. This finding is similar to that of Cai et al. (2010). Note that both structured models allow the dimensions of matrix-variate data to be much larger than the sample size.

10.3.1.2 Sparse Covariance Model

In addition to bandable and tapering methods, the sparse approach is also often considered by constraining many elements of a covariance matrix to be exactly zero. Recently, Leng and Pan (2018) extended the sparse structure to the tensor covariance and proposed a sparse Kronecker-structured estimator by imposing sparsity assumptions on $\mathbf{\Sigma}_1$ and $\mathbf{\Sigma}_2$. In addition, they ensured its positive definiteness by adding a positive definite constraint to the objective function (Xue et al. 2012; Zou et al. 2017). For a more detailed discussion, refer to Sect. 2.6.

Assume that $\mathbf{Y}_i = \mathbf{U}\mathbf{Z}_i\mathbf{V}'$, where $\mathbf{U}$ and $\mathbf{V}$ are square matrices such that $\mathbf{U}\mathbf{U}' = \mathbf{\Sigma}_1$, $\mathbf{V}\mathbf{V}' = \mathbf{\Sigma}_2$, and the entries of $\mathbf{Z}_i$ are i.i.d. random variables with zero mean and unit variance. Based on (10.2), Leng and Pan (2018) obtained a sample estimator of $\mathbf{\Sigma}$:

$$\tilde{\mathbf{\Sigma}}_S = \frac{\widehat{\mathbf{\Sigma}}_2 \otimes \widehat{\mathbf{\Sigma}}_1}{n^{-1}\sum_{i=1}^{n}\|\mathbf{Y}_i\|_F^2},$$

where $\widehat{\mathbf{\Sigma}}_1 = (\widehat{\sigma}_{1,j_1j_2}) := n^{-1}\sum_{i=1}^{n}\mathbf{Y}_i\mathbf{Y}_i'$ and $\widehat{\mathbf{\Sigma}}_2 = (\widehat{\sigma}_{2,j_1'j_2'}) := n^{-1}\sum_{i=1}^{n}\mathbf{Y}_i'\mathbf{Y}_i$. It is straightforward to show that $\tilde{\mathbf{\Sigma}}_S$ is a consistent estimator of $\mathbf{\Sigma}$ when p and q are fixed and $n \to \infty$. In practice, however, p and q can be much larger than n. In order to find covariance matrix estimators, Leng and Pan (2018) proposed to first estimate correlation matrices. To this end, let $\widehat{\mathbf{D}}_1 = \text{diag}^{1/2}(\widehat{\sigma}_{1,11}, \cdots, \widehat{\sigma}_{1,pp})$ and $\widehat{\mathbf{D}}_2 = \text{diag}^{1/2}(\widehat{\sigma}_{2,11}, \cdots, \widehat{\sigma}_{2,qq})$. Then, denote the sample row and column correlation matrices by $\widehat{\mathbf{R}}_1$ and $\widehat{\mathbf{R}}_2$, respectively. As a result, $\widehat{\mathbf{\Sigma}}_1 = \widehat{\mathbf{D}}_1\widehat{\mathbf{R}}_1\widehat{\mathbf{D}}_1$ and $\widehat{\mathbf{\Sigma}}_2 = \widehat{\mathbf{D}}_2\widehat{\mathbf{R}}_2\widehat{\mathbf{D}}_2$.

To estimate correlation matrices, Leng and Pan (2018) proposed the following estimators:

$$\widehat{\mathbf{R}}_{1,LP} = \text{argmin}_{\mathbf{R}\in\mathbb{R}^{p\times p}}\frac{1}{2}\|\mathbf{R} - \widehat{\mathbf{R}}_1\|_F^2 + \lambda_{\mathbf{\Sigma}_1}\|\mathbf{R}\|_1, \; s.t. \; \mathbf{R} \succeq \epsilon\mathbf{I}_p \text{ and } \text{diag}(\mathbf{R}) = \mathbf{1}_p;$$

and

$$\widehat{\mathbf{R}}_{2,LP} = \text{argmin}_{\mathbf{R}\in\mathbb{R}^{q\times q}}\frac{1}{2}\|\mathbf{R} - \widehat{\mathbf{R}}_2\|_F^2 + \lambda_{\mathbf{\Sigma}_2}\|\mathbf{R}\|_1, \; s.t. \; \mathbf{R} \succeq \epsilon\mathbf{I}_q \text{ and } \text{diag}(\mathbf{R}) = \mathbf{1}_q,$$

where $\lambda_{\mathbf{\Sigma}_1}$ and $\lambda_{\mathbf{\Sigma}_2}$ are penalty parameters. In addition, ϵ is an arbitrarily small positive constant that ensures that the resulting estimators are positive definite. Consequently, the estimators of $\mathbf{\Sigma}_1$, $\mathbf{\Sigma}_2$, and $\mathbf{\Sigma}$ are, respectively, given as follows:

$$\widehat{\mathbf{\Sigma}}_{1,LP} = \widehat{\mathbf{D}}_1\widehat{\mathbf{R}}_{1,LP}\widehat{\mathbf{D}}_1, \quad \widehat{\mathbf{\Sigma}}_{2,LP} = \widehat{\mathbf{D}}_2\widehat{\mathbf{R}}_{2,LP}\widehat{\mathbf{D}}_2, \quad \text{and} \quad \widehat{\mathbf{\Sigma}}_{LP} = \frac{\widehat{\mathbf{\Sigma}}_{2,LP} \otimes \widehat{\mathbf{\Sigma}}_{1,LP}}{n^{-1}\sum_{i=1}^{n}\|\mathbf{Y}_i\|_F^2}.$$

Under appropriate sparsity conditions, Leng and Pan (2018) further obtained the non-asymptotic upper bounds for their proposed estimators in terms of both the spectral and Frobenius norms, as p and q go to infinity; see below.

Theorem 10.2 *Let $\mathbf{Y}_i \in \mathbb{R}^{p\times q}$ be i.i.d. random response matrices with the true covariance matrix $\boldsymbol{\Sigma} = \boldsymbol{\Sigma}_2 \otimes \boldsymbol{\Sigma}_1$, where the eigenvalues of $\boldsymbol{\Sigma}_1$ and $\boldsymbol{\Sigma}_2$ are bounded away from zero and infinity. Suppose that $\frac{\log p}{\log q} \in (0, C_3]$ for a finite constant $C_3 > 0$, and, as $p, q \to \infty$, we assume $(\zeta_2 + q)\frac{\log q}{np} \to 0$ and $(\zeta_1 + p)\frac{\log p}{nq} \to 0$, where ζ_1 and ζ_2 are the numbers of non-zero off-diagonal elements in $\boldsymbol{\Sigma}_1$ and $\boldsymbol{\Sigma}_2$, respectively. Then, under certain bounded and moment conditions, the following inequalities hold:*

$$\|\widehat{\boldsymbol{\Sigma}}_{LP} - \boldsymbol{\Sigma}\|_2 \le C_4\left\{\sqrt{(\zeta_2+1)\frac{\log q}{np}} + \sqrt{(\zeta_1+1)\frac{\log p}{nq}}\right\} \quad \textit{and}$$

$$\|\widehat{\boldsymbol{\Sigma}}_{LP} - \boldsymbol{\Sigma}\|_F \le C_4\left\{\sqrt{(\zeta_2+q)\frac{\log q}{n}} + \sqrt{(\zeta_1+p)\frac{\log p}{n}}\right\},$$

with probability $1 - p^{-0.9} - q^{-0.9} - (pq)^{-0.9}$, where C_4 is a positive constant that is independent of p, q and n.

Note that the above non-asymptotic results hold without imposing any distributional assumptions on $\mathbf{Y}_i$. In addition, these results are valid even for a fixed sample size regardless of the dimensionality of $\mathbf{Y}_i$.

10.3.2 Core Shrinkage Covariance Model

In Sect. 10.3.1, we assume that the covariance matrix $\boldsymbol{\Sigma}$ is truly separable. In practice, however, this assumption may not be valid. Hence, the separable covariance estimators might yield poor performance; see Stein (2005), Rougier (2017), and Hoff et al. (2023). To alleviate this assumption, Hoff et al. (2023) proposed a novel matrix decomposition approach for positive definite matrices, named Kronecker-core decomposition (KCD). The main idea of this method is to represent a covariance matrix by a separable matrix, as given in (10.2), together with a "core" matrix. Under this decomposition, the separable matrix enjoys the properties of parsimony and interpretability, while the core matrix is its complement. In sum, KCD comprehensively characterizes covariance matrices that are non-separable.

To introduce their model, Hoff et al. (2023) first defined the Kronecker covariance of $\boldsymbol{\Sigma}$ as $\kappa(\boldsymbol{\Sigma}) = \boldsymbol{\Sigma}_2 \otimes \boldsymbol{\Sigma}_1$, where $\boldsymbol{\Sigma}_1$ and $\boldsymbol{\Sigma}_2$ are positive definite matrices satisfying $\boldsymbol{\Sigma}_1 = E(\mathbf{Y}_i\boldsymbol{\Sigma}_2^{-1}\mathbf{Y}_i')/q$ and $\boldsymbol{\Sigma}_2 = E(\mathbf{Y}_i'\boldsymbol{\Sigma}_1^{-1}\mathbf{Y}_i)/p$. By definition, $\kappa(\boldsymbol{\Sigma})$ represents the separable component of $\boldsymbol{\Sigma}$. Then, they modeled the covariance matrix $\boldsymbol{\Sigma} \in \mathcal{S}_{pq}^{+}$ given below:

$$\mathbf{\Sigma} = \mathbf{HCH}'. \tag{10.6}$$

Note that $\mathbf{H} = \mathbf{H}_2 \otimes \mathbf{H}_1 \in \mathcal{H}_{p,q}$ is a separable matrix, and it satisfies $\mathbf{HH}' = \kappa(\Sigma) = \mathbf{K}$, where $\mathcal{H}_{p,q} = \{\mathbf{H}_2 \otimes \mathbf{H}_1 : \mathbf{H}_1 \in \mathcal{H}_p, \mathbf{H}_2 \in \mathcal{H}_q\}$ is the set of separable matrices, and $\mathcal{H}_p$ and $\mathcal{H}_q$ are the general linear group of nonsingular $p \times p$ and $q \times q$ matrices, respectively. In addition, $\mathbf{C} = \mathbf{H}^{-1}\mathbf{\Sigma}(\mathbf{H}')^{-1} \in \mathbb{R}^{(pq)\times(pq)}$ is the non-separable component of $\mathbf{\Sigma}$, i.e., the core covariance matrix, which satisfies $\kappa(\mathbf{C}) = \mathbf{I}_{pq}$. Thus, the core covariance matrix $\mathbf{C}$ has no across-row or across-column correlation or heteroscedasticity. Consequently, model (10.6) decomposes a covariance matrix into a separable matrix $\mathbf{H}$ and a core matrix $\mathbf{C}$.

Based on model (10.6), Hoff et al. (2023) proposed a covariance shrinkage estimator by assuming that $h(\mathbf{HH}') = \mathbf{H}$, where h is a separable matrix square root function. This estimator essentially shrinks the sample covariance matrix toward the estimator of $\mathbf{K}$ induced by the separable matrix $\mathbf{H}$. To accomplish this task, Hoff et al. (2023) first employed the following objective function to estimate $\mathbf{K}$,

$$\ell(\mathbf{K}) = \log|\mathbf{K}| + tr(\mathbf{K}^{-1}\widehat{\mathbf{\Sigma}}_s),\ s.t.\ \mathbf{K} \in \mathcal{S}_{pq}^{+},$$

and denoted the resulting estimator by $\kappa(\widehat{\mathbf{\Sigma}}_s) = \widehat{\mathbf{K}} = \text{argmin}_{\mathbf{K}}\ell(\mathbf{K})$. Note that $\widehat{\mathbf{K}}$ is symmetric. Let $\widehat{\mathbf{H}} = \widehat{\mathbf{K}}^{1/2}$, which leads to the core covariance matrix estimator of $\mathbf{C}$ in Hoff et al. (2023), that is $\widehat{\mathbf{C}} = \widehat{\mathbf{K}}^{-1/2}\widehat{\mathbf{\Sigma}}_s\widehat{\mathbf{K}}^{-1/2}$. In practice, the dimension of $\widehat{\mathbf{C}}$, pq, can be large. In order to mitigate the estimation error and guarantee the covariance estimator is positive definite, Hoff et al. (2023) proposed the following core shrinkage covariance estimator,

$$\widehat{\mathbf{\Sigma}}_H = \widehat{\mathbf{K}}^{1/2}\widehat{\mathbf{C}}_H\widehat{\mathbf{K}}^{1/2} = (1-w)\widehat{\mathbf{\Sigma}}_s + w\widehat{\mathbf{K}},$$

where $\widehat{\mathbf{C}}_H = (1-w)\widehat{\mathbf{C}} + w\mathbf{I}_{pq}$ is the shrunken version of $\widehat{\mathbf{C}}$, and $w \in [0, 1]$ is a weight parameter. Since $\widehat{\mathbf{C}}_H$ linearly shrinks $\widehat{\mathbf{C}}$ toward the identity matrix, the sample covariance matrix $\widehat{\mathbf{\Sigma}}_s$ linearly shrinks toward the separable covariance $\widehat{\mathbf{K}}$. In addition, the core shrinkage covariance estimator is positive definite even when the sample size is much smaller than the dimension of the covariance matrix. As demonstrated by Hoff et al. (2023), when the sample covariance matrix $\widehat{\mathbf{\Sigma}}_s$ converges in probability to $\mathbf{\Sigma}$, $\widehat{\mathbf{\Sigma}}_H$, is consistent. This result also holds by replacing w by its estimator in Hoff et al. (2023).

10.3.3 Tensor Factor Model

In addition to the separable and core shrinkage approaches, another popular method for estimating tensor covariances is based on tensor factor models (Chen et al. 2022; Chang et al. 2023; Chen et al. 2024a; Chen and Lam 2024; Chen et al. 2024b; Han et al. 2024a; Han et al. 2024b). The main idea behind these models is to generalize factor models for vector data in Chap. 8 to those for tensor data. This generalization can be done by utilizing Tucker or CP decompositions (Kolda and Bader 2009) to construct low-rank structures.

Before presenting the general form of tensor factor models, we first revisit the matrix-variate factor model proposed by Chen and Fan (2023) for the second-order tensor time series data $\mathbf{X}_t = (x_{t,j_1 j_2}) \in \mathbb{R}^{p\times q}$, which was introduced in Sect. 8.5. The model is given by:

$$\mathbf{X}_t = \mathbf{A}\mathbf{F}_t\mathbf{B}' + \mathbf{E}_t, \tag{10.7}$$

for $t = 1, \cdots, T$, where $\mathbf{F}_t \in \mathbb{R}^{m_1\times m_2}$ is the latent factor matrix with $m_1 \ll p$ and $m_2 \ll q$, $\mathbf{A}$ and $\mathbf{B}$ are $p \times m_1$ and $q \times m_2$ loading matrices, respectively, and $\mathbf{E}_t$ is a $p \times q$ noise matrix. When matrices $\mathbf{X}_t$ are stacked along the third mode (time), we obtain a third-order tensor $\mathcal{X} = (\mathbf{X}_1, \mathbf{X}_2, \cdots, \mathbf{X}_T) \in \mathbb{R}^{p\times q\times T}$. In this setting, the model (10.7) can be re-expressed as a noisy Tucker decomposition

$$\mathcal{X} = \mathcal{F} \times_1 \mathbf{A} \times_2 \mathbf{B} \times_3 \mathbf{I}_T + \mathcal{E}, \tag{10.8}$$

where $\times_k$ is the k-mode product as introduced in Sect. 10.1.2, $\mathcal{F} = (\mathbf{F}_1, \mathbf{F}_2, \cdots, \mathbf{F}_T) \in \mathbb{R}^{m_1\times m_2\times T}$, and $\mathcal{E} = (\mathbf{E}_1, \mathbf{E}_2, \cdots, \mathbf{E}_T) \in \mathbb{R}^{p\times q\times T}$ is a third-order tensor. By assuming that $\mathbf{E}_t$ is uncorrelated with $\mathbf{F}_t$, we then obtain the covariance matrix of $\text{vec}(\mathbf{X}_t)$,

$$\mathbf{\Sigma} = (\mathbf{B} \otimes \mathbf{A})\mathbf{\Sigma}_F(\mathbf{B} \otimes \mathbf{A})' + \mathbf{\Sigma}_E,$$

where $\mathbf{\Sigma}_F = \text{Cov}\{\text{vec}(\mathbf{F}_t)\}$ and $\mathbf{\Sigma}_E = \text{Cov}\{\text{vec}(\mathbf{E}_t)\}$. To estimate $\mathbf{\Sigma}$, Chen and Fan (2023) estimated $\mathbf{A}$, $\mathbf{B}$, and $\mathbf{F}_t$. In addition, they established the limiting distributions of $\widehat{\mathbf{A}}$ and $\widehat{\mathbf{B}}$, and the rate of convergence of $\widehat{\mathbf{F}}_t$, under general conditions. Thus, we can estimate $\mathbf{\Sigma}$ by $\widehat{\mathbf{\Sigma}} = (\widehat{\mathbf{B}} \otimes \widehat{\mathbf{A}})\widehat{\mathbf{\Sigma}}_F(\widehat{\mathbf{B}} \otimes \widehat{\mathbf{A}})' + \widehat{\mathbf{\Sigma}}_E$, where $\widehat{\mathbf{\Sigma}}_F$ is the sample covariance matrix of $\widehat{\mathbf{F}}_t$, and $\widehat{\mathbf{\Sigma}}_E$ is the sample covariance matrix of $\text{vec}(\widehat{\mathbf{E}}_\mathbf{t}) = \text{vec}(\mathbf{X}_t - \widehat{\mathbf{A}}\widehat{\mathbf{F}}_t\widehat{\mathbf{B}}')$.

It should be noted that the factor model proposed by Chen and Fan (2023) is equivalent to a noisy Tucker decomposition as expressed in (10.8). In addition to Tucker decomposition, Chen et al. (2024a) adapted Han and Zhang's (2023) CP decomposition and produced a unique set of one-dimensional latent factors. Specifically, when defining $\mathcal{X}_t \in \mathbb{R}^{I_1\times I_2\times\cdots\times I_N}$ for $t = 1, \cdots, T$, a general tensor factor model with a CP low-rank structure is

$$\mathcal{X}_t = \sum_{r=1}^{G} \lambda_{rt}\mathbf{A}_1^{(r)} \otimes \cdots \otimes \mathbf{A}_N^{(r)} + \mathcal{E}_t, \tag{10.9}$$

where λ_{rt} represents the r-th element of a latent factor vector $\boldsymbol{\lambda}_t = (\lambda_{1t}, \cdots, \lambda_{Gt})^\top \in \mathbb{R}^G$, $\mathbf{A}_k^{(r)}$ is an I_k-dimensional loading vector with $||\mathbf{A}_k^{(r)}||_2 = 1$, for $k = 1, \cdots, N$ and $r = 1, \cdots, G$, and $\mathcal{E}_t \in \mathbb{R}^{I_1\times I_2\times\cdots\times I_N}$ is an N-th order noise tensor. When $N = 2$, model (10.9) has the same structure as model (10.7) with $\mathbf{A} = (\mathbf{A}_1^{(1)}, \cdots, \mathbf{A}_1^{(G)}) \in \mathbb{R}^{I_1 \times G}$, $\mathbf{B} = (\mathbf{A}_2^{(1)}, \cdots, \mathbf{A}_2^{(G)}) \in \mathbb{R}^{I_2 \times G}$, and $\mathbf{F}_t = \mathrm{diag}\{\lambda_{1t}, \cdots, \lambda_{Gt}\} \in \mathbb{R}^{G \times G}$. Based on model (10.9), Chen et al. (2024a) proposed the estimators of the tensor covariance of $\mathcal{X}_t$ and factor loading vectors. They also demonstrated the consistency and asymptotic normality of the estimated factor loading vectors under relaxed assumptions.

10.4 Empirical Examples

In the previous section, we have introduced tensor-based covariance matrices. To demonstrate the effectiveness of tensor methods, we present two empirical examples on speech recognition and object tracking.

10.4.1 Speech Recognition

Speech is the primary and most natural and efficient way for human beings to communicate and exchange information; see Gaikwad et al. (2010). Hence, enabling machines to interpret and understand human speech has the potential to significantly enhance work efficiency and improve living standards. To achieve this goal, speech recognition has become increasingly popular among researchers. Many methods have been developed to improve recognition accuracy; see a useful reference book of Rao and Manjunath (2017).

Among those methods, Hoff et al. (2023) applied quadratic discriminant analysis (QDA), in conjunction with their proposed core shrinkage estimator (CSE) of covariance matrices, to speech recognition. Specifically, they focused on a classification task for spoken-word audio involving training a model to identify and categorize Q different command words such as "yes" and "no" from 20,600 1-second-long audio WAV files. To this end, let $\mathbf{X}_i^{(\tau)} \in \mathbb{R}^{p \times q}$ represent the feature matrix for the i-th audio sample of the τ-th command word, where $\tau = 1, \cdots, Q$ and $i = 1, \cdots, n_\tau$. That is, each of the command words has its own set of n_τ audio files. In addition, for each τ, the n_τ audio samples were divided into a training set of size $n_{\tau 1}$ and a testing set of size $n_{\tau 2}$, where $n_\tau = n_{\tau 1} + n_{\tau 2}$. Accordingly, the total training and testing sample sizes are $N_1 = \sum_{\tau=1}^{Q} n_{\tau 1}$ and $N_2 = \sum_{\tau=1}^{Q} n_{\tau 2}$, respectively. Using the training data, Hoff et al. (2023) computed the sample means ($\widehat{\boldsymbol{\mu}}^{(\tau)}$) and the core shrinkage estimates of covariance matrices ($\widehat{\boldsymbol{\Sigma}}_H^{(\tau)}$) for each of the Q words.

After obtaining the sample means and sample covariance matrices, Hoff et al. (2023) then employed the QDA to classify the audio samples in the testing data. Specifically, for an audio sample in the testing data with the feature matrix $\mathbf{X}^* \in \mathbb{R}^{p \times q}$, the score obtained from the category $\tau \in \{1, \cdots, Q\}$ is defined below,

$$\ell_\tau(\mathbf{X}^*) = (\text{vec}(\mathbf{X}^*) - \widehat{\boldsymbol{\mu}}^{(\tau)})'\widehat{\boldsymbol{\Sigma}}_H^{(\tau)-1}(\text{vec}(\mathbf{X}^*) - \widehat{\boldsymbol{\mu}}^{(\tau)}) + \log|\widehat{\boldsymbol{\Sigma}}_H^{(\tau)}|.$$

When the frequencies of the different classes are equal, the classification rule assigns a new audio sample with feature matrix $\mathbf{X}^*$ to the class with the lowest score.

Based on the above general setting, Hoff et al. (2023) used the data from Warden (2017) to compare their method with two competing approaches. The competing methods are the QDA with the sample covariance matrix and the QDA with an estimated covariance matrix obtained from a separable structure. A detailed description of this dataset can be found in Warden (2018). In this example, we have $Q = 10$, $p = 99$, and $q = 13$. For $\tau = 1, \cdots, Q$, n_τ ranged from 1,987 to 2,103. In addition, the training sample size and the testing sample size are set to $N_1 = 19,600$ and $N_2 = 100$, respectively. According to the confusion matrix, QDA in conjunction with CSE yields substantially higher accuracy than both of the competing approaches.

10.4.2 Object Tracking

Object tracking plays a crucial role in image data analysis, with applications in diverse fields such as robot vision, visual surveillance, traffic management, vehicle navigation, sports analytics, and human–computer interaction. Numerous methods have been developed for object tracking. For a comprehensive overview, one can refer to survey article by Balaji and Karthikeyan (2017) or by Tiwari and Singhai (2017). In this study, we focus on tensor-based covariance approaches for assessing the difference between the covariance matrices of the target and candidate objects in object tracking. These approaches are effective at modeling correlations in multimodal data and able to capture complex intrinsic structures in image data (Tuzel et al. 2006; Li and Sun 2011; Qin et al. 2014).

Let $\mathbf{O} = (o_{i_1 i_2}) \in \mathbb{R}^{I_1 \times I_2}$ denote the observed image of a target object over the image region $I_1 \times I_2$ such as $W \times H$. For each pixel $o_{i_1 i_2}$, define the third-order tensor $X_{i_1 i_2} \in \mathbb{R}^{I_3}$ as its feature vector, which can consist of colors, gradients, and infrared, among others. Thus, an object can be naturally represented by a third-order tensor $\mathcal{X} \in \mathbb{R}^{I_1 \times I_2 \times I_3}$. Based on the $I_1 I_2$ pixel points, we construct the sample covariance matrix of the I_3 features as

$$\widehat{\boldsymbol{\Sigma}}_X = \frac{1}{I_1 I_2 - 1} \sum_{i_1=1}^{I_1} \sum_{i_2=1}^{I_2} (X_{i_1 i_2} - \bar{X})(X_{i_1 i_2} - \bar{X})' = \frac{1}{I_1 I_2 - 1} \mathbf{X}^{*\prime}\mathbf{X}^*,$$

where $\bar{X} = \frac{1}{I_1 I_2} \sum_{i_1=1}^{I_1} \sum_{i_2}^{I_2} X_{i_1 i_2} \in \mathbb{R}^{I_3}$ is the sample mean of the I_3 features, $\mathbf{X}^* = (X_{11}^*, \cdots, X_{I_1 1}^*, \cdots, X_{1 I_2}^*, \cdots, X_{I_1 I_2}^*)' \in \mathbb{R}^{I_1 I_2 \times I_3}$, and $X_{i_1 i_2}^* = X_{i_1 i_2} - \bar{X}$ for $i_1 = 1, \cdots, I_1$ and $i_2 = 1, \cdots, I_2$. Accordingly, we obtain the target covariance matrix, $\widehat{\boldsymbol{\Sigma}}_X^T \in \mathbb{R}^{I_3 \times I_3}$, computed at the target object window, and the candidate covariance matrix, $\widehat{\boldsymbol{\Sigma}}_X^C \in \mathbb{R}^{I_3 \times I_3}$, calculated at the candidate regions.

To analyze object tracking with tensor covariance matrices, we employ the covariance distance measure proposed by Förstner and Moonen (2003) to assess the difference between the covariance matrices of the target and candidate objects. That is,

$$\operatorname{dist}\left(\widehat{\boldsymbol{\Sigma}}_X^T, \widehat{\boldsymbol{\Sigma}}_X^C\right) = \left\{\sum_{i_3=1}^{I_3} \log^2 \gamma_{i_3}\left(\widehat{\boldsymbol{\Sigma}}_X^T, \widehat{\boldsymbol{\Sigma}}_X^C\right)\right\}^{1/2}, \tag{10.10}$$

where $\gamma_{i_3}(\widehat{\boldsymbol{\Sigma}}_X^T, \widehat{\boldsymbol{\Sigma}}_X^C)$ is the generalized eigenvalue obtained by solving $|\gamma \widehat{\boldsymbol{\Sigma}}_X^T - \widehat{\boldsymbol{\Sigma}}_X^C| = 0$ for $i_3 = 1, \cdots, I_3$. By this measure, we conduct a search process across the entire image to identify the optimal candidate object. The goal is to find the region that results in the minimum covariance distance. The region that exhibits the best match is then utilized to determine the location of the target object within the image.

It is expected that calculating the distance measure (10.10) becomes computationally intractable when the value of I_3 gets large. To alleviate this computational burden, Li and Sun (2011) employed the principal component approach to obtain a low-rank approximation of the covariance matrices. Specifically, they conducted a spectral decomposition on $\widehat{\boldsymbol{\Sigma}}_X^T$, such that $\widehat{\boldsymbol{\Sigma}}_X^T = \mathbf{U}_T \boldsymbol{\Lambda} \mathbf{U}_T'$, where $\mathbf{U}_T \in \mathbb{R}^{I_3 \times I_3}$ represents an orthogonal matrix, and $\boldsymbol{\Lambda}$ is the diagonal matrix whose diagonal elements correspond to the eigenvalues of $\widehat{\boldsymbol{\Sigma}}_X^T$. To determine the most relevant components to approximate $\widehat{\boldsymbol{\Sigma}}_X^T$, the following methods can be considered: the scree plot (Jolliffe 2002), variance explained criteria (Li and Sun 2011), or the component selection criterion (Sclove 2021). Let the number of the selected components be $\bar{d}$, where $\bar{d} \ll I_3$. We then construct the principal component score matrix, $\tilde{\mathbf{X}} = \mathbf{X}\tilde{\mathbf{U}} \in \mathbb{R}^{I_1 I_2 \times \bar{d}}$, where the columns of $\tilde{\mathbf{U}} \in \mathbb{R}^{I_3 \times \bar{d}}$ is the first $\bar{d}$ columns of $\mathbf{U}_T$.

Based on the principal component score matrix and assuming that $\tilde{\mathbf{X}}$ has been centralized, Li and Sun (2011) obtained the target covariance matrix with dimension $\bar{d} \times \bar{d}$ given below:

$$\tilde{\boldsymbol{\Sigma}}_X^T = \frac{1}{I_1 I_2 - 1} \tilde{\mathbf{X}}' \tilde{\mathbf{X}}.$$

Analogously, the candidate covariance matrix $\tilde{\boldsymbol{\Sigma}}_X^C \in \mathbb{R}^{\bar{d} \times \bar{d}}$ can also be obtained. Then, employing the following distance measure as defined in (10.10), they were able to assess the difference between these two covariance matrices,

$$\operatorname{dist}\left(\tilde{\boldsymbol{\Sigma}}_X^T, \tilde{\boldsymbol{\Sigma}}_X^C\right) = \left\{\sum_{j=1}^{\bar{d}} \log^2 \tilde{\gamma}_j\left(\tilde{\boldsymbol{\Sigma}}_X^T, \tilde{\boldsymbol{\Sigma}}_X^C\right)\right\}^{1/2},$$

where $\tilde{\gamma}_j(\tilde{\boldsymbol{\Sigma}}_X^T, \tilde{\boldsymbol{\Sigma}}_X^C)$ is the generalized eigenvalue obtained by solving the equation $|\gamma\tilde{\boldsymbol{\Sigma}}_X^T - \tilde{\boldsymbol{\Sigma}}_X^C| = 0$ for $j = 1, \cdots, \bar{d}$. This measure allows us to effectively select the optimal candidate object and determine the location of the target object.

As discussed in Sect. 10.1.1, there are three mode-k ($k = 1, 2, 3$) unfolding matrices, $\mathbf{X}_{(1)}$, $\mathbf{X}_{(2)}$ and $\mathbf{X}_{(3)}$, induced by the third-order tensor $\mathcal{X}$. It can be seen that the above discussion is only for the model-3 unfolding matrix $\mathbf{X}_{(3)} \in \mathbb{R}^{I_3 \times I_1 I_2}$. To further enhance the accuracy of object tracking, Li and Sun (2011) proposed an ensemble distance measure by integrating the three distance measures computed from the unfolding matrices, $\mathbf{X}_{(1)}$, $\mathbf{X}_{(2)}$, and $\mathbf{X}_{(3)}$. To mitigate the effects of environmental changes such as noise, illumination, background clutter, occlusion, and pose variation during the tracking process, they introduced a model-updating procedure for a long-term and stable tracking.

According to the above general setting, Li and Sun (2011) compared the tensor-based covariance tracking (TCT) method with the conventional covariance tracking (CCT) approach in three datasets. The first dataset consists of a dog sequence, where the primary objective is to detect the dog within the frames. In this example, $I_1 = 44$, $I_2 = 56$ and $I_3 = 9$. The second dataset is a street sequence with $I_1 = 36$, $I_2 = 55$, and $I_3 = 9$. The third dataset is known as the PETS2001 sequence and was recorded on campus with $I_1 = 49$, $I_2 = 25$, and $I_3 = 9$. Note that the nine components in the feature factor $X(i_1, i_2)$ are

$$\Big(X_r(i_1, i_2), X_g(i_1, i_2), X_b(i_1, i_2), \frac{\partial X_r(i_1, i_2)}{\partial i_1}, \frac{\partial X_r(i_1, i_2)}{\partial i_2},$$

$$\frac{\partial X_g(i_1, i_2)}{\partial i_1}, \frac{\partial X_g(i_1, i_2)}{\partial i_2}, \frac{\partial X_b(i_1, i_2)}{\partial i_1}, \frac{\partial X_b(i_1, i_2)}{\partial i_2}\Big)',$$

where $X_r(i_1, i_2)$, $X_g(i_1, i_2)$, and $X_b(i_1, i_2)$ denote the values of three color channels, respectively, at pixel (i_1, i_2), and $\frac{\partial z}{\partial t}$ denotes the partial derivative of z versus t.

The experimental results showed that, for both the first and the third datasets, the TCT method yields more accurate tracking than the CCT method. In the second dataset, as a black van gradually occludes the white van, the CCT method encounters difficulties. Accordingly, the resulting position errors from CCT increase, and the tracking eventually fails at approximately frame 130. In contrast, the TCT method is able to continuously track the white van even under this adverse condition. In sum, the tensor-based covariance tracking is superior to the conventional covariance tracking.

References

Abadi, M., Barham, P., Chen, J. M., Chen, Z. F., Davis, A., Dean, J., Devin, M., Ghemawat, S., Irving, G., Isard, M., et al. (2016). TensorFlow: a system for large-scale machine learning. *Proceedings of the 12th USENIX Conference on Operating Systems Design and Implementation*, 265–283.

Abbe, E. (2018). Community detection and stochastic block models: recent developments. *Journal of Machine Learning Research* 18, 1–86.

Aggarwal, C. C. (2018). *Neural Networks and Deep Learning: A Textbook*. Cham: Springer International Publishing.

Ahn, S. C. and Horenstein, A. R. (2013). Eigenvalue ratio test for the number of factors. *Econometrica* 81, 1203–1227.

Aitken, A. C. (1936). On least squares and linear combination of observations. *Proceedings of the Royal Society of Edinburgh* 55, 42–48.

Akaike, H. (1973). Information theory and an extension of the maximum likelihood principle. *Proceedings of the 2nd International Symposium on Information Theory*, 267–281.

Akaike, H. (1987). Factor analysis and AIC. *Psychometrika* 52, 317–332.

Allen, G. I. and Tibshirani, R. (2010). Transposable regularized covariance models with an application to missing data imputation. *The Annals of Applied Statistics* 4, 764–790.

Alpaydin, E. (2020). *Introduction to Machine Learning*. Cambridge: MIT Press.

Alzubaidi, L., Zhang, J. L., Humaidi, A. J., Al-Dujaili, A., Duan, Y., Al-Shamma, O., Santamaría, José, Fadhel, M. A., Al-Amidie, M., and Farhan, L. (2021). Re-view of deep learning: concepts, CNN architectures, challenges, applications, future directions. *Journal of Big Data* 8, 1–74.

Amemiya, T. (1985). *Advanced Econometrics*. Cambridge: Harvard University Press.

Amini, M. R., Feofanov, V., Pauletto, L., Hadjadj, L., Devijver, É., and Maximov, Y. (2025). Self-training: A survey. *Neurocomputing* 616, 128904.

Anderson, T. W. (1973). Asymptotically efficient estimation of covariance matrices with linear structure. *The Annals of Statistics* 1, 135–141.

Anderson, T. W. (2003). *An Introduction to Multivariate Statistical Analysis* New York: Wiley.

Anderson, T. W. and Amemiya, Y. (1988). The asymptotic normal distribution of estimators in factor analysis under general conditions. *The Annals of Statistics* 16, 759–771.

Andreini, P., Izzo, C., and Ricco, G. (2020). Deep dynamic factor models. *arXiv preprint arXiv:2007.11887.*

Anselin, L. (1988). *Spatial Econometrics: Methods and Models*. Dordrecht: Springer Netherlands.

Archakov, I. and Hansen, P. R. (2021). A new parametrization of correlation matrices. *Econometrica* 89, 1699–1715.

W. Lan, C.-L. Tsai, *Covariance Analysis and Beyond*,
https://doi.org/10.1007/978-3-032-08796-6

Avagyan, V. (2022). Precision matrix estimation using penalized generalized sylvester matrix equation. *Test* 31, 950–967.

Avella-Medina, M., Battey, H., Fan, J., and Li, Q. F. (2018). Robust estimation of high-dimensional covariance and precision matrices. *Biometrika* 105, 271–284.

Ba, J., Kiros, J. R., and Hinton, G. (2016). Layer normalization. *Advances in Nips 2016 Deep Learning Symposium. arXiv preprint arXiv:1607.06450.*

Bacciu, D., Errica, F., Micheli, A., and Podda, M. (2020). A gentle introduction to deep learning for graphs. *Neural Networks* 129, 203–221.

Bai, J. (2003). Inferential theory for factor models of large dimensions. *Econometrica* 71, 135–171.

Bai, J. (2009). Panel data models with interactive fixed effects. *Econometrica* 77, 1229–1279.

Bai, J. and Li, K. P. (2012). Statistical analysis of factor models of high dimension. *The Annals of Statistics* 40, 436–465.

Bai, J., Li, K. P., and Lu, L. N. (2016). Estimation and inference of FAVAR models. *Journal of Business & Economic Statistics* 34, 620–641.

Bai, J. and Liao, Y. (2016). Efficient estimation of approximate factor models via penalized maximum likelihood. *Journal of Econometrics* 191, 1–18.

Bai, J. and Ng, S. (2002). Determining the number of factors in approximate factor models. *Econometrica* 70, 191–221.

Bai, J. and Ng, S. (2006). Confidence intervals for diffusion index forecasts and inference for factor-augmented regressions. *Econometrica* 74, 1133–1150.

Bai, J. and Ng, S. (2007). Determining the number of primitive shocks in factor models. *Journal of Business & Economic Statistics* 25, 52–60.

Bai, J. and Ng, S. (2008a). Forecasting economic time series using targeted predictors. *Journal of Econometrics* 146, 304–317.

Bai, J. and Ng, S. (2008b). Large dimensional factor analysis. *Foundations and Trends in Econometrics* 3, 89–163.

Bai, J. and Ng, S. (2019). Rank regularized estimation of approximate factor models. *Journal of Econometrics* 212, 78–96.

Bai, J. and Ng, S. (2023). Approximate factor models with weaker loadings. *Journal of Econometrics* 235, 1893–1916.

Bai, J. and Shi, S. (2011). Estimating high dimensional covariance matrices and its applications. *Annals of Economics and Finance* 12, 199–215.

Bai, Y., Chen, F., Wang, H., Xiong, C., and Mei, S. (2023). Transformers as statisticians: provable in-context learning with in-context algorithm selection. *Proceedings of the 37th International Conference on Neural Information Processing Systems*, 57125–57211.

Bai, Z. D. (1999). Methodologies in spectral analysis of large dimensional random matrices, a review. *Statistica Sinica* 9, 611–677.

Bai, Z. D., Jiang, D., Yao, J. F., and Zheng, S. (2009). Corrections to LRT on large-dimensional covariance matrix by RMT. *The Annals of Statistics* 37, 3822–3840.

Bai, Z. D. and Yin, Y. Q. (1993). Limit of the smallest eigenvalue of a large dimensional sample covariance matrix. *The Annals of Probability* 21, 1275–1294.

Baik, J., Ben Arous, G., and Péché, S. (2005). Phase transition of the largest eigenvalue for nonnull complex sample covariance matrices. *The Annals of Probability* 33, 1643–1697.

Baik, J. and Silverstein, J. W. (2006). Eigenvalues of large sample covariance matrices of spiked population models. *Journal of Multivariate Analysis* 97, 1382–1408.

Balaji, S. R. and Karthikeyan, S. (2017). A survey on moving object tracking using image processing. *2017 11th International Conference on Intelligent Systems and Control (ISCO)*, 469–474.

Baltagi, B. H. (2008). *Econometric Analysis of Panel Data*. New York: John Wiley & Sons.

Banerjee, O., El Ghaoui, L., and d'Aspremont, A. (2008). Model selection through sparse maximum likelihood estimation for multivariate Gaussian or binary data. *Journal of Machine Learning Research* 9, 485–516.

Banerjee, S., Carlin, B. P., and Gelfand, A. E. (2015). *Hierarchical Modeling and Analysis for Spatial Data*. Boca Raton: CRC Press.

Bao, Z., Ding, X., Wang, J., and Wang, K. (2022). Statistical inference for principal components of spiked covariance matrices. *The Annals of Statistics* 50, 1144–1169.

Barhoumi, K., Darné, O., and Ferrara, L. (2013). Dynamic factor models: a review of the literature. *Journal of Business Cycle Measurement and Analysis* 2, 73–107.

Barigozzi, M., He, Y., Li, Lingxiao., and Trapani, L. (2023). Robust estimation of large factor models for tensor-valued time series. *arXiv preprint arXiv:2303.18163.*

Barigozzi, M. and Luciani, M. (2019). Quasi maximum likelihood estimation and inference of large approximate dynamic factor models via the EM algorithm. *arXiv preprint arXiv:1910.03821.*

Barratt, S. and Boyd, S. (2023). Covariance prediction via convex optimization. *Optimization and Engineering* 24, 2045–2078.

Bartholomew, D. J. (2005). History of factor analysis: a statistical perspective. *Encyclopedia of Statistics in Behavioral Science*. Hoboken: John Wiley & Sons.

Bartlett, M. S. (1950). Tests of significance in factor analysis. *British Journal of Psychology* 3, 77–85.

Bartlett, P. L., Montanari, A., and Rakhlin, A. (2021). Deep learning: a statistical viewpoint. *Acta Numerica* 30, 87–201.

Basu, S. and Michailidis, G. (2015). Regularized estimation in sparse high-dimensional time series models. *The Annals of Statistics* 43, 1535–1567.

Battey, H. (2017). Eigen structure of a new class of covariance and inverse covariance matrices. *Bernoulli* 23, 3166–3177.

Bellman, R. (1997). *Introduction to Matrix Analysis*. Philadelphia: Society for Industrial and Applied Mathematics.

Bernard, E. (2021). *Introduction to Machine Learning* Champaign: Wolfram Media.

Beyhum, J. and Striaukas, J. (2024). Testing for sparse idiosyncratic components in factor-augmented regression models. *Journal of Econometrics* 244, 105845.

Bhatt, V., Kumar, S., and Saini, S. (2021). Tucker decomposition and applications. *Materials Today: Proceedings* 46, 10787–10792.

Bhattacharjee, M. and Bose, A. (2014). Consistency of large dimensional sample covariance matrix under weak dependence. *Statistical Methodology* 20, 11–26.

Bi, D., Han, X., Nie, A., and Yang, Y. R. (2022). Spiked eigenvalues of high-dimensional sample autocovariance matrices: CLT and applications. *arXiv preprint arXiv:2201.03181.*

Bi, X., Qu, A., and Shen, X. (2018). Multilayer tensor factorization with applications to recommender systems. *The Annals of Statistics* 46, 3308–3333.

Bi, X., Tang, X. W., Yuan, Y. B., Zhang, Y. Q., and Qu, A. (2021). Tensors in statistics. *Annual Review of Statistics and Its Application* 8, 345–368.

Bi, Y., Lu, Y., Long, Z., Zhu, C., and Liu, Y. P. (2022). Tensor decompositions: computations, applications, and challenges. *Tensors for Data Processing*, 1–30.

Bickel, P. J. and Chen, A. (2009). A nonparametric view of network models and Newman–Girvan and other modularities. *Proceedings of the National Academy of Sciences* 106, 21068–21073.

Bickel, P. J. and Levina, E. (2008a). Covariance regularization by thresholding. *The Annals of Statistics* 36, 2577–2604.

Bickel, P. J. and Levina, E. (2008b). Regularized estimation of large covariance matrices. *The Annals of Statistics* 36,199–227.

Bickel, P. J. and Sarkar, P. (2016). Hypothesis testing for automated community detection in networks. *Journal of the Royal Statistical Society: Series B (Statistical Methodology)* 78, 253–273.

Bien, J., Bunea, F., and Xiao, L. (2016). Convex banding of the covariance matrix. *Journal of the American Statistical Association* 111, 834–845.

Bishop, Y. M. M., Fienberg, S. E., and Holland, P. W. (1975). *Discrete Multivariate Analysis: Theory and Practice*. Cambridge: MIT Press.

Bontempi, G., Le Borgne, Y. A., and de Stefani, J. (2017). A dynamic factor machine learning method for multi-variate and multi-step-ahead forecasting. *2017 IEEE International Conference on Data Science and Advanced Analytics (DSAA)*, 222–231.

Bose, A. and Bhattacharjee, M. (2018). *Large Covariance and Autocovariance Matrices* Boca Raton: CRC Press.

Boser, B., Guyon, I. M., and Vapnik, V. N. (1992). A training algorithm for optimal margin classifiers. *Proceedings of the Fifth Annual Workshop on Computational Learning Theory*, 144–152.

Bouveyron, C., Celeux, G., Murphy, T. B., and Raftery, A. E. (2019). *Model-Based Clustering and Classification for Data Science: With Applications in R*. Cambridge: Cambridge University Press.

Box, G. E. P., Jenkins, G. M., Reinsel, G. C., and Ljung, G. M. (2015). *Time Series Analysis: Forecasting and Control*. San Francisco: Wiley.

Bravais, A. (1844). *Analyse Mathematique Sur Les Probabilités Des Erreurs De Situation D'un Point*. Paris: Imprimerie Royale.

Breiman, L. (1995). Better subset regression using the nonnegative garrote. *Technometrics* 37, 373–384.

Breiman, L. (1996). Bagging predictors. *Machine Learning* 24, 123–140.

Breiman, L. (2001). Random forests. *Machine Learning* 45, 5–32.

Breiman, L., Friedman, J., Olshen, R. A., and Stone, C. J. (1984). *Classification and Regression Trees*. Belmont: Wadsworth International Group.

Breiman, L. and Spector, P. (1992). Submodel selection and evaluation in regression. The X-random case. *International Statistical Review* 60, 291–319.

Breusch, T. S. and Pagan, A. R. (1979). A Simple Test for Heteroscedasticity and Random Coefficient Variation. *Econometrica* 47, 1287–1294.

Brobbey, A., Wiebe, S., Nettel-Aguirre, A., Josephson, C. B., Williamson, T., Lix, L. M., and Sajobi, T. T. (2022). Repeated measures discriminant analysis using multivariate generalized estimation equations. *Statistical Methods in Medical Research* 31, 646–657.

Brown, H. and Prescott, R. (2015). *Applied Mixed Models in Medicine*. Chichester: Wiley.

Brown, L. D. (1990). An ancillarity paradox which appears in multiple linear regression. *The Annals of Statistics* 18, 471–493.

Bu, X., Peng, J., Yan, J., Tan, T., and Zhang, Z. X. (2021). Gaia: A transfer learning system of object detection that fits your needs. *Proceedings of the IEEE/CVF Conference on Computer Vision and Pattern Recognition* 274–283.

Bühlmann, P. and van de Geer, S. (2011). *Statistics for High-Dimensional Data: Methods, Theory and Applications*. Heidelberg: Springer.

Buja, A., Brown, L., Berk, R., George, E., Pitkin, E., Traskin, M., Zhang, Kai., and Zhao, L. (2019). Models as approximations I: consequences illustrated with linear regression. *Statistical Science* 34, 523–544.

Burnham, K. P. and Anderson, D. R. (2003). *Model Selection and Multimodel Inference: A Practical Information-Theoretic Approach*. New York: Springer.

Burnham, K. P. and Anderson, D. R. (2004). Multimodel inference: understanding AIC and BIC in model selection. *Sociological Methods & Research* 33, 261–304.

Cai, T. T. (2017). Global testing and large-scale multiple testing for high-dimensional covariance structures. *Annual Review of Statistics and Its Application* 4, 423–446.

Cai, T. T. and Kim, D. (2024). Transfer learning for covariance matrix estimation: optimality and adaptivity. Technical Report. University of Pennsylvania.

Cai, T. T. and Liu, W. D. (2011). Adaptive thresholding for sparse covariance matrix estimation. *Journal of the American Statistical Association* 106, 672–684.

Cai, T. T., Liu, W. D., and Luo, X. (2011). A constrained ℓ_1 minimization approach to sparse precision matrix estimation. *Journal of the American Statistical Association* 106, 594–607.

Cai, T. T., Liu, W. D., and Xia, Y. (2013). Two-sample covariance matrix testing and support recovery in high-dimensional and sparse settings. *Journal of the American Statistical Association* 108, 265–277.

Cai, T. T., Ren, Z., and Zhou, H. H. (2013). Optimal rates of convergence for estimating Toeplitz covariance matrices. *Probability Theory and Related Fields* 156, 101–143.

Cai, T. T., Zhang, C. H., and Zhou, H. H. (2010). Optimal rates of convergence for covariance matrix estimation. *The Annals of Statistics* 38, 2118–2144.

Candes, E. and Tao, T. (2007). The Dantzig selector: Statistical estimation when $p$is much larger than n. *The Annals of Statistics* 35, 2313–2351.

Carroll, R. J. and Ruppert, D. (1988). *Transformation and Weighting in Regression* New York: Chapman & Hall.

Ceulemans, E. and Kiers, H. A. L. (2006). Selecting among three-mode principal component models of different types and complexities: a numerical convex hull based method. *British Journal of Mathematical and Statistical Psychology* 59, 133–150.

Chamberlain, G. and Rothschild, M. (1983). Arbitrage, factor structure, and mean-variance analysis on large asset markets. *Econometrica* 51, 1281–1304.

Chang, J., He, J., Yang, L., and Yao, Q. (2023). Modelling matrix time series via a tensor CP-decomposition. *Journal of the Royal Statistical Society: Series B (Statistical Methodology)* 85, 127–148.

Chapelle, O., Schölkopf, B., and Zien, A. (2006). *Semi-Supervised Learning*. Cambridge: MIT Press.

Chen, B., Han, Y. F., and Yu, Q. (2024). Estimation and inference for CP tensor factor models. *arXiv preprint arXiv:2406.17278.*

Chen, E. Y. and Fan, J. (2023). Statistical inference for high-dimensional matrix-variate factor models. *Journal of the American Statistical Association* 118, 1038–1055.

Chen, E. Y., Fan, J., and Zhu, X. N. (2023). Community network auto-regression for high-dimensional time series. *Journal of Econometrics* 235, 1239–1256.

Chen, E. Y., Xia, D., Cai, C., and Fan, J. (2024). Semi-parametric tensor factor analysis by iteratively projected singular value decomposition. *Journal of the Royal Statistical Society: Series B (Statistical Methodology)* 86, 793–823.

Chen, J. H. and Chen, Z. H. (2008). Extended bayesian information criteria for model selection with large model spaces. *Biometrika* 95, 759–771.

Chen, J. H. and Chen, Z. H. (2012). Extended BIC for small-n-large-Psparse GLM. *Statistica Sinica* 22, 555–574.

Chen, J. H. and Shao, J. (1993). Iterative weighted least squares estimators. *The Annals of Statistics* 21, 1071–1092.

Chen, R., Yang, D., and Zhang, C. H. (2022). Factor models for high-dimensional tensor time series. *Journal of the American Statistical Association* 117, 94–116.

Chen, S. X., Guo, B., and Qiu, Y. (2023). Testing and signal identification for two-sample high-dimensional covariances via multi-level thresholding. *Journal of Econometrics* 235, 1337–1354.

Chen, S. X., Zhang, L. X., and Zhong, P. S. (2010). Tests for high-dimensional covariance matrices. *Journal of the American Statistical Association* 105, 810–819.

Chen, W. and Lam, C. (2024). Rank and factor loadings estimation in time series tensor factor model by pre-averaging. *The Annals of Statistics* 52, 364–391.

Chen, X. H., Xu, M., and Wu, W. B. (2013). Covariance and precision matrix estimation for high-dimensional time series. *The Annals of Statistics* 41, 2994–3021.

Chen, Yilun., Wiesel, A., and Hero III, A. O. (2011). Robust shrinkage estimation of high-dimensional covariance matrices. *IEEE Transactions on Signal Processing* 59, 4097–4107.

Chen, Youlin., Kolar, M., and Tsay, R. S. (2021). Tensor canonical correlation analysis with convergence and statistical guarantees. *Journal of Computational and Graphical Statistics* 30, 728–744.

Cheng, X. and Hansen, B. E. (2015). Forecasting with factor-augmented regression: a frequentist model averaging approach. *Journal of Econometrics* 186, 280–293.

Chew, V. (1970). Covariance matrix estimation in linear models. *Journal of the American Statistical Association* 65, 173–181.

Chiou, J. M., Ma, Yanyuan., and Tsai, C. L. (2012). Functional random effect time-varying coefficient model for longitudinal data. *Stat* 1, 75–89.

Chiu, T. Y. M., Leonard, T., and Tsui, K. W. (1996). The matrix-logarithmic covariance model. *Journal of the American Statistical Association* 91, 198–210.

Choi, D. S., Wolfe, P. J., and Airoldi, E. M. (2012). Stochastic blockmodels with a growing number of classes. *Biometrika* 99, 273–284.

Choi, I. (2012). Efficient estimation of factor models. *Econometric Theory* 28, 274–308.

Choi, I. and Jeong, H. (2019). Model selection for factor analysis: some new criteria and performance comparisons. *Econometric Reviews* 38, 577–596.

Choi, J., Zou, H., and Oehlert, G. (2010). A penalized maximum likelihood approach to sparse factor analysis. *Statistics and Its Interface* 3, 429–436.

Chung, F. R. K. (1997). *Spectral Graph Theory*. Providence: American Mathematical Society.

Chung, K. L. (2001). *A Course in Probability Theory* San Diego: Elsevier Science.

Cichocki, A., Mandic, D., De Lathauwer, L., Zhou, G. X., Zhao, Q., Caiafa, C., and Phan, H. A. (2015). Tensor decompositions for signal processing applications: from two-way to multiway component analysis. *IEEE Signal Processing Magazine* 32, 145–163.

Claeskens, G. and Hjort, N. L. (2008). *Model Selection and Model Averaging* Cambridge: Cambridge University Press.

Clarke, B., Fokoue, E., and Zhang, H. H. (2009). *Principles and Theory for Data Mining and Machine Learning*. New York: Springer.

Cliff, A. D. and Ord, J. K. (1981). *Spatial Processes: Models and Applications*. London: Pion.

Cook, R. D. (2018). *An Introduction to Envelopes: Dimension Reduction for Efficient Estimation in Multivariate Statistics*. Hoboken: John Wiley & Sons.

Cook, R. D. and Weisberg, S. (1983). Diagnostics for heteroscedasticity in regression. *Biometrika* 70, 1–10.

Coroneo, L. and Iacone, F. (2020). Comparing predictive accuracy in small samples using fixed-smoothing asymptotics. *Journal of Applied Econometrics* 35, 391–409.

Cox, D. R. and Hinkley, D. V. (1974). *Theoretical Statistics* London: Chapman and Hall.

Cox, D. R. and Wermuth, N. (2014). *Multivariate Dependencies: Models, Analysis and Interpretation*. New York: CRC Press.

Cressie, N. A. C. (1993). *Statistics for Spatial Data* New York: Wiley.

Cristianini, N. and Shawe-Taylor, J. (2000). *An Introduction to Support Vector Machines and Other Kernel-Based Learning Methods*. Cambridge: Cambridge University Press.

Daniel, K., Grinblatt, M., Titman, S., and Wermers, R. (1997). Measuring mutual fund performance with characteristic-based benchmarks. *The Journal of Finance* 52, 1035–1058.

Daniele, M., Pohlmeier, W., and Zagidullina, A. (2025). A sparse approximate factor model for high-dimensional covariance matrix estimation and portfolio selection. *Journal of Financial Econometrics* 23, nbae017.

Daumé III, H. (2007). Frustratingly easy domain adaptation. *Proceedings of the 45th Annual Meeting of the Association of Computational Linguistics* 256–263.

Davidian, M. and Carroll, R. J. (1987). Variance function estimation. *Journal of the American Statistical Association* 82, 1079–1091.

Davidson, E. H., Rast, J. P., Oliveri, P., Ransick, A., Calestani, C., Yuh, C. H., Mi-nokawa, T., Amore, G., Hinman, V., Arenas-Mena, C., et al. (2002). A genomic regulatory network for development. *Science* 295, 1669–1678.

Davino, C., Furno, M., and Vistocco, D. (2013). *Quantile Regression: Theory and Applications*. Chichester: Wiley.

Davis, C. S. (2002). *Statistical Methods for the Analysis of Repeated Measurements* New York: Springer.

Davison, A. C. and Hinkley, D. V. (1997). *Bootstrap Methods and Their Application* Cambridge: Cambridge University Press.

Dawid, A. P. (1981). Some matrix-variate distribution theory: notational considerations and a Bayesian application. *Biometrika* 68, 265–274.

De Lathauwer, L., De Moor, B., and Vandewalle, J. (2000a). A multilinear singular value decomposition. *SIAM Journal on Matrix Analysis and Applications* 21, 1253–1278.

De Lathauwer, L., De Moor, B., and Vandewalle, J. (2000b). On the best rank-1 and rank-$(R_1\ R_2, \ldots R_n)$ approximation of higher-order tensors. *SIAM Journal on Matrix Analysis and Applications* 21, 1324–1342.

de Stefani, J., Le Borgne, Y. A., Caelen, O., Hattab, D., and Bontempi, G. (2019). Batch and incremental dynamic factor machine learning for multivariate and multi-step-ahead forecasting. *International Journal of Data Science and Analytics* 7, 311–329.

Demidenko, E. (2013). *Mixed Models: Theory and Applications with R* Hoboken: John Wiley & Sons.

DeMiguel, V., Garlappi, L., Nogales, F. J., and Uppal, R. (2009). A generalized approach to portfolio optimization: improving performance by constraining portfolio norms. *Management Science* 55, 798–812.

Dempster, A. P. (1972). Covariance selection. *Biometrics* 28, 157–175.

Deng, X. and Tsui, K. W. (2013). Penalized covariance matrix estimation using a matrix-logarithm transformation. *Journal of Computational and Graphical Statistics* 22, 494–512.

Deng, Y., Tang, X. W., and Qu, A. (2023). Correlation tensor decomposition and its application in spatial imaging data. *Journal of the American Statistical Association* 118, 440–456.

Dennis, J. E. and Schnabel, R. B. (1996). *Numerical Methods for Unconstrained Optimization and Nonlinear Equations* Philadelphia: Society for Industrial and Applied Mathematics.

Derksen, H. and Makam, V. (2021). Maximum likelihood estimation for matrix normal models via quiver representations. *SIAM Journal on Applied Algebra and Geometry* 5, 338–365.

Devlin, J., Chang, M. W., Lee, K., and Toutanova, K. (2019). BERT: pre-training of deep bidirectional transformers for language understanding. *Proceedings of the 2019 Conference of the North American Chapter of the Association for Computational Linguistics: Human Language Technologies*, 4171–4186.

Diebold, F. X. (2015). Comparing predictive accuracy, twenty years later: a personal perspective on the use and abuse of Diebold–Mariano tests. *Journal of Business & Economic Statistics* 33, 1–9.

Diebold, F. X. and Mariano, R. S. (2002). Comparing predictive accuracy. *Journal of Business & Economic Statistics* 20, 134–144.

Dieci, L. (1996). Considerations on computing real logarithms of matrices, Hamiltonian logarithms, and skew-symmetric logarithms. *Linear Algebra and Its Applications* 244, 35–54.

Diggle, P. J., Heagerty, P. J., Liang, K. Y., and Zeger, S. L. (2002). *Analysis of Longitudinal Data*. New York: Oxford University Press.

Ding, J., Tarokh, V., and Yang, Y. (2018). Model selection techniques: an overview. *IEEE Signal Processing Magazine* 35, 16–34.

Ding, S. and Cook, R. D. (2018). Matrix variate regressions and envelope models. *Journal of the Royal Statistical Society: Series B (Statistical Methodology)* 80, 387–408.

Ding, X. and Yang, F. (2021). Spiked separable covariance matrices and principal components. *The Annals of Statistics* 49, 1113–1138.

Dobriban, E. and Wager, S. (2018). High-dimensional asymptotics of prediction: ridge regression and classification. *The Annals of Statistics* 46, 247–279.

Donoho, D. L., Gavish, M., and Johnstone, I. M. (2018). Optimal shrinkage of eigen-values in the spiked covariance model. *The Annals of Statistics* 46, 1742–1778.

Dorta, G., Vicente, S., Agapito, L., Campbell, N. D. F., and Simpson, I. (2018). Structured uncertainty prediction networks. *IEEE/CVF Conference on Computer Vision and Pattern Recognition*, 5477–5485.

Doz, C. and Fuleky, P. (2020). Dynamic factor models. *Macroeconomic Forecasting in the Era of Big Data: Theory and Practice* 52, 27–64.

Doz, C., Giannone, D., and Reichlin, L. (2011). A two-step estimator for large approximate dynamic factor models based on Kalman filtering. *Journal of Econometrics* 164, 188–205.

Doz, C., Giannone, D., and Reichlin, L. (2012). A quasi–maximum likelihood approach for large, approximate dynamic factor models. *Review of Economics and Statistics* 94, 1014–1024.

Drton, M., Kuriki, S., and Hoff, P. (2021). Existence and uniqueness of the Kronecker covariance MLE. *The Annals of Statistics* 49, 2721–2754.

Drton, M. and Maathuis, M. H. (2017). Structure learning in graphical modeling. *Annual Review of Statistics and Its Application* 4, 365–393.

Dubey, S. R., Singh, S. K., and Chaudhuri, B. B. (2022). Activation functions in deep learning: a comprehensive survey and benchmark. *Neurocomputing* 503, 92–108.

Durbin, J. and Koopman, S. J. (2012). *Time Series Analysis by State Space Methods* Oxford: Oxford University Press.

Dutilleul, P. (1999). The MLE algorithm for the matrix normal distribution. *Journal of Statistical Computation and Simulation* 64, 105–123.

Edgeworth, F. Y. (1893). Statistical correlation between social phenomena. *Journal of the Royal Statistical Society* 56, 670–675.

Edwards, D. (2000). *Introduction to Graphical Modelling* New York: Springer.

Efron, B. (1979). Bootstrap methods: another look at the jackknife. *The Annals of Statistics* 7, 1–26.

Efron, B. (2014). Estimation and accuracy after model selection. *Journal of the American Statistical Association* 109, 991–1007.

El-Nouby, A., Touvron, H., Caron, M., Bojanowski, P., Douze, M., Joulin, A., Laptev, I., Neverova, N., Synnaeve, G., Verbeek, J., and Jégou, H. (2021). Xcit: crosscovariance image transformers. *Proceedings of the 35th International Conference on Neural Information Processing Systems* 34, 20014–20027.

Elman, J. L. (1990). Finding structure in time. *Cognitive Science* 14, 179–211.

Esmaeili Nezhad, A. and Samimi, M. H. (2024). A review of the applications of machine learning in the condition monitoring of transformers. *Energy Systems* 15, 463–493.

Everitt, B. and Hothorn, T. (2011). *An Introduction to Applied Multivariate Analysis with R*. New York: Springer.

Fama, E. F. and French, K. R. (1993). Common risk factors in the returns on stocks and bonds. *Journal of Financial Economics* 33, 3–56.

Fama, E. F. and French, K. R. (2015). A five-factor asset pricing model. *Journal of Financial Economics* 116, 1–22.

Fan, J. and Gijbels, I. (1996). *Local Polynomial Modelling and Its Applications: Monographs on Statistics and Applied Probability* London: Chapman and Hall.

Fan, J., Huang, T., and Li, R. (2007). Analysis of longitudinal data with semiparametric estimation of covariance function. *Journal of the American Statistical Association* 102, 632–641.

Fan, J., Ke, Y., and Wang, K. Z. (2020). Factor-adjusted regularized model selection. *Journal of Econometrics* 216, 71–85.

Fan, J., Li, K. P., and Liao, Y. (2021). Recent developments in factor models and applications in econometric learning. *Annual Review of Financial Economics* 13, 401–430.

Fan, J. and Li, R. (2001). Variable selection via nonconcave penalized likelihood and its oracle properties. *Journal of the American Statistical Association* 96, 1348–1360.

Fan, J., Li, R., Zhang, C. H., and Zou, H. (2020). *Statistical Foundations of Data Science* New York: CRC Press.

Fan, J., Liao, Y., and Liu, H. (2016). An overview of the estimation of large covariance and precision matrices. *The Econometrics Journal* 19, C1–C32.

Fan, J., Liao, Y., and Mincheva, M. (2011). High dimensional covariance matrix estimation in approximate factor models. *The Annals of Statistics* 39, 3320–3356.

Fan, J., Liu, H., and Wang, W. C. (2018). Large covariance estimation through elliptical factor models. *The Annals of Statistics* 46, 1383–1414.

Fan, J., Lou, Z., and Yu, M. (2024). Are latent factor regression and sparse regression adequate? *Journal of the American Statistical Association* 119, 1076–1088.

Fan, J. and Lv, J. (2008). Sure independence screening for ultrahigh dimensional feature space. *Journal of the Royal Statistical Society: Series B (Statistical Methodology)* 70, 849–911.

Fan, J. and Lv, J. (2010). A selective overview of variable selection in high dimensional feature space. *Statistica Sinica* 20, 101–148.

Fan, J., Ma, C., and Zhong, Y. (2021). A selective overview of deep learning. *Statistical Science* 36, 264–290.

Fan, J., Xue, L., and Yao, J. W. (2017). Sufficient forecasting using factor models. *Journal of Econometrics* 201, 292–306.

Fan, J. and Yao, Q. (1998). Efficient estimation of conditional variance functions in stochastic regression. *Biometrika* 85, 645–660.

Fan, J. and Zhang, W. (1999). Statistical estimation in varying coefficient models. *The Annals of Statistics* 27, 1491–1518.

(2008). Statistical methods with varying coefficient models. *Statistics and Its Interface* 1, 179–195.

Fan, X., Lan, W., Zou, T., and Tsai, C. L. (2024a). Covariance model with general linear structure and divergent parameters. *Journal of Business & Economic Statistics* 42, 36–48.

Fan, X., Lan, W., Zou, T., and Tsai, C. L. (2024b). Mutual influence regression model. *Statistica Sinica* 34, 1723–1743.

Fang, K. T. and Zhang, Y. T. (1990). *Generalized Multivariate Analysis* Berlin: Springer-Verlag.

Fang, Y., Wang, B., and Feng, Y. (2016). Tuning-parameter selection in regularized estimations of large covariance matrices. *Journal of Statistical Computation and Simulation* 86, 494–509.

Feldman, M. J. (2025). Spectral properties of elementwise-transformed spiked matrices. *SIAM Journal on Mathematics of Data Science* 7, 542–571.

Finch, W. H. (2020). *Exploratory Factor Analysis* Thousand Oaks: Sage Publications.

Fisher, R. A. (1915). Frequency distribution of the values of the correlation coefficient in samples from an indefinitely large population. *Biometrika* 10, 507–521.

Fisher, R. A. (1925). *Statistical Methods for Research Workers*. Edinburgh: Oliver and Boyd.

Fitzmaurice, G. M., Davidian, M., Verbeke, G., and Molenberghs, G. (2008). *Longitudinal Data Analysis* New York: CRC Press.

Fitzmaurice, G. M., Laird, N. M., and Ware, J. H. (2011). *Applied Longitudinal Analysis*. Hoboken: Wiley.

Fletcher, R. (1987). *Practical Methods of Optimization* Chichester: Wiley.

Förstner, W. and Moonen, B. (2003). A metric for covariance matrices. *Geodesy-The Challenge of the 3rd Millennium*. Berlin: Springer, 299–309.

Fotheringham, A. S., Brunsdon, C., and Charlton, M. (2002). *Geographically Weighted Regression: The Analysis of Spatially Varying Relationships*. Chichester: Wiley.

Frees, E. W. (2004). *Longitudinal and Panel Data: Analysis and Applications in the Social Sciences*. Cambridge: Cambridge University Press.

Freijeiro-González, L., Febrero-Bande, M., and González-Manteiga, W. (2022). A critical review of Lasso and its derivatives for variable selection under dependence among covariates. *International Statistical Review* 90, 118–145.

Freund, Y. and Schapire, R. E. (1997). A decision-theoretic generalization of on-line learning and an application to boosting. *Journal of Computer and System Sciences* 55, 119–139.

Friedman, J., Hastie, T., and Tibshirani, R. (2008). Sparse inverse covariance estimation with the graphical Lasso. *Biostatistics* 9, 432–441.

Furrer, R. and Bengtsson, T. (2007). Estimation of high-dimensional prior and posterior covariance matrices in Kalman filter variants. *Journal of Multivariate Analysis* 98, 227–255.

Gaikwad, S. K., Gawali, B. W., and Yannawar, P. (2010). A review on speech recognition technique. *International Journal of Computer Applications* 10, 16–24.

Galecki, A. and Burzykowski, T. (2013). *Linear Mixed-Effects Models Using R: A Step- by-Step Approach*. New York: Springer.

Galton, F. (1889). I. Co-relations and their measurement, chiefly from anthropometric data. *Proceedings of the Royal Society of London* 45, 135–145.

Gao, X., Pu, D. Q., Wu, Y., and Xu, H. (2012). Tuning parameter selection for penalized likelihood estimation of Gaussian graphical model. *Statistica Sinica* 22, 1123–1146.

Gao, Y., Zhang, Z. Y., Cai, Z., Zhu, X. N., Zou, T., and Wang, H. S. (2025). Penalized sparse covariance regression with high dimensional covariates. *Journal of Business & Economic Statistics* 43, 615–626.

Gao, Z. X., Ma, Y. Y., Wang, H. S., and Yao, Q. (2019). Banded spatio-temporal autoregressions. *Journal of Econometrics* 208, 211–230.

Geisser, S. (1975). The predictive sample reuse method with applications. *Journal of the American Statistical Association* 70, 320–328.
Geweke, J. (1977). The dynamic factor analysis of economic time series. *Latent Variables in Socio-Economic Models*. Amsterdam: North-Holland, 365–383.
Ghosh, M., Reid, N., and Fraser, D. A. S. (2010). Ancillary statistics: a review. *Statistica Sinica* 20, 1309–1332.
Giacomini, R. and White, H. (2006). Tests of conditional predictive ability. *Econometrica* 74, 1545–1578.
Girard, D. A. (1998). Asymptotic comparison of (partial) cross-validation, GCV and randomized GCV in nonparametric regression. *The Annals of Statistics* 26, 315–334.
Gitelman, D. (2022). Tensor methods for network analysis. PhD thesis. Princeton: Princeton University.
Glymour, C., Zhang, Kun., and Spirtes, P. (2019). Review of causal discovery methods based on graphical models. *Frontiers in Genetics* 10, 524.
Goncalves, S. and Perron, B. (2014). Bootstrapping factor-augmented regression models. *Journal of Econometrics* 182, 156–173.
Goodfellow, I., Bengio, Y., and Courville, A. (2016). *Deep Learning*. New York: MIT Press.
Govaers, F. (2019). Object detection in tensor decomposition based multi target tracking. *2019 22th International Conference on Information Fusion (FUSION)* 1–8.
Gray, R. M. (2006). Toeplitz and circulant matrices: a review. *Foundations and Trends in Communications and Information Theory* 2, 155–239.
Greenaway-McGrevy, R., Han, C., and Sul, D. (2012). Asymptotic distribution of factor augmented estimators for panel regression. *Journal of Econometrics* 169, 48–53.
Greene, W. H. (2018). *Econometric Analysis* New York: Pearson.
Greenewald, K., Zelnio, E., and Hero III, A. O. (2016). Robust SAR STAP via Kronecker decomposition. *IEEE Transactions on Aerospace and Electronic Systems* 52, 2612–2625.
Greenwell, B. M. (2022). *Tree-based Methods for Statistical Learning in R*. New York: Chapman and Hall/CRC.
Grenander, U. and Szego, G. (1958). *Toeplitz Forms and Their Applications*. Berkeley: University of California Press.
Grinblatt, M., Titman, S., and Wermers, R. (1995). Momentum investment strategies, portfolio performance, and herding: a study of mutual fund behavior. *American Economic Review* 85, 1088–1105.
Guggenberger, P., Kleibergen, F., and Mavroeidis, S. (2023). A test for Kronecker product structure covariance matrix. *Journal of Econometrics* 233, 88–112.
Guo, X. and Tang, C. Y. (2021). Specification tests for covariance structures in high-dimensional statistical models. *Biometrika* 108, 335–351.
Haddad, J. N. (2004). On the closed form of the covariance matrix and its inverse of the causal ARMA process. *Journal of Time Series Analysis* 25, 443–448.
Hair, J. F., Babin, B. J., Black, W. C., and Anderson, R. E. (2019). *Multivariate Data Analysis*. Hampshire: Cengage.
Hale, E. and Prater-Bennette, A. (2021). Comparison of CP and Tucker tensor decomposition algorithms. *Big Data III: Learning, Analytics, and Applications*, 117300D.
Hall, P. and Heyde, C. C. (1980). *Martingale Limit Theory and Its Application* New York: Academic Press.
Hamilton, J. D. (1994). *Time Series Analysis*. Princeton: Princeton University Press.
Hamilton, W. L. (2020). *Graph Representation Learning* San Rafael: Morgan & Claypool Publishers.
Hamilton, W. L., Ying, R., and Leskovec, J. (2017). Inductive representation learning on large graphs. *Proceedings of the 31st International Conference on Neural Information Processing Systems*, 1025–1035.
Han, Y. F., Chen, R., Yang, D., and Zhang, C. H. (2024). Tensor factor model estimation by iterative projection. *The Annals of Statistics* 52, 2641–2667.

Han, Y. F., Yang, D., Zhang, C. H., and Chen, R. (2024). CP factor model for dynamic tensors. *Journal of the Royal Statistical Society: Series B (Statistical Methodology)* 86, 1383–1413.

Han, Y. F. and Zhang, C. H. (2023). Tensor principal component analysis in high dimensional CP models. *IEEE Transactions on Information Theory* 69, 1147–1167.

Hanson, D. L. and Wright, F. T. (1971). A bound on tail probabilities for quadratic forms in independent random variables. *The Annals of Mathematical Statistics* 42, 1079–1083.

Hao, L. and Naiman, D. Q. (2007). *Quantile Regression* Thousand Oaks: SAGE Publications.

Hardin, J. W. and Hilbe, J. M. (2012). *Generalized Estimating Equations*. New York: CRC Press.

Härdle, W. K. and Simar, L. (2015). *Applied Multivariate Statistical Analysis*. Berlin: Springer International Publishing.

Harrison, M. J. and McCabe, B. P. M. (1975). Autocorrelation with heteroscedasticity: a note on the robustness of the Durbin-Watson, Geary and Henshaw tests. *Biometrika* 62, 214–216.

Harvey, A. C. (1990). *Forecasting, Structural Time Series Models and the Kalman Filter*. Cambridge: Cambridge University Press.

Hastie, T., Montanari, A., Rosset, S., and Tibshirani, R. J. (2022). Surprises in high-dimensional ridgeless least squares interpolation. *The Annals of Statistics* 50, 949–986.

Hastie, T. and Tibshirani, R. (1993). Varying-coefficient models. *Journal of the Royal Statistical Society: Series B (Statistical Methodology)* 55, 757–779.

Hastie, T., Tibshirani, R., and Friedman, J. (2009). *The Elements of Statistical Learning: Data Mining, Inference, and Prediction*. New York: Springer.

Hastie, T., Tibshirani, R., and Wainwright, M. (2015). *Statistical Learning with Sparsity: The Lasso and Generalizations*. Boca Raton: Chapman & Hall/CRC.

Hatch, J. P. (1980). The effects of operant reinforcement schedules on the modification of human heart rate. *Psychophysiology* 17, 559–567.

Haykin, S. (2009). *Neural Networks and Learning Machines* New York: Pearson.

He, J. and Chen, S. X. (2016). Testing super-diagonal structure in high dimensional covariance matrices. *Journal of Econometrics* 194, 283–297.

He, K., Zhang, X., Ren, S., and Sun, J. (2016). Deep residual learning for image recognition. *IEEE Conference on Computer Vision and Pattern Recognition*, 770–778.

He, X., Pan, X., Tan, K. M., and Zhou, W. X. (2023). Smoothed quantile regression with large-scale inference. *Journal of Econometrics* 232, 367–388.

He, Y., Li, Q. S., Hu, Q. Q., and Liu, L. (2022). Transfer learning in high-dimensional semiparametric graphical models with application to brain connectivity analysis. *Statistics in Medicine* 41, 4112–4129.

Hearne III, E. M., Clark, G. M., and Hatch, J. P. (1983). A test for serial correlation in univariate repeated-measures analysis. *Biometrics* 39, 237–243.

Hedeker, D. and Gibbons, R. D. (2006). *Longitudinal Data Analysis* Hoboken: Wiley.

Hinton, G. and Salakhutdinov, R. R. (2006). Reducing the dimensionality of data with neural networks. *Science* 313, 504–507.

Hochreiter, S. and Schmidhuber, J. (1997). Long short-term memory. *Neural Computation* 9, 1735–1780.

Hoeding, W. (1963). Probability inequalities for sums of bounded random variables. *Journal of the American Statistical Association* 58, 13–30.

Hoff, P. (2011). Separable covariance arrays via the Tucker product, with applications to multivariate relational data. *Bayesian Analysis* 6, 179–196.

Hoff, P. (2015). Multilinear tensor regression for longitudinal relational data. *The Annals of Applied Statistics* 9, 1169–1193.

Hoff, P., McCormack, A., and Zhang, A. R. (2023). Core shrinkage covariance estimation for matrix-variate data. *Journal of the Royal Statistical Society: Series B (Statistical Methodology)* 85, 1659–1679.

Hoff, P. and Niu, X. (2012). A covariance regression model. *Statistica Sinica* 22, 729–753.

Højsgaard, S., Edwards, D., and Lauritzen, S. (2012). *Graphical Models with R*. New York: Springer.

Holland, P. W., Laskey, K. B., and Leinhardt, S. (1983). Stochastic blockmodels: first steps. *Social Networks* 5, 109–137.

Hoover, D. R., Rice, J. A., Wu, C. O., and Yang, L. P. (1998). Nonparametric smoothing estimates of time-varying coefficient models with longitudinal data. *Biometrika* 85, 809–822.

Hosmer, D. W. and Lemeshow, S. (2000). *Applied Logistic Regression* New York: Wiley.

Hotelling, H. (1936). Relations between two sets of variates. *Biometrika* 28, 321–377.

Hou, K., Xue, C., and Zhang, Lu (2020). Replicating anomalies. *The Review of Financial Studies* 33, 2019–2133.

Hou, Z., Li, W., Tao, R., and Du, Q. (2021). Three-order tucker decomposition and reconstruction detector for unsupervised hyperspectral change detection. *IEEE Journal of Selected Topics in Applied Earth Observations and Remote Sensing* 14, 6194–6205.

Hsiao, C. (2022). *Analysis of Panel Data* Cambridge: Cambridge University Press.

Hu, J., Li, W. M., Liu, Z., and Zhou, W. (2019). High-dimensional covariance matrices in elliptical distributions with application to spherical test. *The Annals of Statistics* 47, 527–555.

Hu, J. C. and Szymczak, S. (2023). A review on longitudinal data analysis with random forest. *Briefings in Bioinformatics* 24, bbad002.

Hu, J. W., Zhang, J. F., Qin, H., Yan, T., and Zhu, J. (2021). Using maximum entry-wise deviation to test the goodness of fit for stochastic block models. *Journal of the American Statistical Association* 116, 1373–1382.

Huang, D., Lan, W., Zhang, H. H., and Wang, H. S. (2019). Least squares estimation of spatial autoregressive models for large-scale social networks. *Electronic Journal of Statistics* 13, 1135–1165.

Huang, J., Chai, J., and Cho, S. (2020). Deep learning in finance and banking:A literature review and classification. *Frontiers of Business Research in China* 14, 1–24.

Huang, J. Z., Liu, N., Pourahmadi, M., and Liu, L. X. (2006). Covariance matrix selection and estimation via penalised normal likelihood. *Biometrika* 93, 85–98.

Hubert, M., Rousseeuw, P. J., and Van Aelst, S. (2008). High-breakdown robust multivariate methods. *Statistical Science* 23, 92–119.

Hurvich, C. M., Simonoff, J. S., and Tsai, C. L. (1998). Smoothing parameter selection in nonparametric regression using an improved Akaike information criterion. *Journal of the Royal Statistical Society: Series B (Statistical Methodology)* 60, 271–293.

Hurvich, C. M. and Tsai, C. L. (1989). Regression and Time Series Model Selection in Small Samples. *Biometrika* 76, 297–307.

Hurvich, C. M. and Tsai, C. L. (1990). The impact of model selection on inference in linear regression. *The American Statistician* 44, 214–217.

Hurvich, C. M. and Tsai, C. L. (1993). A corrected Akaike information criterion for vector autoregressive model selection. *Journal of Time Series Analysis* 14, 271–279.

Huynh, K. and Lenhard, G. (2022). Asymmetric autoencoders for factor-based covariance matrix estimation. *Proceedings of the Third ACM International Conference on AI in Finance* 403–410.

Ionescu, C., Vantzos, O., and Sminchisescu, C. (2015). Matrix backpropagation for deep networks with structured layers. *Proceedings of the IEEE International Conference on Computer Vision*, 2965–2973.

Islam, S., Elmekki, H., Elsebai, A., Bentahar, J., Drawel, N., Rjoub, G., and Pedrycz, W. (2024). A comprehensive survey on applications of transformers for deep learning tasks. *Expert Systems with Applications* 241, 122666.

Izenman, A. J. (2008). *Modern Multivariate Statistical Techniques* New York: Springer.

Jagannathan, R. and Ma, T. (2003). Risk reduction in large portfolios: Why imposing the wrong constraints helps. *The Journal of Finance* 58, 1651–1683.

James, G., Witten, D., Hastie, T., and Tibshirani, R. (2013). *An Introduction to Statistical Learning: with Applications in R*. New York: Springer.

Janiszewska, M. (2022). Structures of the covariance matrix: an overview. *Biometrical Letters* 59, 141–157.

Janková, J. and van de Geer, S. (2017). Honest confidence regions and optimality in high-dimensional precision matrix estimation. *Test* 26, 143–162.

Jennrich, R. I. and Schluchter, M. D. (1986). Unbalanced repeated-measures models with structured covariance matrices. *Biometrics* 42, 805–820.

Jiang, H. (2021). *Machine Learning Fundamentals: A Concise Introduction* Cambridge: Cambridge University Press.

Jiang, J. (2019). *Robust Mixed Model Analysis* Singapore: World Scientific Publishing Company.

Jiang, J. and Nguyen, T. (2021). *Linear and Generalized Linear Mixed Models and Their Applications* New York: Springer.

Jin, J. (2012). Fast community detection by score. *The Annals of Statistics* 43, 57–89.

Jockers, M. L. and Thalken, R. (2020). *Text Analysis with R: For Students of Literature* Cham: Springer International Publishing.

John, S. (1971). Some optimal multivariate tests. *Biometrika* 58, 123–127.

Johnson, R. A. and Wichern, D. W. (2007). *Applied Multivariate Statistical Analysis* Upper Saddle River: Pearson Prentice Hall.

Johnstone, I. M. (2001). On the distribution of the largest eigenvalue in principal components analysis. *The Annals of Statistics* 29, 295–327.

Johnstone, I. M. and Lu, A. Y. (2009). On consistency and sparsity for principal components analysis in high dimensions. *Journal of the American Statistical Association* 104, 682–693.

Johnstone, I. M. and Paul, D. (2018). PCA in high dimensions: an orientation. *Proceedings of the IEEE* 106, 1277–1292.

Jolliffe, I. T. (2002). *Principal Component Analysis* New York: Springer.

Jolliffe, I. T., Trendafilov, N. T., and Uddin, M. (2003). A modified principal component technique based on the Lasso. *Journal of Computational and Graphical Statistics* 12, 531–547.

Jones, B. and West, M. (2005). Covariance decomposition in undirected Gaussian graphical models. *Biometrika* 92, 779–786.

Jonsson, D. (1982). Some limit theorems for the eigenvalues of a sample covariance matrix. *Journal of Multivariate Analysis* 12, 1–38.

Jöreskog, K. G. (1967). Some contributions to maximum likelihood factor analysis. *Psychometrika* 32, 443–482.

Juodis, A. and Sarafidis, V. (2022). A linear estimator for factor-augmented fixed-T panels with endogenous regressors. *Journal of Business & Economic Statistics* 40, 1–15.

Kalman, R. E. (1960). A new approach to linear filtering and prediction problems. *Transactions of the ASME–Journal of Basic Engineering* 82, 35–45.

Kamath, U., Graham, K., and Emara, W. (2022). *Transformers for Machine Learning: A Deep Dive* New York: CRC Press.

Kan, R. and Zhou, G. F. (2007). Optimal portfolio choice with parameter uncertainty. *Journal of Financial and Quantitative Analysis* 42, 621–656.

Kan, R. and Zhou, G. F. (2017). Modeling non-normality using multivariate t: implications for asset pricing. *China Finance Review International* 7, 2–32.

Karoglou, M. (2010). Breaking down the non-normality of stock returns. *The European Journal of Finance* 16, 79–95.

Karoui, N. E. (2008). Operator norm consistent estimation of large-dimensional sparse covariance matrices. *The Annals of Statistics* 36, 2717–2756.

Kaya, O. and Robert, Y. (2019). Computing dense tensor decompositions with optimal dimension trees. *Algorithmica* 81, 2092–2121.

Ke, Y., Minsker, S., Ren, Z., Sun, Q., and Zhou, W. X. (2019). User-friendly covariance estimation for heavy-tailed distributions. *Statistical Science* 34, 454–471.

Ke, Z. T., Ji, P., Jin, J., and Li, W. S. (2024). Recent advances in text analysis. *Annual Review of Statistics and Its Application* 11, 347–372.

Khamaru, K. and Mazumder, R. (2019). Computation of the maximum likelihood estimator in low-rank factor analysis. *Mathematical Programming* 176, 279–310.

Khan, S., Xu, G., Chan, R., and Yan, H. (2017). An online spatio-temporal tensor learning model for visual tracking and its applications to facial expression recognition. *Expert Systems with Applications* 90, 427–438.

Kiers, H. A. L. (2000). Towards a standardized notation and terminology in multiway analysis. *Journal of Chemometrics* 14, 105–122.

Kilian, L. and Lütkepohl, H. (2017). *Structural Vector Autoregressive Analysis*. Cambridge: Cambridge University Press.

Kim, T. K., Wong, S., and Cipolla, R. (2007). Tensor canonical correlation analysis for action classification. *IEEE Conference on Computer Vision and Pattern Recognition*, 1–8.

Kingma, D. P. and Ba, J. (2017). Adam: A method for stochastic optimization. *arXiv preprint arXiv:1412.6980*.

Kipf, T. N. and Welling, M. (2017). Semi-supervised classification with graph convolutional networks. *5th International Conference on Learning Representations, ICLR 2017*. OpenReview.net.

Kirillov, A., He, K., Girshick, R., Rother, C., and Dollár, P. (2019). Panoptic segmentation. *IEEE/CVF Conference on Computer Vision and Pattern Recognition*, 9404–9413.

Kleinbaum, D. G. and Klein, M. (2010). *Logistic Regression: A Self-Learning Text* New York: Springer.

Koenker, R. (2005). *Quantile Regression*. Cambridge: Cambridge University Press.

Koenker, R. (2017). Quantile regression: 40 years on. *Annual Review of Economics* 9, 155–176.

Koenker, R. and Bassett, G. (1978). Regression quantiles. *Econometrica* 46, 33–50.

Koenker, R. and Xiao, Z. (2006). Quantile autoregression. *Journal of the American Statistical Association* 101, 980–990.

Kolda, T. G. and Bader, B. W. (2009). Tensor decompositions and applications. *SIAM Review* 51, 455–500.

Kollo, T. and Rosen, D. (2005). *Advanced Multivariate Statistics with Matrices*. Dordrecht: Springer Netherlands.

Konishi, S. and Kitagawa, G. (2008). *Information Criteria and Statistical Modeling* New York: Springer.

Kostopoulos, G., Karlos, S., Kotsiantis, S., and Ragos, O. (2018). Semi-supervised regression: a recent review. *Journal of Intelligent & Fuzzy Systems* 35, 1483–1500.

Krizhevsky, A., Sutskever, I., and Hinton, G. (2017). ImageNet classification with deep convolutional neural networks. *Commun. ACM* 60, 84–90.

Kumar, S., Roy, P. P., Dogra, D. P., and Kim, B. G. (2023). A comprehensive review on sentiment analysis: tasks, approaches and applications. *arXiv preprint arXiv:2311.11250*.

Kuo, C. J. and Madni, A. M. (2023). Green learning: introduction, examples and outlook. *Journal of Visual Communication and Image Representation* 90, 103685.

Lam, C. (2020). High-dimensional covariance matrix estimation. *WIREs Computational Statistics* 12, e1485.

Lam, C. and Fan, J. (2009). Sparsistency and rates of convergence in large covariance matrix estimation. *The Annals of Statistics* 37, 4254–4278.

Lam, C. and Yao, Q. (2012). Factor modeling for high-dimensional time series: inference for the number of factors. *The Annals of Statistics* 40, 694–726.

Lan, W., Chen, X. R., Zou, T., and Tsai, C. L. (2022). Imputations for high missing rate data in covariates via semi-supervised learning approach. *Journal of Business & Economic Statistics* 40, 1282–1290.

Lan, W., Fang, Z., Wang, H. S., and Tsai, C. L. (2018). Covariance matrix estimation via network structure. *Journal of Business & Economic Statistics* 36, 359–369.

Lan, W., Lei, B., Feng, L., and Tsai, C. L. (2024). Maximum-subsampling test of equal predictive ability. *Journal of Business & Economic Statistics* 42, 1344–1355.

Lan, W., Luo, R., Tsai, C. L., Wang, H. S., and Yang, Y. (2015). Testing the diagonality of a large covariance matrix in a regression setting. *Journal of Business & Economic Statistics* 33, 76–86.

Lauritzen, S. (1996). *Graphical Models* Oxford: Clarendon Press.

Lawley, D. N. and Maxwell, A. E. (1962). Factor analysis as a statistical method. *Journal of the Royal Statistical Society Series D: The Statistician* 12, 209–229.

Lazer, D., Pentland, A., Adamic, L., Aral, S., Barabási, A. L., Brewer, D., Christakis, N., Contractor, N., Fowler, J., Gutmann, M., et al. (2009). Life in the network: the coming age of computational social. *Science* 323, 721–723.

LeCun, Y., Bengio, Y., and Hinton, G. (2015). Deep learning. *Nature* 521, 436–444.

LeCun, Y., Boser, B., Denker, J. S., Henderson, D., Howard, R. E., Hubbard, W., and Jackel, L. D. (1989). Backpropagation applied to handwritten zip code recognition. *Neural Computation* 1, 541–551.

LeCun, Y., Bottou, L., Bengio, Y., and Haffner, P. (1998). Gradient-based learning applied to document recognition. *Proceedings of the IEEE* 86, 2278–2324.

Ledoit, O. and Wolf, M. (2002). Some hypothesis tests for the covariance matrix when the dimension is large compared to the sample size. *The Annals of Statistics* 30, 1081–1102.

Ledoit, O. and Wolf, M. (2004). A well-conditioned estimator for large-dimensional covariance matrices. *Journal of Multivariate Analysis* 88, 365–411.

Lee, C. and Wilkinson, D. J. (2019). A review of stochastic block models and extensions for graph clustering. *Applied Network Science* 4, 1–50.

Lee, J., Lee, I., and Kang, J. (2019). Self-attention graph pooling. *International Conference on Machine Learning* 97, 3734–3743.

Lee, L. F. (2002). Consistency and efficiency of least squares estimation for mixed regressive, spatial autoregressive models. *Econometric Theory* 18, 252–277.

Lee, L. F. (2004). Asymptotic distributions of quasi-maximum likelihood estimators for spatial autoregressive models. *Econometrica* 72, 1899–1925.

Lee, L. F. (2023). *Spatial Econometrics: Spatial Autoregressive Models*. Singapore: World Scientific Publishing Company.

Lee, L. F. and Liu, X. D. (2010). Efficient GMM estimation of high order spatial autoregressive models with autoregressive disturbances. *Econometric Theory* 26, 187–230.

Lee, S. H. and Choi, S. (2007). Two-dimensional canonical correlation analysis. *IEEE Signal Processing Letters* 14, 735–738.

Lehmann, E. L. (1998). *Elements of Large-Sample Theory* New York: Springer.

Lei, J. and Rinaldo, A. (2015). Consistency of spectral clustering in stochastic block models. *The Annals of Statistics* 43, 215–237.

Leng, C. and Pan, G. (2018). Covariance estimation via sparse Kronecker structures. *Bernoulli* 24, 3833–3863.

Leng, C. and Tang, C. Y. (2012). Sparse matrix graphical models. *Journal of the American Statistical Association* 107, 1187–1200.

Leonard, T. and Hsu, J. S. J. (1992). Bayesian inference for a covariance matrix. *The Annals of Statistics* 20, 1669–1696.

LeSage, J. and Pace, R. K. (2009). *Introduction to Spatial Econometrics* New York: CRC Press.

Lewis, M., Liu, Y. H., Goyal, N., Ghazvininejad, M., Mohamed, A., Levy, O., Stoyanov, V., and Zettlemoyer, L. (2020). BART: denoising sequence-to-sequence pre-training for natural language generation, translation, and comprehension. *Proceedings of the 58th Annual Meeting of the Association for Computational Linguistics* 7871–7880.

Li, D. G., Tosasukul, J., and Zhang, W. (2020). Nonlinear factor-augmented predictive regression models with functional coefficients. *Journal of Time Series Analysis* 41, 367–386.

Li, H. M. and Sewell, D. K. (2021). A comparison of estimators for the network autocorrelation model based on observed social networks. *Social Networks* 66, 202–210.

Li, Jie. and Song, Y. (2024). Incremental transfer learning for spatial autoregressive model with linear constraints. *Spatial Statistics* 61, 100833.

Li, Jingyang., Cai, T. T., Xia, D., and Zhang, A. R. (2024). Federated PCA and estimation for spiked covariance matrices: optimal rates and efficient algorithm. *arXiv preprint arXiv:2411.15660*.

Li, Jun. and Chen, S. X. (2012). Two sample tests for high-dimensional covariance matrices. *The Annals of Statistics* 40, 908–940.

Li, K. C. (1991). Sliced inverse regression for dimension reduction. *Journal of the American Statistical Association* 86, 316–327.

Li, P. H. and Sun, Qi. (2011). Tensor-based covariance matrices for object tracking. *2011 IEEE International Conference on Computer Vision Workshops (ICCV Workshops)*, 1681–1688.

Li, P. H., Xie, J., Wang, Q. L., and Gao, Z. L. (2018). Towards faster training of global covariance pooling networks by iterative matrix square root normalization. *IEEE Conference on Computer Vision and Pattern Recognition* 947–955.

Li, P. H., Xie, J., Wang, Q. L., and Zuo, W. M. (2017). Is second-order information helpful for large-scale visual recognition? *Proceedings of the IEEE International Conference on Computer Vision*, 2070–2078.

Li, Q. and Racine, J. S. (2007). *Nonparametric Econometrics: Theory and Practice* Princeton: Princeton University Press.

Li, S., Cai, T. T., and Li, H. Z. (2023). Transfer learning in large-scale gaussian graphical models with false discovery rate control. *Journal of the American Statistical Association* 118, 2171–2183.

Li, X. S., Xu, D., Zhou, H., and Li, Lexin. (2018). Tucker tensor regression and neuroimaging analysis. *Statistics in Biosciences* 10, 520–545.

Li, X. T., Wang, F. F., Lan, W., and Wang, H. S. (2023). Subnetwork estimation for spatial autoregressive models in large-scale networks. *Electronic Journal of Statistics* 17, 1768–1805.

Li, Z. W., Liu, F., Yang, W., Peng, S., and Zhou, Jun. (2022). A survey of convolutional neural networks: analysis, applications, and prospects. *IEEE Transactions on Neural Networks and Learning Systems* 33, 6999–7019.

Lian, H. (2011). Shrinkage tuning parameter selection in precision matrices estimation. *Journal of Statistical Planning and Inference* 141, 2839–2848.

Liang, K. Y. and Zeger, S. L. (1986). Longitudinal data analysis using generalized linear models. *Biometrika* 73, 13–22.

Liang, Y., Coelho, C. A., and von Rosen, T. (2022). Hypothesis testing in multivariate normal models with block circular covariance structures. *Biometrical Journal* 64, 557–576.

Lin, D. Y. and Ying, Z. (2001). Semiparametric and nonparametric regression analysis of longitudinal data. *Journal of the American Statistical Association* 96, 103–126.

Lin, M., Chen, Q., and Yan, S. (2014). Network in network. *2ndălnternational ConferenceăonăLearning Representations, ICLR 2014*

Lin, T., Wang, Yuxin., Liu, X. Y., and Qiu, X. (2022). A survey of transformers. *AI Open* 3, 111–132.

Lin, Tsung. Yi., Dollár, P., Girshick, R., He, K., Hariharan, B., and Belongie, S. (2017). Feature pyramid networks for object detection. *IEEE Conference on Computer Vision and Pattern Recognition*, 936–944.

Lin, Tsung. Yu. and Maji, S. (2017). Improved bilinear pooling with CNNs. *Proceedings of the British Machine Vision Conference (BMVC)* 117.1–117.12.

Lin, Tsung. Yu., RoyChowdhury, A., and Maji, S. (2015). Bilinear CNN models for fine-grained visual recognition. *Proceedings of the IEEE International Conference on Computer Vision*, 1449–1457.

Linton, O. and McCrorie, J. R. (1995). Differentiation of an exponential matrix function. *Econometric Theory* 11, 1182–1185.

Little, M. A., McSharry, P., Hunter, E., Spielman, J., and Ramig, L. (2009). Suitability of dysphonia measurements for telemonitoring of Parkinson's disease. *IEEE Transactions on Biomedical Engineering* 56, 1015–1022.

Liu, A., Feng, B., Xue, B., Wang, B. X., Wu, B. C., Lu, C., Zhao, C. G., Deng, C., Zhang, C. Y., Ruan, C., et al. (2024). Deepseek-V3 technical report. *arXiv preprint arXiv:2412.19437*.

Liu, Bing (2020). *Sentiment Analysis: Mining Opinions, Sentiments, and Emotions* Cambridge: Cambridge University Press.

Liu, C., Wu, H. Y., Wang, J. Y., and Wang, M. (2020). A unified fourth-order tensor-based smart community system. *Sensors* 20, 5990.

Liu, H. and Wang, L. (2017). TIGER: a tuning-insensitive approach for optimally estimating gaussian graphical models. *Electronic Journal of Statistics* 11, 241–294.

Liu, H., Wang, L., and Zhao, T. (2014). Sparse covariance matrix estimation with eigenvalue constraints. *Journal of Computational and Graphical Statistics* 23, 439–459.

Liu, W. D. and Luo, X. (2015). Fast and adaptive sparse precision matrix estimation in high dimensions. *Journal of Multivariate Analysis* 135, 153–162.

Liu, X. F., Feng, Y. P., Liu, X. M., and Huang, B. (2015). CP decomposition and its application in noise reduction and multiple sources identification. *2015 International Conference on Computer Science and Intelligent Communication* 47–51.

Liu, Xiao. (2023). Deep learning in marketing: a review and research agenda. *Artificial Intelligence in Marketing* 20, 239–271.

Liu, Z. M., Wang, Yixuan., Vaidya, S., Ruehle, F., Halverson, J., Soljačić, M., Hou, T. Y., and Tegmark, M. (2025). KAN: Kolmogorov-Arnold Networks. *The Thirteenth International Conference on Learning Representations, ICLR 2025*

Locascio, J. J. and Atri, A. (2011). An overview of longitudinal data analysis methods for neurological research. *Dementia and Geriatric Cognitive Disorders Extra* 1, 330–357.

Lounici, K. (2012). High-dimensional covariance matrix estimation with missing observations. *Bernoulli* 20, 1029–1058.

Lu, H., Plataniotis, K. N., and Venetsanopoulos, A. (2013). *Multilinear Subspace Learning: Dimensionality Reduction of Multidimensional Data*. Hoboken: CRC Press.

Lu, J., Han, F., and Liu, Han. (2021). Robust scatter matrix estimation for high dimensional distributions with heavy tail. *IEEE Transactions on Information Theory* 67, 5283–5304.

Luo, Y., Tao, D., Ramamohanarao, K., Xu, C., and Wen, Y. (2015). Tensor canonical correlation analysis for multi-view dimension reduction. *IEEE Transactions on Knowledge and Data Engineering* 27, 3111–3124.

Lütkepohl, H. (2007). *New Introduction to Multiple Time Series Analysis*. Berlin: Springer.

Lyche, T. (2020). *and Matrix Factorizations* Cham: Springer International Publishing.

Ma, S. J., Li, R., and Tsai, C. L. (2017). Variable screening via quantile partial correlation. *Journal of the American Statistical Association* 112, 650–663.

Ma, S. J., Su, L., and Zhang, Yichong. (2021). Determining the number of communities in degree-corrected stochastic block models. *Journal of Machine Learning Research* 22, 1–63.

Ma, Y. Y., Pan, R., Zou, T., and Wang, H. S. (2020). A naive least squares method for spatial autoregression with covariates. *Statistica Sinica* 30, 653–672.

Ma, Yao., Wang, S. H., Aggarwal, C. C., and Tang, J. L. (2019). Graph convolutional networks with eigenpooling. *Proceedings of the 25th ACM SIGKDD International Conference on Knowledge Discovery & Data Mining*, 723–731.

Maas, A. L., Hannun, A. Y., and Ng, A. Y. (2013). Rectifier nonlinearities improve neural network acoustic models. *Proceedings of the 30th International Conference on Machine Learning*

MacQueen, J. (1967). Some methods for classification and analysis of multivariate observations. *Proceedings of the Fifth Berkeley Symposium on Mathematical Statistics and Probability*, 281–297.

Magnus, J. R. and Neudecker, H. (2019). *Matrix Differential Calculus with Applications in Statistics and Econometrics*. New York: Wiley.

Mallows, C. L. (1973). Some comments on Cp. *Technometrics* 15, 661–675.

Manceur, A. M. and Dutilleul, P. (2013). Maximum likelihood estimation for the tensor normal distribution: algorithm, minimum sample size, and empirical bias and dispersion. *Journal of Computational and Applied Mathematics* 239, 37–49.

Mardia, K. V. and Goodall, C. R. (1993). Spatial-temporal analysis of multivariate environmental monitoring data. *Multivariate Environmental Statistics* 6, 347–385.

Mardia, K. V., Kent, J. T., and Bibby, J. M. (1979). *Multivariate Analysis* New York: Academic Press.

Marin, J. M. and Dhorne, T. (2002). Linear Toeplitz covariance structure models with optimal estimators of variance components. *Linear Algebra and Its Applications* 354, 195–212.

Markowitz, H. (1952). Portfolio selection. *The Journal of Finance* 7, 77–91.

Maronna, R. A. (1976). Robust M-estimators of multivariate location and scatter. *Annals of Statistics* 4, 51–67.

Masak, T., Sarkar, S., and Panaretos, V. M. (2023). Separable expansions for covariance estimation via the partial inner product. *Biometrika* 110, 225–247.

Massacci, D. and Kapetanios, G. (2024). Forecasting in factor augmented regressions under structural change. *International Journal of Forecasting* 40, 62–76.

McCullagh, P. and Nelder, J. A. (1989). *Generalized Linear Models* New York: Chapman and Hall.

McCulloch, C. E., Searle, S. R., and Neuhaus, J. M. (2008). *Generalized, Linear, and Mixed Models* New York: Wiley.

McCulloch, W. S. and Pitts, W. (1943). A logical calculus of the ideas immanent in nervous activity. *The Bulletin of Mathematical Biophysics* 5, 115–133.

Mcmurry, T. L. and Politis, D. N. (2015). High-dimensional autocovariance matrices and optimal linear prediction. *Electronic Journal of Statistics* 9, 753–788.

McQuarrie, A. D. R. and Tsai, C. L. (1998). *Regression and Time Series Model Selection.* Singapore: World Scientific.

Mei, S. Y., Fei, W., and Zhou, S. (2011). Gene ontology based transfer learning for protein subcellular localization. *BMC Bioinformatics* 12, 44.

Meinshausen, N. and Bühlmann, P. (2006). High-dimensional graphs and variable selection with the Lasso. *The Annals of Statistics* 34, 1436–1462.

Meng, X., Zhang, J. F., and Li, Y. (2024). Statistical inference on high dimensional gaussian graphical regression models. *arXiv preprint arXiv:2411.01588.*

Mikolov, T., Chen, K., Corrado, G., and Dean, J. (2013). Efficient estimation of word representations in vector space. *1st International Conference on Learning Representations, ICLR 2013*

Minh, H. Q. and Murino, V. (2017). *Covariances in Computer Vision and Machine Learning.* Cham: Morgan & Claypool Publishers.

Misra, D. (2019). Mish: a self regularized non-monotonic activation function. *arXiv preprint arXiv:1908.08681.*

Mokhtari, F., Laurienti, P. J., Rejeski, W. J., and Ballard, G. (2019). Dynamic functional magnetic resonance imaging connectivity tensor decomposition: a new approach to analyze and interpret dynamic brain connectivity. *Brain Connectivity* 9, 95–112.

Nakatsuji, M., Zhang, Q. P., Lu, X., Makni, B., and Hendler, J. A. (2017). Semantic social network analysis by cross-domain tensor factorization. *IEEE Transactions on Computational Social Systems* 4, 207–217.

Nalisnick, E., Smyth, P., and Tran, D. (2023). A brief tour of deep learning from a statistical perspective. *Annual Review of Statistics and Its Application* 10, 219–246.

Nelder, J. A. and Wedderburn, R. W. M. (1972). Generalized linear models. *Journal of the Royal Statistical Society Series A: Statistics in Society* 135, 370–384.

Neter, J. and Wasserman, W. (1996). *Applied Linear Statistical Models* Chicago: Irwin.

Neudecker, H. and Wesselman, A. M. (1990). The asymptotic variance matrix of the sample correlation matrix. *Linear Algebra Applic* 127, 589–599.

Neuman, E. J. and Mizruchi, M. S. (2010). Structure and bias in the network autocorrelation model. *Social Networks* 32, 290–300.

Newman, M. E. J. and Girvan, M. (2004). Finding and evaluating community structure in networks. *Physical Review E* 69, 026113.

Newton, H. J. (1988). *TIMESLAB: A Time Series Analysis Laboratory* Pacific Grove: Wadsworth & Brooks/Cole.

Nguyen, T., Lyu, B., Ishwar, P., Scheutz, M., and Aeron, S. (2022). Joint covariate-alignment and concept-alignment: a framework for domain generalization. *IEEE 32nd International Workshop on Machine Learning for Signal Processing (MLSP)*, 1–6.

Nielsen, M. (2015). *Neural Networks and Deep Learning* San Francisco: Determination Press.

Noci, L., Li, C. N., Li, M. F., He, B., Hofmann, T., Maddison, C. J., and Roy, D. M. (2023). The shaped transformer: attention models in the infinite depth-and-width limit. *Proceedings of the 37th International Conference on Neural Information Processing Systems*, 54250–54281.

Noschese, S., Pasquini, L., and Reichel, L. (2013). Tridiagonal Toeplitz matrices: properties and novel applications. *Numerical Linear Algebra with Applications* 20, 302–326.

Novikov, A., Podoprikhin, D., Osokin, A., and Vetrov, D. (2015). Tensorizing neural networks. *Proceedings of the 29th International Conference on Neural Information Processing Systems*, 442–450.

Nowicki, K. and Snijders, T. A. B. (2001). Estimation and prediction for stochastic blockstructures. *Journal of the American Statistical Association* 96, 1077–1087.

Olive, D. J. (2017). *Robust Multivariate Analysis*. Cham: Springer International Publishing.

Ord, J. K. (1975). Estimation methods for models of spatial interaction. *Journal of the American Statistical Association* 70, 120–126.

Pan, S. J. and Yang, Q. (2010). A survey on transfer learning. *IEEE Transactions on Knowledge and Data Engineering* 22, 1345–1359.

Paul, D. (2007). Asymptotics of sample eigenstructure for a large dimensional spiked covariance model. *Statistica Sinica* 17, 1617–1642.

Paul, S. and Nath, S. (2024). Spatial autoregressive model with measurement error in covariates. *arXiv preprint arXiv:2402.04593*.

Pearson, K. (1900). On the criterion that a given system of deviations from the probable in the case of a correlated system of variables is such that it can be reasonably supposed to have arisen from random sampling. *The London, Edinburgh, and Dublin Philosophical Magazine and Journal of Science* 50, 157–175.

Pennington, J., Socher, R., and Manning, C. D. (2014). Glove: Global vectors for word representation. *Proceedings of the 2014 Conference on Empirical Methods in Natural Language Processing (EMNLP)*, 1532–1543.

Pesaran, M. H. (2006). Estimation and inference in large heterogeneous panels with a multifactor error structure. *Econometrica* 74, 967–1012.

Petersen, A. and Müller, H. G. (2019). Fréchet regression for random objects with Euclidean predictors. *The Annals of Statistics* 47, 691–719.

Pinheiro, J. C. and Bates, D. M. (1996). Unconstrained parametrizations for variancecovariance matrices. *Statistics and Computing* 6, 289–296.

Poignard, B. and Terada, Y. (2020). Statistical analysis of sparse approximate factor models. *Electronic Journal of Statistics* 14, 3315–3365.

Potthoff, R. F. and Roy, S. N. (1964). A generalized multivariate analysis of variance model useful especially for growth curve problems. *Biometrika* 51, 313–326.

Pourahmadi, M. (1999). Joint mean-covariance models with applications to longitudinal data: unconstrained parameterisation. *Biometrika* 86, 677–690.

Pourahmadi, M. (2011). Covariance estimation: the GLM and regularization perspectives. *Statistical Science* 26, 369–387.

Pourahmadi, M. (2013). *High-Dimensional Covariance Estimation: With High-Dimensional Data* Hoboken: John Wiley & Sons.

Preacher, K. J., Zhang, G., Kim, C., and Mels, G. (2013). Choosing the optimal number of factors in exploratory factor analysis: a model selection perspective. *Multivariate Behavioral Research* 48, 28–56.

Preisser, J. S. and Qaqish, B. F. (1999). Robust regression for clustered data with application to binary responses. *Biometrics* 55, 574–579.

Puchkin, N. and Rakhuba, M. (2024). Dimension-free structured covariance estimation. *Proceedings of Thirty Seventh Conference on Learning Theory* 4276–4306.

Qi, F., Čerňanová, V., and Semenov, Y. S. (2019). Some tridiagonal determinants related to central Delannoy numbers, the Chebyshev polynomials, and the Fibonacci polynomials. *UPB Scientific Bulletin, Series A: Applied Mathematics and Physics* 81, 123–136.

Qin, L., Snoussi, H., and Abdallah, F. (2014). Object tracking using adaptive covariance descriptor and clustering-based model updating for visual surveillance. *Sensors* 14, 9380–9407.

Qiu, Y. and Chen, S. X. (2012). Test for bandedness of high-dimensional covariance matrices and bandwidth estimation. *The Annals of Statistics* 40, 1285–1314.

Qiu, Y. and Chen, S. X. (2015). Bandwidth selection for high-dimensional covariance matrix estimation. *Journal of the American Statistical Association* 110, 1160–1174.

Qiu, Y. and Liyanage, J. S. (2019). Threshold selection for covariance estimation. *Biometrics* 75, 895–905.

Qu, A. and Li, R. (2006). Quadratic inference functions for varying-coefficient models with longitudinal data. *Biometrics* 62, 379–391.

Qu, A., Lindsay, B. G., and Li, Bing. (2000). Improving generalised estimating equations using quadratic inference functions. *Biometrika* 87, 823–836.

Qu, A. and Song, P. X. K. (2004). Assessing robustness of generalised estimating equations and quadratic inference functions. *Biometrika* 91, 447–459.

Rabanser, S., Shchur, O., and Günnemann, S. (2017). Introduction to tensor decompositions and their applications in machine learning. *arXiv preprint arXiv:1711.10781.*

Radford, A., Narasimhan, K., Salimans, T., and Sutskever, I. (2018). Improving language understanding by generative pre-training. https://cdn.openai.com/research-covers/language-unsupervised/language_understanding_paper.pdf. OpenAI technical report.

Rajwade, A., Rangarajan, A., and Banerjee, A. (2012). Image denoising using the higher order singular value decomposition. *IEEE Transactions on Pattern Analysis and Machine Intelligence* 35, 849–862.

Ramachandran, P., Zoph, B., and Le, Q. V. (2018). Searching for activation functions. *arXiv preprint arXiv:1710.05941.*

Ramsay, J. O. and Dalzell, C. J. (1991). Some tools for functional data analysis. *Journal of the Royal Statistical Society: Series B (Statistical Methodology)* 53, 539–561.

Ramsay, J. O. and Silverman, B. W. (2005). *Functional Data Analysis* New York: Springer.

Rangapuram, S. S., Seeger, M., Gasthaus, J., Stella, L., Wang, Y. Y., and Januschowski,T. (2018). Deep state space models for time series forecasting. *Proceedings of the 32nd International Conference on Neural Information Processing Systems*, 7796–7805.

Rao, C. R. (1965). *Linear Statistical Inference and Its Applications* New York: Wiley.

Rao, C. R. (1970). Estimation of heteroscedastic variances in linear models. *Journal of the American Statistical Association* 65, 161–172.

Rao, C. R. (1972). Estimation of variance and covariance components in linear models. *Journal of the American Statistical Association* 67, 112–115.

Rao, K. S. and Manjunath, K. E. (2017). *Speech Recognition Using Articulatory and Excitation Source Features*. Cham: Springer International Publishing.

Ravikumar, P., Wainwright, M. J., Raskutti, G., and Yu, Bin. (2011). High-dimensional covariance estimation by minimizing ℓ_1-penalized log-determinant divergence. *Electronic Journal of Statistics* 5, 935–980.

Rebala, G., Ravi, A., and Churiwala, S. (2019). *An Introduction to Machine Learning* Cham: Springer International Publishing.

Redmon, J., Divvala, S., Girshick, R., and Farhadi, A. (2015). You only look once: unified, real-time object detection. *IEEE Conference on Computer Vision and Pattern Recognition*, 779–788.

Reid, N. (1995). The roles of conditioning in inference. *Statistical Science* 10, 138–157.

Ren, M., Zhen, Y., and Wang, J. H. (2024). Transfer learning for tensor Gaussian graphical models. *Journal of Machine Learning Research* 25, 1–40.

Ren, Y., Li, Z., Zhu, X. N., Gao, Y., and Wang, H. S. (2024). Distributed estimation and inference for spatial autoregression model with large scale networks. *Journal of Econometrics* 238, 105629.

Ricci, M. M. G. and Levi-Civita, T. (1900). Méthodes de calcul différentiel absolu et leurs applications. *Mathematische Annalen* 54, 125–201.

Rodman, L. and Shalom, T. (1992). On inversion of symmetric Toeplitz matrices. *SIAM Journal on Matrix Analysis and Applications* 13, 530–549.

Rohe, K., Chatterjee, S., and Yu, Bin. (2011). Spectral clustering and the high-dimensional stochastic blockmodel. *The Annals of Statistics* 39, 1878–1915.

Ronneberger, O., Fischer, P., and Brox, T. (2015). U-Net: convolutional networks for biomedical image segmentation. *Medical Image Computing and Computer-Assisted Intervention – MICCAI 2015*, 234–241.

Roś, B., Bijma, F., Munck, J. C. de, and de Gunst, M. C. M. (2016). Existence and uniqueness of the maximum likelihood estimator for models with a Kronecker product covariance structure. *Journal of Multivariate Analysis* 143, 345–361.

Rosenblatt, F. (1958). The perceptron: a probabilistic model for information storage and organization in the brain. *Psychological Review* 65, 386–408.

Rosset, S. and Tibshirani, R. J. (2020). From fixed-X to Random-X regression: bias-variance decompositions, covariance penalties, and prediction error estimation. *Journal of the American Statistical Association* 115, 138–151.

Rothman, A. J. (2012). Positive definite estimators of large covariance matrices. *Biometrika* 99, 733–740.

Rothman, A. J., Bickel, P. J., Levina, E., and Zhu, J. (2008). Sparse permutation invariant covariance estimation. *Electronic Journal of Statistics* 2, 494–515.

Rothman, A. J., Levina, E., and Zhu, J. (2009). Generalized thresholding of large covariance matrices. *Journal of the American Statistical Association* 104, 177–186.

Rougier, J. (2017). A representation theorem for stochastic processes with separable co-variance functions, and its implications for emulation. *arXiv preprint arXiv:1702.05599*.

Rousseeuw, P. J. (1987). Silhouettes: a graphical aid to the interpretation and validation of cluster analysis. *Journal of Computational and Applied Mathematics* 20, 53–65.

Rousseeuw, P. J. and Leroy, A. M. (2005). *Robust Regression and Outlier Detection* New York: John Wiley & Sons.

Ruder, S. (2016). An overview of gradient descent optimization algorithms. *arXiv preprint arXiv:1609.04747*.

Rumelhart, D. E., Hinton, G., and Williams, R. J. (1986). Learning representations by back-propagating errors. *Nature* 323, 533–536.

Rummery, G. A. and Niranjan, M. (1994). *Online Q-learning Using Connectionist Systems* Cambridge: University of Cambridge.

Ruppert, D., Wand, M. P., and Carroll, R. J. (2003). *Semiparametric Regression*. Cambridge: Cambridge university Press.

Saito, D., Suzuki, S., and Minematsu, N. (2019). Speech representation based on tensor factor analysis and its application to speaker recognition and language identification. *2019 Asia-Pacific Signal and Information Processing Association Annual Summit and Conference (APSIPA ASC)*, 402–406.

Saito, D., Yamamoto, K., Minematsu, N., and Hirose, K. (2011). One-to-many voice conversion based on tensor representation of speaker space. *Twelfth Annual Coference of the International Speech Communication Association* 653–656.

Sargent, T. J. and Sims, C. A. (1977). Business cycle modeling without pretending to have too much a priori economic theory. *ăNew Methods in Business Cycle Research* 1, 145–168.

Schelldorfer, J., Meier, L., and Bühlmann, P. (2014). GLMMLasso: an algorithm for high-dimensional generalized linear mixed models using ℓ_1-penalization. *Journal of Computational and Graphical Statistics* 23, 460–477.

Schmidhuber, J. (2015). Deep learning in neural networks: an overview. *Neural Networks* 61, 85–117.

Schmidhuber, J. (2022). Annotated history of modern ai and deep learning. *arXiv preprint arXiv:2212.11279*

Schott, J. R. (2007). A test for the equality of covariance matrices when the dimension is large relative to the sample sizes. *Computational Statistics & Data Analysis* 51, 6535–6542.

Schott, J. R. (2017). *Matrix Analysis for Statistics* Hoboken: Wiley.

Schwarz, G. (1978). Estimating the dimension of a model. *The Annals of Statistics* 6, 461–464.

Sclove, S. L. (1987). Application of model-selection criteria to some problems in multivariate analysis. *Psychometrika* 52, 333–343.

Sclove, S. L. (2021). Using model selection criteria to choose the number of principal components. *Journal of Statistical Theory and Applications* 20, 450–461.

Seber, G. A. F. (2004). *Multivariate Observations* Hoboken: Wiley.

Seber, G. A. F. (2008). *A Matrix Handbook For Statisticians* New York: John Wiley & Sons.

Seber, G. A. F. and Lee, A. J. (2003). *Linear Regression Analysis* Hoboken: Wiley.

Sentürk, D. and Müller, H. G. (2010). Functional varying coefficient models for longitudinal data. *Journal of the American Statistical Association* 105, 1256–1264.

Shaha, M. and Pawar, M. (2018). Transfer learning for image classification. *Second International Conference On Electronics, Communication and Aerospace Technology (ICECA)*. IEEE, 656–660.

Shalev-Shwartz, S. and Ben-David, S. (2014). *Understanding Machine Learning: From Theory to Algorithms*. Cambridge: Cambridge university Press.

Shao, J. (1997). An asymptotic theory for linear model selection. *Statistica Sinica* 7, 221–242.

Sharpe, W. F. (1964). Capital asset prices: a theory of market equilibrium under conditions of risk. *The Journal of Finance* 19, 425–442.

Sharpe, W. F. (1966). Mutual fund performance. *The Journal of Business* 39, 119–138.

Shen, X. and Huang, H. C. (2020). Discussion of "From fixed-X to random-X regression: bias-variance decompositions, covariance penalties, and prediction error estimation". *Journal of the American Statistical Association* 115, 152–156.

Sheppard, C. (2017). *Tree-based Machine Learning Algorithms: Decision Trees, Random Forests, and Boosting* Austin: CreateSpace Independent Publishing Platform.

Sherman, M. (2011). *Spatial Statistics and Spatio-Temporal Data: Covariance Functions and Directional Properties* Chichester: Wiley.

Sherman, P. J. (2023). On the eigenstructure of the AR (1) covariance. *2023 IEEE Statistical Signal Processing Workshop (SSP)*, 6–10.

Shi, J., Liu, Xu., Li, Y. X., Yu, C., and Han, Y. S. (2022). Does supply chain network centrality affect stock price crash risk? Evidence from Chinese listed manufacturing companies. *International Review of Financial Analysis* 80, 102040.

Shin, H. C., Roth, H. R., Gao, M., Lu, L., Xu, Z. Y., Nogues, I., Yao, J. H., Mollura, D., and Summers, R. M. (2016). Deep convolutional neural networks for computeraided detection: CNN architectures, dataset characteristics and transfer learning. *IEEE Transactions on Medical Imaging* 35, 1285–1298.

Shu, H. and Nan, B. (2019). Estimation of large covariance and precision matrices from temporally dependent observations. *The Annals of Statistics* 47, 1321–1350.

Shumway, R. H. and Stoffer, D. S. (2017). *Time Series Analysis and Its Applications: With R Examples*. Cham: Springer International Publishing.

Sidiropoulos, N. D. and Bro, R. (2000). On the uniqueness of multilinear decomposition of N-way arrays. *Journal of Chemometrics: A Journal of the Chemometrics Society* 14, 229–239.

Sigillito, V., Wing, S., Hutton, L., and Baker, K. (1989). Ionosphere. UCI Machine Learning Repository. https://doi.org/10.24432/C5W01B

Sihag, S., Mateos, G., McMillan, C., and Ribeiro, A. (2022). coVariance Neural Networks. *Proceedings of the 36th International Conference on Neural Information Processing System*, 17003–17016.

Simonoff, J. S. (1996). *Smoothing Methods in Statistics* New York: Springer.

Singer, J. D. and Willett, J. B. (2003). *Applied Longitudinal Data Analysis: Modeling Change and Event Occurrence* Oxford: Oxford University Press.

Singh, D., Febbo, P. G., Ross, K., Jackson, D. G., Manola, J., Ladd, C., Tamayo, P., Renshaw, A. A., D'Amico, A. V., Richie, J. P., et al. (2002). Gene expression correlates of clinical prostate cancer behavior. *Cancer Cell* 1, 203–209.

Skinner, B. F. (1958). Reinforcement today. *American Psychologist* 13, 94–99.

Smirnov, O. and Anselin, L. (2001). Fast maximum likelihood estimation of very large spatial autoregressive models: a characteristic polynomial approach. *Computational Statistics & Data Analysis* 35, 301–319.

Soloveychik, I. and Trushin, D. (2016). Gaussian and robust Kronecker product covariance estimation: Existence and uniqueness. *Journal of Multivariate Analysis* 149, 92–113.

Son, S., Park, Y., Cho, M., and Kang, U. (2022). DAO-CP: data-adaptive online CP decomposition for tensor stream. *PLOS One* 17, e0267091.

Srivastava, M. S. and Yanagihara, H. (2010). Testing the equality of several covariance matrices with fewer observations than the dimension. *Journal of Multivariate Analysis* 101, 1319–1329.

Srivastava, V. K. and Giles, D. E. A. (1987). *Seemingly Unrelated Regression Equations Models: Estimation and Inference*. New York: Marcel Dekker.

Stein, M. L. (2005). Space–time covariance functions. *Journal of the American Statistical Association* 100, 310–321.

Steinwart, I. and Christmann, A. (2008). *Support Vector Machines* New York: Springer Science & Business Media.

Stock, J. H. and Watson, M. W. (2002a). Forecasting using principal components from a large number of predictors. *Journal of the American Statistical Association* 97, 1167–1179.

Stock, J. H. and Watson, M. W. (2002b). Macroeconomic forecasting using diffusion indexes. *Journal of Business & Economic Statistics* 20, 147–162.

Stock, J. H. and Watson, M. W. (2005). Implications of dynamic factor models for VAR analysis. NBER Working Papers. National Bureau of Economic Research.

Stock, J. H. and Watson, M. W. (2016). Dynamic factor models, factor-augmented vector autoregressions, and structural vector autoregressions in macroeconomics. *Handbook of Macroeconomics* Saint Louis: Elsevier, 415–525.

Stone, M. (1974). Cross-validatory choice and assessment of statistical predictions. *Journal of the Royal Statistical Society: Series B (Statistical Methodology)* 36, 111–133.

Su, L. and Wang, X. (2017). On time-varying factor models: estimation and testing. *Journal of Econometrics* 198, 84–101.

Sun, W. W., Hao, B., and Li, Lexin. (2021). Tensors in modern statistical learning. *Wiley StatsRef: Statistics Reference Online* 1–25.

Sun, Y., Wang, X. G., and Tang, X. O. (2014). Deep learning face representation from predicting 10,000 classes. *IEEE Conference on Computer Vision and Pattern Recognition*, 1891–1898.

Sun, Y. V. and Hu, Y. (2016). Integrative analysis of multi-omics data for discovery and functional studies of complex human diseases. *Advances in Genetics* 93, 147–190.

Sutskever, I., Vinyals, O., and Le, Q. V. (2014). Sequence to sequence learning with neural networks. *Proceedings of the 27th International Conference on Neural Information Processing Systems*, 3104–3112.

Sutton, R. S. and Barto, A. G. (1998). *Reinforcement Learning: An Introduction* Cambridge: MIT Press.

Szatrowski, T. H. (1980). Necessary and sufficient conditions for explicit solutions in the multivariate normal estimation problem for patterned means and covariances. *The Annals of Statistics* 8, 802–810.

Szatrowski, T. H. (1982). Testing and estimation in the block compound symmetry problem. *Journal of Educational Statistics* 7, 3–18.

Taigman, Y., Yang, M., Ranzato, M., and Wolf, L. (2014). DeepFace: Closing the gap to human-level performance in face verification. *IEEE Conference on Computer Vision and Pattern Recognition*, 1701–1708.

Taunk, K., De, S., Verma, S., and Swetapadma, A. (2019). A brief review of nearest neighbor algorithm for learning and classification. *2019 International Conference on Intelligent Computing and Control Systems (ICCS)* 1255–1260.

Thanh Noi, P. and Kappas, M. (2018). Comparison of random forest, k-nearest neighbor, and support vector machine classifiers for land cover classification using Sentinel-2 imagery. *Sensors* 18, 18.

Tho, Z. Y., Ding, D., Hui, F. K., Welsh, A., and Zou, T. (2023). On the robust estimation of spatial autoregressive models. *Econometrics and Statistics*. In Press.

Thorndike, R. L. (1953). Who belongs in the family? *Psychometrika* 18, 267–276.

Tian, Y. and Feng, Y. (2023). Transfer learning under high-dimensional generalized linear models. *Journal of the American Statistical Association* 118, 2684–2697.

Tibshirani, R. (1996). Regression shrinkage and selection via the Lasso. *Journal of the Royal Statistical Society: Series B (Statistical Methodology)* 58, 267–288.

–(2011). Regression shrinkage and selection via the Lasso: a retrospective. *Journal of the Royal Statistical Society: Series B (Statistical Methodology)* 73, 273–282.

Tibshirani, R., Saunders, M., Rosset, S., Zhu, J., and Knight, K. (2005). Sparsity and smoothness via the fused Lasso. *Journal of the Royal Statistical Society: Series B (Statistical Methodology)* 67, 91–108.

Timm, N. H. (2002). *Applied Multivariate Analysis* New York: Springer.

Timmerman, M. E. and Kiers, H. A. L. (2000). Three-mode principal components analysis: Choosing the numbers of components and sensitivity to local optima. *British Journal of Mathematical and Statistical Psychology* 53, 1–16.

Tiwari, M. and Singhai, R. (2017). A review of detection and tracking of object from image and video sequences. *International Journal of Computational Intelligence Research* 13, 745–765.

Tong, Y. L. (1990). *The Multivariate Normal Distribution*. New York: Springer.

Torrey, L. and Shavlik, J. (2010). Transfer learning. *Handbook of Research on Machine Learning Applications and Trends: Algorithms, Methods, and Techniques*, 242–264.

Tsai, C. L. (1986). Score test for the first-order autoregressive model with heteroscedasticity. *Biometrika* 73, 455–460.

Tsay, R. S. (2010). *Analysis of Financial Time Series* Cambridge: Wiley.

Tsay, R. S. (2013). *Multivariate Time Series Analysis: With R and Financial Applications* Hoboken: Wiley.

Tsay, R. S. (2014). *An Introduction to Analysis of Financial Data with R* Somerset: Wiley.

Tsay, R. S. (2024). Matrix-variate time series analysis: a brief review and some new developments. *International Statistical Review* 92, 246–262.

Tsiligkaridis, T. and Hero, A. O. (2013). Covariance estimation in high dimensions via Kronecker product expansions. *IEEE Transactions on Signal Processing* 61, 5347–5360.

Tsukuma, H. and Kubokawa, T. (2020). *Shrinkage Estimation for Mean and Covariance Matrices*. Singapore: Springer Nature.

Tucker, L. R. (1966). Some mathematical notes on three-mode factor analysis. *Psychometrika* 31, 279–311.

Turki, T., Wei, Z., and Wang, J. T. L. (2017). Transfer learning approaches to improve drug sensitivity prediction in multiple myeloma patients. *IEEE Access* 5, 7381–7393.

Tuzel, O., Porikli, F., and Meer, P. (2006). Region covariance: a fast descriptor for detection and classification. *ECCV*, 589–600.

Twisk, J. W. R. (2023). *Applied Longitudinal Data Analysis for Medical Science: A Practical Guide*. Cambridge: Cambridge University Press.

Tyler, D. E. (1987). A distribution-free M-estimator of multivariate scatter. *Annals of Statistics* 15, 234–251.

Uppuluri, V. R. R. and Carpenter, J. A. (1969). The inverse of a matrix occurring in first-order moving-average models. *Sankhya: The Indian Journal of Statistics, Series A* 31, 79–82.

van der Vaart, A. W. (1998). *Asymptotic Statistics* Cambridge: Cambridge University Press.

Vannieuwenhoven, N., Vandebril, R., and Meerbergen, K. (2012). A new truncation strategy for the higher-order singular value decomposition. *SIAM Journal on Scientific Computing* 34, A1027–A1052.

Vapnik, V. N. (1995). *The Nature of Statistical Learning Theory* New York: Springer.

Vapnik, V. N. (1998). *Statistical Learning Theory* New York: John Willey & Sons.

Varoquaux, G., Gramfort, A., Poline, J., and Thirion, B. (2010). Brain covariance selection: better individual functional connectivity models using population prior. *Proceedings of the 23rd International Conference on Neural Information Processing Systems* 2334–2342.

Vaswani, A., Shazeer, N., Parmar, N., Uszkoreit, J., Jones, L., Gomez, A. N., Kaiser, L., and Polosukhin, I. (2017). Attention is all you need. *Proceedings of the 31st International Conference on Neural Information Processing Systems* 6000–6010.

Velickovic, P., Cucurull, G., Casanova, A., Romero, A., Lio, P., and Bengio, Y. (2018). Graph attention networks. *International Conference on Learning Representations (ICLR)*

Verbeke, G. and Molenberghs, G. (2001). *Linear Mixed Models for Longitudinal Data* New York: Springer.

Vidal, R., Ma, Yi., and Sastry, S. (2016). *Generalized Principal Component Analysis* New York: Springer.

Wachter, K. W. (1978). The strong limits of random matrix spectra for sample matrices of independent elements. *The Annals of Probability* 6, 1–18.

Wang, C., Yang, J., Miao, B., and Cao, L. (2013). Identity tests for high dimensional data using RMT. *Journal of Multivariate Analysis* 118, 128–137.

Wang, Di, Zheng, Y., Lian, H., and Li, G. D. (2022). High-dimensional vector autoregressive time series modeling via tensor decomposition. *Journal of the American Statistical Association* 117, 1338–1356.

Wang, Dong., Liu, X. L., and Chen, R. (2019). Factor models for matrix-valued high-dimensional time series. *Journal of Econometrics* 208, 231–248.

Wang, F. (2022). Maximum likelihood estimation and inference for high dimensional generalized factor models with application to factor-augmented regressions. *Journal of Econometrics* 229, 180–200.

Wang, H. S. (2009). Forward regression for ultra-high dimensional variable screening. *Journal of the American Statistical Association* 104, 1512–1524.

Wang, H. S. (2012). Factor profiled sure independence screening. *Biometrika* 99, 15–28.

Wang, H. S., Li, Bo., and Leng, C. (2009). Shrinkage tuning parameter selection with a diverging number of parameters. *Journal of the Royal Statistical Society: Series B (Statistical Methodology)* 71, 671–683.

Wang, J. D., Lan, C., Liu, C., Ouyang, Y., Qin, T., Lu, W., Chen, Y. Q., Zeng, W., and Yu, P. S. (2023). Generalizing to unseen domains: a survey on domain generalization. *IEEE Transactions on Knowledge and Data Engineering* 35, 8052–8072.

Wang, J. L., Chiou, J. M., and Müller, H. G. (2016). Functional data analysis. *Annual Review of Statistics and Its Application* 3, 257–295.

Wang, Q., Ma, Yue., Zhao, K., and Tian, Y. J. (2022). A comprehensive survey of loss functions in machine learning. *Annals of Data Science* 9, 187–212.

Wang, Q. L., Li, P. H., and Zhang, Lei. (2017). G2DeNet: global gaussian distribution embedding network and its application to visual recognition. *IEEE Conference on Computer Vision and Pattern Recognition*, 2730–2739.

Wang, Q. L., Xie, J., Zuo, W. M., Zhang, Lei., and Li, P. H. (2021). Deep CNNs meet global covariance pooling: better representation and generalization. *IEEE Transactions on Pattern Analysis and Machine Intelligence* 43, 2582–2597.

Wang, Q. L., Zhang, Li., Wu, B., Ren, D., Li, P. H., Zuo, W. M., and Hu, Q. (2020). What deep CNNs benefit from global covariance pooling: An optimization perspective. *Proceedings of the IEEE/CVF Conference on Computer Vision and Pattern Recognition*, 10771–10780.

Wang, S., Xie, C., and Kang, X. (2023). A novel robust estimation for high-dimensional precision matrices. *Statistics in Medicine* 42, 656–675.

Wang, T. and Singh, S. (2024). KAN based Autoencoders for factor models. *arXiv preprint arXiv:2408.02694*.

Wang, W., Neuman, E. J., and Newman, D. A. (2014). Statistical power of the social network autocorrelation model. *Social Networks* 38, 88–99.

Wang, W. C. and Fan, J. (2017). Asymptotics of empirical eigenstructure for high dimensional spiked covariance. *Annals of Statistics* 45, 1342–1374.

Wang, X. C. and Zhang, B. (2024). Target selection in shrinkage estimation of covariance matrix: a structural similarity approach. *Statistics & Probability Letters* 208, 110048.

Wang, Z., Lei, B., Liu, Bin., and Yin, G. (2023). Structure-aware covariance estimation using graph neural networks. *Available at SSRN 4670342*

Warden, P. (2017). Speech commands dataset streaming test version 2. http://download.tensorflow.org/data/speech_commands_streaming_test_v0.02.tar.gz

Warden, P. (2018). Speech commands: A dataset for limited-vocabulary speech recognition. *arXiv preprint arXiv:1804.03209*.

Wasserman, L. (2004). *All of Statistics: A Concise Course in Statistical Inference*. New York: Springer.

Wasserman, S. and Faust, K. (1994). *Social Network Analysis: Methods and Applications* Cambridge: Cambridge University Press.

Watkins, C. J. (1989). Learning from delayed rewards. PhD thesis. Cambridge: King's College.

Watkins, C. J. and Dayan, P. (1992). Q-learning. *Machine learning* 8, 279–292.

Wedderburn, R. W. M. (1974). Quasi-likelihood functions, generalized linear models, and the Gauss-Newton method. *Biometrika* 61, 439–447.

Weiss, K., Khoshgoftaar, T. M., and Wang, D. D. (2016). A survey of transfer learning. *Journal of Big Data* 3, 1–40.

Weiss, R. E. (2005). *Modeling Longitudinal Data* New York: Springer.

Werner, K., Jansson, M., and Stoica, P. (2008). On estimation of covariance matrices with Kronecker product structure. *IEEE Transactions on Signal Processing* 56, 478–491.

West, B. T., Welch, K. B., and Galecki, A. T. (2014). *Linear Mixed Models: A Practical Guide Using Statistical Software* Boca Raton: Chapman & Hall/CRC.

White, H. (1980a). A heteroskedasticity-consistent covariance matrix estimator and a direct test for heteroskedasticity. *Econometrica* 48, 817–838.

White, H. (1980b). Using least squares to approximate unknown regression functions. *International Economic Review* 21, 149–170.

White, H. (1981). Consequences and detection of misspecified nonlinear regression models. *Journal of the American Statistical Association* 76, 419–433.

White, H. (2001). *Asymptotic Theory for Econometricians* San Diego: Academic Press.

Whittaker, J. (1991). *Graphical Models in Applied Multivariate Statistics* New York:Wiley.

Whittle, P. (1954). On stationary processes in the plane. *Biometrika* 41, 434–449.

Wille, A., Zimmermann, P., Vranová, E., Fürholz, A., Laule, O., Bleuler, S., Hennig, L., Prelić, A., von Rohr, P., Thiele, L., Zitzler, E., Gruissem, W., and Bühlmann, P. (2004). Sparse graphical Gaussian modeling of the isoprenoid gene network in Arabidopsis thaliana. *Genome Biology* 5, R92.

Williams, P. M. (1996). Using neural networks to model conditional multivariate densities. *Neural Computation* 8, 843–854.

Wishart, J. (1928). The generalised product moment distribution in samples from a normal multivariate population. *Biometrika* 20, 32–52.

Witten, D., Tibshirani, R., and Hastie, T. (2009). A penalized matrix decomposition, with applications to sparse principal components and canonical correlation analysis. *Biostatistics* 10, 515–534.

Wooldridge, J. M. (2010). *Econometric Analysis of Cross Section and Panel Data* Cambridge: MIT Press.

Wu, A. (2021). Domain generalization via domain-based covariance minimization. *arXiv preprint arXiv:2110.06298.*

Wu, C. F. J. (1986). Jackknife, bootstrap and other resampling methods in regression analysis. *The Annals of Statistics* 14, 1261–1295.

Wu, C. O., Chiang, C. T., and Hoover, D. R. (1998). Asymptotic confidence regions for kernel smoothing of a varying-coefficient model with longitudinal data. *Journal of the American Statistical Association* 93, 1388–1402.

Wu, C. O. and Tian, X. (2018). *Nonparametric Models for Longitudinal Data: With Implementation in R.* Boca Raton: CRC Press.

Wu, H. L. and Zhang, J. T. (2006). *Nonparametric Regression Methods for Longitudinal Data Analysis: Mixed-Effects Modeling Approaches* Hoboken: Wiley.

Wu, W. B. and Wu, Y. N. (2016). Performance bounds for parameter estimates of high-dimensional linear models with correlated errors. *Electronic Journal of Statistics* 10, 352–379.

Wu, Y. C., Fan, J., and Müller, H. G. (2010). Varying-coefficient functional linear regression. *Bernoulli* 16, 730–758.

Wu, Y. H., Schuster, M., Chen, Z. F., Le, Q. V., Norouzi, M., Macherey, W., Krikun, M., Cao, Y., Gao, Q., Macherey, K., et al. (2016). Google's neural machine translation system: Bridging the gap between human and machine translation. *arXiv preprint arXiv:1609.08144.*

Wu, Y. J., Lan, W., Feng, L., and Tsai, C. L. (2025). A goodness-of-fit test for sparse networks. *arXiv preprint arXiv:2503.11990.*

Wu, Y. J., Lan, W., Zou, T., and Tsai, C. L. (2022). Inward and outward network influence analysis. *Journal of Business & Economic Statistics* 40, 1617–1628.

Wu, Z., Pan, S. R., Chen, F. W., Long, G., Zhang, C. Q., and Yu, P. S. (2021). A comprehensive survey on graph neural networks. *IEEE Transactions on Neural Networks and Learning Systems* 32, 4–24.

Xu, B., Wang, N., Chen, T., and Li, M. (2015). Empirical evaluation of rectified activations in convolutional network. *arXiv preprint arXiv:1505.00853*.

Xu, H. S. and Li, H. Z. (2025). Test of partial effects for Fréchet regression on Bures-Wasserstein manifolds. *arXiv preprint arXiv:2506.23487*.

Xu, J., Sun, X., Zhang, Zhiyuan., Zhao, G., and Lin, J. (2019). Understanding and improving layer normalization. *Proceedings of the 33rd International Conference on Neural Information Processing Systems*, 4381–4391.

Xu, K., Hu, W. H., Leskovec, J., and Jegelka, S. (2019). How powerful are graph neural networks? *7th International Conference on Learning Representations, ICLR 2019*.

Xu, S., Yuan, C., and Guo, J. (2024). Quasi maximum likelihood estimation for large-dimensional matrix factor models. *Journal of Business & Economic Statistics* 43, 439–453.

Xue, L., Ma, S. Q., and Zou, H. (2012). Positive-definite ℓ_1-penalized estimation of large covariance matrices. *Journal of the American Statistical Association* 107, 1480–1491.

Yan, Y. and Cheng, T. (2022). Factor-augmented forecasting regressions with threshold effects. *The Econometrics Journal* 25, 134–154.

Yang, Q., Zhang, Yu., Dai, W., and Pan, S. J. (2020). *Transfer Learning*. Cambridge: Cambridge University Press.

Yang, X. X., Zheng, X., and Chen, J. Q. (2021). Testing high-dimensional covariance matrices under the elliptical distribution and beyond. *Journal of Econometrics* 221, 409–423.

Yang, Y., Zhou, Jie., and Pan, J. (2021). Estimation and optimal structure selection of high-dimensional Toeplitz covariance matrix. *Journal of Multivariate Analysis* 184, 104739.

Yang, Y. H. (2005). Can the strengths of AIC and BIC be shared? A conflict between model indentification and regression estimation. *Biometrika* 92, 937–950.

Yao, F., Müller, H. G., and Wang, J. L. (2005). Functional data analysis for sparse longitudinal data. *Journal of the American Statistical Association* 100, 577–590.

Yao, J. F., Zheng, S., and Bai, Z. D. (2015). *Large Sample Covariance Matrices and High-Dimensional Data Analysis*. Cambridge: Cambridge University Press.

Yeh, I. C. (2007). Modeling slump flow of concrete using second-order regressions and artificial neural networks. *Cement and Concrete Composites* 29, 474–480.

Yeh, I. C. (2008). Modeling slump of concrete with fly ash and superplasticizer. *Computers and Concrete* 5, 559–572.

Yin, J. and Li, H. Z. (2012). Model selection and estimation in the matrix normal graphical model. *Journal of Multivariate Analysis* 107, 119–140.

Yin, Y. Q., Bai, Z. D., and Krishnaiah, P. R. (1988). On the limit of the largest eigenvalue of the large dimensional sample covariance matrix. *Probability Theory and Related Fields* 78, 509–521.

Ying, R., You, J., Morris, C., Ren, X., Hamilton, W. L., and Leskovec, J. (2018). Hierarchical graph representation learning with differentiable pooling. *Proceedings of the 32nd International Conference on Neural Information Processing System*, 4805–4815.

Yu, Bing., Yin, H. T., and Zhu, Z. (2018). Spatio-temporal graph convolutional networks: A deep learning framework for traffic forecasting. *Proceedings of the 27th International Joint Conference on Artificial Intelligence* 3634–3640.

Yu, D., Deng, L., and Seide, F. (2013). The deep tensor neural network with applications to large vocabulary speech recognition. *IEEE Transactions on Audio, Speech, and Language Processing* 21, 388–396.

Yu, R., Chen, R., Xiao, H., and Han, Y. F. (2024). Dynamic matrix factor models for high dimensional time series. *arXiv preprint arXiv:2407.05624*.

Yuan, M. (2010). High dimensional inverse covariance matrix estimation via linear programming. *Journal of Machine Learning Research* 11, 2261–2286.

Yuan, M. and Lin, Y. (2007). Model selection and estimation in the Gaussian graphical model. *Biometrika* 94, 19–35.

Yuan, Y., Deng, Y., Zhang, Y. Q., and Qu, A. (2020). Deep learning from a statistical perspective. *Stat* 9, e294.

Zagidullina, A. (2021). *High-Dimensional Covariance Matrix Estimation: An Introduction to Random Matrix Theory*. Cham: Springer International Publishing.

Zaremba, W., Sutskever, I., and Vinyals, O. (2015). Recurrent neural network regularization. *3rd International Conference on Learning Representations, ICLR 2015*

Zeger, S. L. and Diggle, P. J. (1994). Semiparametric models for longitudinal data with application to CD4 cell numbers in HIV seroconverters. *Biometrics* 50, 689–699.

Zeger, S. L. and Liang, K. Y. (1992). An overview of methods for the analysis of longitudinal data. *Statistics in Medicine* 11, 1825–1839.

Zeger, S. L., Liang, K. Y., and Albert, P. S. (1988). Models for longitudinal data: a generalized estimating equation approach. *Biometrics* 44, 1049–1060.

Zellner, A. (1962). An efficient method of estimating seemingly unrelated regressions and tests for aggregation bias. *Journal of the American Statistical Association* 57, 348–368.

Zeng, H., Zhong, W., and Xu, X. (2024). Transfer Learning for Spatial Autoregressive Models with Application to U.S. Presidential Election Prediction. *arXiv preprint arXiv:2405.15600*.

Zhang, A. Y. and Zhou, H. H. (2016). Minimax rates of community detection in stochastic block models. *The Annals of Statistics* 44, 2252–2280.

Zhang, C. H. (2010). Nearly unbiased variable selection under minimax concave penalty. *The Annals of Statistics* 38, 894–942.

Zhang, D. N. (2021). Robust estimation of the mean and covariance matrix for high dimensional time series. *Statistica Sinica* 31, 797–820.

Zhang, D. N. and Wu, W. B. (2017). Gaussian approximation for high dimensional time series. *The Annals of Statistics* 45, 1895–1919.

Zhang, D. W., Lin, X., Raz, J., and Sowers, M. (1998). Semiparametric stochastic mixed models for longitudinal data. *Journal of the American Statistical Association* 93, 710–719.

Zhang, Q. and Wang, H. S. (2011). On BIC'S selection consistency for discriminant analysis. *Statistica Sinica* 21, 731–740.

Zhang, T. and Zou, H. (2014). Sparse precision matrix estimation via Lasso penalized D-trace loss. *Biometrika* 101, 103–120.

Zhang, Yi. and Schneider, J. (2010). Learning multiple tasks with a sparse matrixnormal penalty. *Proceedings of the 24th International Conference on Neural Information Processing Systems*, 2550–2558.

Zhang, Yichi., Shen, W., and Kong, D. (2023). Covariance estimation for matrix-valued data. *Journal of the American Statistical Association* 118, 2620–2631.

Zhang, Z. W., Cui, P., and Zhu, W. (2022). Deep learning on graphs: A survey. *IEEE Transactions on Knowledge and Data Engineering* 34, 249–270.

Zhao, B., Ma, C., and Kolar, M. (2024). Trans-GLasso: A transfer learning approach to precision matrix estimation. *arXiv preprint arXiv:2411.15624*.

Zhao, S. D., Cai, T. T., and Li, H. Z. (2014). Direct estimation of differential networks. *Biometrika* 101, 253–268.

Zhao, W. X., Zhou, K., Li, Junyi., Tang, T., Wang, X. L., Hou, Y., Min, Y., Zhang, B. C., Zhang, J. J., Dong, Z., et al. (2023). A survey of large language models. *arXiv preprint arXiv:2303.18223*.

Zheng, S., Chen, Z., Cui, H., and Li, R. (2019). Hypothesis testing on linear structures of high dimensional covariance matrix. *The Annals of Statistics* 47, 3300–3334.

Zheng, Y. H., Xu, Z. S., and Xiao, A. (2023). Deep learning in economics: a systematic and critical review. *Artificial Intelligence Review* 56, 9497–9539.

Zhong, L. (2024). Comparison of Q-learning and SARSA reinforcement learning models on cliff walking problem. *Proceedings of the 2023 International Conference on Data Science, Advanced Algorithm and Intelligent Computing (DAI 2023)* Atlantis Press, 207–213.

Zhong, P. S., Lan, W., Song, P. X. K., and Tsai, C. L. (2017). Tests for covariance structures with high-dimensional repeated measurements. *The Annals of Statistics* 45, 1185–1213.

Zhong, P. S., Li, R., and Santo, S. (2019). Homogeneity tests of covariance matrices with high-dimensional longitudinal data. *Biometrika* 106, 619–634.

Zhou, H., Li, Lexin., and Zhu, H. T. (2013). Tensor regression with applications in neuroimaging data analysis. *Journal of the American Statistical Association* 108, 540–552.

Zhou, Jie., Cui, G., Hu, S., Zhang, Zhengyan., Yang, C., Liu, Z. Y., Wang, L. F., Li, C. C., and Sun, M. (2020). Graph neural networks: A review of methods and applications. *AI open*, 57–81.

Zhou, Jing., Li, D., Pan, R., and Wang, H. S. (2020). Network GARCH model. *Statistica Sinica* 30, 1723–1740.

Zhou, Jing., Tu, Y., Chen, Y., and Wang, H. S. (2017). Estimating spatial autocorrelation with sampled network data. *Journal of Business & Economic Statistics* 35, 130–138.

Zhou, K. Y., Liu, Z. W., Qiao, Y., Xiang, T., and Loy, C. C. (2023). Domain generalization: a survey. *IEEE Transactions on Pattern Analysis and Machine Intelligence* 45, 4396–4415.

Zhou, Z., Xing, J., Zhang, M., and Hu, W. M. (2018). Online multi-target tracking with tensor-based high-order graph matching. *2018 24th International Conference on Pattern Recognition (ICPR)*, 1809–1814.

Zhu, H. T., Li, R., and Kong, L. (2012). Multivariate varying coefficient model for functional responses. *The Annals of Statistics* 40, 2634–2666.

Zhu, L., Lei, J., Devlin, B., and Roeder, K. (2017). Testing high-dimensional covariance matrices, with application to detecting schizophrenia risk genes. *The Annals of Applied Atatistics* 11, 1810–1831.

Zhu, X. J. and Goldberg, A. B. (2009). *Introduction to Semi-Supervised Learning* San Rafael: Morgan & Claypool.

Zhu, X. N., Cai, Z., and Ma, Yanyuan. (2022). Network functional varying coefficient model. *Journal of the American Statistical Association* 117, 2074–2085.

Zhu, X. N., Chang, X., Li, R., and Wang, H. S. (2019). Portal nodes screening for large scale social networks. *Journal of Econometrics* 209, 145–157.

Zhu, X. N., Huang, D., Pan, R., and Wang, H. S. (2020). Multivariate spatial autore-gressive model for large scale social networks. *Journal of Econometrics* 215, 591–606.

Zhu, X. N., Pan, R., Li, G. D., Liu, Y. W., and Wang, H. S. (2017). Network vector autoregression. *The Annals of Statistics* 45, 1096–1123.

Zhu, X. N., Wang, W. N., Wang, H. S., and Härdle, W. K. (2019). Network quantile autoregression. *Journal of Econometrics* 212, 345–358.

Zhuang, F., Qi, Z., Duan, K., Xi, D., Zhu, Y., Zhu, H. S., Xiong, H., and He, Q. (2021). A comprehensive survey on transfer learning. *Proceedings of the IEEE* 109, 43–76.

Zou, H. (2006). The adaptive Lasso and its oracle properties. *Journal of the American Statistical Association* 101, 1418–1429.

Zou, H. and Hastie, T. (2005). Regularization and variable selection via the elastic net. *Journal of the Royal Statistical Society: Series B (Statistical Methodology)* 67, 301–320.

Zou, H. and Li, R. (2008). One-step sparse estimates in nonconcave penalized likelihood models. *The Annals of Statistics* 36, 1509–1533.

Zou, T., Lan, W., Li, R., and Tsai, C. L. (2022). Inference on covariance-mean regression. *Journal of Econometrics* 230, 318–338.

Zou, T., Lan, W., Li, R., and Tsai, C. L. (2025). Fixed and random covariance regression analyses. *The Annals of Statistics* 53, 1587–1612.

Zou, T., Lan, W., Wang, H. S., and Tsai, C. L. (2017). Covariance regression analysis. *Journal of the American Statistical Association* 112, 266–281.

Zou, T., Luo, R., Lan, W., and Tsai, C. L. (2020). Covariance regression model for non-normal data. *Handbook of Financial Econometrics, Mathematics, Statistics, and Machine Learning*. World Scientific, 3933–3945.

Zou, T., Luo, R., Lan, W., and Tsai, C. L. (2021). Network influence analysis. *Statistica Sinica* 31, 1727–1748.

Zwiernik, P., Uhler, C., and Richards, D. (2017). Maximum likelihood estimation for linear Gaussian covariance models. *Journal of the Royal Statistical Society: Series B (Statistical Methodology)* 79, 1269–1292.

Author Index

W. Lan, C.-L. Tsai, *Covariance Analysis and Beyond*,
https://doi.org/10.1007/978-3-032-08796-6

M

N

T

Z

Subject Index

W. Lan, C.-L. Tsai, *Covariance Analysis and Beyond*,
https://doi.org/10.1007/978-3-032-08796-6

MIX
Papier aus verantwortungsvollen Quellen
Paper from responsible sources
FSC® C105338
FSC
www.fsc.org

If you have any concerns about our products,
you can contact us on
ProductSafety@springernature.com

In case Publisher is established outside the EU,
the EU authorized representative is:
Springer Nature Customer Service Center GmbH
Europaplatz 3, 69115 Heidelberg, Germany

Printed by Libri Plureos GmbH
in Hamburg, Germany